Get the eBook FREE!

(PDF, ePub, Kindle, and liveBook all included)

We believe that once you buy a book from us, you should be able to read it in any format we have available. To get electronic versions of this book at no additional cost to you, purchase and then register this book at the Manning website.

Go to https://www.manning.com/freebook and follow the instructions to complete your pBook registration.

That's it!
Thanks from Manning!

Geometry for Programmers

Geometry for Programmers

OLEKSANDR KALENIUK

SHELTER ISLAND

For online information and ordering of this and other Manning books, please visit www.manning.com. The publisher offers discounts on this book when ordered in quantity. For more information, please contact

Special Sales Department
Manning Publications Co.
20 Baldwin Road
PO Box 761
Shelter Island, NY 11964
Email: orders@manning.com

Manning Publications Co.
20 Baldwin Road
PO Box 761
Shelter Island, NY 11964

Development editor: Elesha Hyde
Technical development editor: Jerry Kuch
Review editor: Adriana Sabo
Production editor: Kathleen Rossland
Copy editor: Keir Simpson
Proofreader: Jason Everett
Technical proofreader: Alain Couniot
Typesetter: Dennis Dalinnik
Cover designer: Marija Tudor

ISBN: 9781633439603
Printed in the United States of America

brief contents

contents

preface

I wasn't even going to write this book. Well, I do enjoy speaking at programmers' events, and I do like writing about geometry, especially online, where I can add interactive widgets instead of still illustrations. But a whole book? Aren't there enough geometry books already? Does anybody read them anyway?

But then I had a chat with Michael Stephens, an associate publisher at Manning, and he convinced me that first of all, people do read books—even the hard ones, even books on the most sophisticated or obscure topics. Second, there's a specific interest in a book explaining geometrical concepts to practicing programmers. Programmers will read a book about the geometry but there's a catch: it has to be a good book.

What makes a good "Something for Programmers" book? Three things:

- *The book has to be applicable.* One should be able to easily try out the explained concepts in practice.
- *The book has to be approachable.* Modern geometry relies heavily on algebra and calculus, and contrary to the popular belief from the 1970s, not every computer programmer is automatically good at math.
- *The book has to be concise.* The programmer's time is valuable.

To make the book applicable, I gathered a list of practical examples from my own practice and from popular interview questions. Although I write mostly in C++ professionally, I present the source-code samples in Python solely for ease of trying them.

To make the book approachable, I delegated all the tedious math work to SymPy, a Python library that does your algebra and calculus for you. I also introduced some algebra and calculus in the book, but briefly. You don't have to spend months in training to become good at moving *x*s and *y*s around; SymPy, which you can learn in 15 minutes, will take care of that.

To make the book concise, or to let myself be brief with the explanations, I gathered a list of interactive tutorials online and placed references to them all over the book. Now if you find my version of the explanation for something too brief, you can usually find a more detailed version in a nearby reference. Don't get me wrong: I believe in brief explanations, but I believe in interactive experiences more.

So the book you're looking at is the product of several major design decisions, some of which may appear to be controversial. A hardcore mathematician, for example, might object to using SymPy instead of pen and paper, because that would be "cheating." A hardcore programmer, on the other hand, might object to using Python as a language of examples because mathematical code belongs to Fortran and C++. I recognize these objections as being well justified. The goal of this book, however, is not to be a flawless piece of art, but to help you to get good at geometry. And sometimes, as you'll see in chapter 11, cutting corners is a valid, desirable strategy for achieving your goal.

acknowledgments

First, I want to thank all the people from Manning who were involved with the book, but especially Elesha Hyde, my wonderful development editor, and Jerry Kuch, and Alain Couniot, who did technical editing and proofing. Thank you for believing in this book, for your everlasting enthusiasm and dedication, and for the hard work that followed.

I also want to thank James Cooper, Tim Biddington, and Volodymyr Anokhin, who were incredibly helpful with their feedback. Thank you for both your support and your honesty. Also, I thank all the reviewers: Aidan Twomey, Andrey Solodov, Anton Herzog, Christian Sutton, Christopher Kardell, Darrin Bishop, David Moskowitz, Dipkumar Patel, Fernando García Sedano, George Thomas, Hal Moroff, Hilde Van Gysel, Jean Francois Morrin, Jedidiah Clemons-Johnson, Jens Hansen, Jeremy Chen, James J. Byleckie, John Guthrie, Jort Rodenburg, Jose San Leandro, Justin J. Vincent, Kent Spillner, Laud Bentil, Maciej Jurkowski, Manoj Reddy, Mark Graham, Mary Anne Thygesen, Matt Gukowsky, Maxim Volgin, Maxime Boillot, Megan Jones, Oliver Korten, Patrick Regan, V. V. Phansalkar, Ranjit Sahai, Rich Yonts, Rodney Weis, Ruben Gonzalez Rubio, Sadhana Ganapathiraju, Samuel Bosch, Sanjeev Kilarapu, Sergio Govoni, Sriram Macharla, Sujith Surendranathan Pillai, Tam Thanh Nguyen, Teresa Abreu, Theo Despoudis, Tim van Deurzen, Vidhya Vinay, Werner Nindl, and Zbigniew Curylo. Your suggestions helped make this book better.

I want to thank my friends, family members, and colleagues for their patience and understanding. For completely unrelated reasons, this period has been a difficult one for all of us, and I'm deeply sorry if I couldn't be with you when you needed me most.

Last but not least, I want to thank the armed forces of Ukraine for keeping me, my friends, my family, and my colleagues safe and alive during the past year. Yes, there's a saying that the greatest books are written by dead people, but I don't think that this rule applies to technical literature.

about this book

Geometry for Programmers explains the geometry behind video games, computer-aided design (CAD) applications, 3D-printing software, geoinformation systems, virtual and augmented reality, computer vision, computed tomography, and so on. It concerns curves, surfaces, 3D objects, object spatial positioning, object spatial transformations, and several models to represent geometrical data, each with its pros and cons.

Who should read this book

You should read this book if you work with a game engine, a CAD framework, or a 3D-printing software development kit and want to know the math behind your tools of the trade to use them more efficiently. If you want to develop such tools, you should read much more than one book, but this one might be a good start.

How this book is organized

The book consists of 12 chapters:

- Chapter 1 shows the relationship between geometry and programming: which programming domains rely on geometry and which part of geometry applies to programming. The chapter showcases SymPy as a tool to do our math for us and provides a brief but good-enough-to-get-started tutorial.
- Chapter 2 teaches the basic language of applied geometry, including terms such as *near-degenerate triangle, nonmanifold mesh,* and *continuous transformation.*
- Chapter 3 discusses linear systems as though they are systems of lines, planes, and hyperplanes. The chapter shows the conditions in which such systems have

computable solutions and provides practical guidance on choosing the best algorithm for the job.

- Chapter 4 explains how the most common geometric transformation—such as rotation, scale, and translation—are generalized and extended by projective transformation. The chapter also introduces the concept of homogeneous coordinates and explains how using them simplifies programming in practice.
- Chapter 5 shows how calculus is connected to the geometric properties of curves and surfaces. The chapter also explains how to use these properties to program curves and surfaces.
- Chapter 6 introduces polynomials as a kind of digital clay—a building material for arbitrary functions with requested properties. The chapter also discusses the downsides of polynomial modeling and ways to mitigate them.
- Chapter 7 teaches how to craft custom curves with splines. The chapter explains the general idea of splines and proposes a few well-known approaches, such as Bézier splines and NURBS. It also gives you the necessary tool set to design your own polynomial splines with the specific properties you want.
- Chapter 8 expands on the discussion of projective transformations in chapter 4, introducing nonlinear transformations. The chapter also shows how to build and modify surfaces with such transformations.
- Chapter 9 introduces the basics of vector algebra from the geometric perspective. The chapter explains how vector products work, how they generalize, and how to use their geometric properties to solve real-world problems.
- Chapter 10 teaches about signed distance functions as one possible way to represent geometrical data. The chapter presents both the benefits and the drawbacks, explaining how to create a signed distance function from a contour of a triangle mesh and how to do Boolean operations, offsets, and transformations on these functions as though they were geometric objects.
- Chapter 11 discusses more popular ways to represent geometric bodies. Like the preceding chapter, chapter 11 discuss benefits, drawbacks, common operations, and conversion techniques for boundary representations and triangle meshes.
- Chapter 12 showcases modeling bodies with 3D images and voxels. The chapter also discusses common operations for this representation, along with conversion techniques.

The last three chapters should give you enough guidance to choose the best representation for your practical problem, whichever it might be.

The book progresses from purely theoretical material to the data representations and algorithms used in real-world applications. This progression is gradual, and later chapters rely heavily on earlier ones. I highly advise reading the chapters in their natural order, although if you're already familiar with calculus, vector algebra, or linear systems, you can safely skip the pertinent chapters; they propose only alternative perspectives on the topics, not unique and exclusive material.

About the code

This book contains many examples of source code both in numbered listings and inline with normal text. In both cases, source code is formatted in a `fixed-width font like this` to separate it from ordinary text.

In many cases, the original source code has been reformatted; we've added line breaks and reworked indentation to accommodate the available page space in the book. In rare cases, even this was not enough, and listings include line-continuation markers (➥). Additionally, comments in the source code were often removed from the listings when the code is described in the text. Code annotations accompany many of the listings, highlighting important concepts.

As the book progresses from theory to applications, the code examples progress from SymPy snippets to full-fledged algorithms:

- The SymPy snippets don't produce any curves or surfaces. They produce formulas, or, alternatively, other snippets of code, this time in a language of your choice. More on that later.
- The algorithmic examples do produce curves and surfaces, and as such, they need visualization. The specific details of visualization implementation aren't important from the geometric standpoint, so in the book itself, all the visualization code is omitted. It exists only in the code examples that you download separately.

All the code is in Python. There are no specific dependencies on modern Python features or obscure libraries, so any generic 3.x Python will do. There are no specific hardware requirements, either.

You can get executable snippets of code from the liveBook (online) version of this book at https://livebook.manning.com/book/geometry-for-programmers. The complete code for the examples in the book is available for download from the Manning website at https://www.manning.com/books/geometry-for-programmers and from GitHub at https://github.com/akalenuk/geometry-for-programmers-code.

liveBook discussion forum

Purchase of *Geometry for Programmers* includes free access to liveBook, Manning's online reading platform. Using liveBook's exclusive discussion features, you can attach comments to the book globally or to specific sections or paragraphs. It's a snap to make notes for yourself, ask and answer technical questions, and receive help from the author and other users. To access the forum, go to https://livebook.manning.com/book/geometry-for-programmers/discussion. You can also learn more about Manning's forums and the rules of conduct at https://livebook.manning.com/discussion.

Manning's commitment to our readers is to provide a venue where meaningful dialogue between individual readers and between readers and the author can take place. It isn't a commitment to any specific amount of participation on the part of the author, whose contribution to the forum remains voluntary (and unpaid). We suggest that you try asking the author some challenging questions lest his interest stray! The forum and the archives of previous discussions will be accessible on the publisher's website as long as the book is in print.

about the author

OLEKSANDR KALENIUK is the creator of Words and Buttons Online, a collection of interactive tutorials on math and programming. He works for Materialise NV as a senior software engineer specializing in geometric algorithms.

about the cover illustration

The figure on the cover of *Geometry for Programmers* is "Femme de l'isle de Naxia," or "Woman of Naxos Island," taken from a collection by Jacques Grasset de Saint-Sauveur, published in 1788. Each illustration is finely drawn and colored by hand.

In those days, it was easy to identify where people lived and what their trade or station in life was by their dress alone. Manning celebrates the inventiveness and initiative of the computer business with book covers based on the rich diversity of regional culture centuries ago, brought back to life by pictures from collections such as this one.

1 Getting started

This chapter covers

- Which programming domains rely on geometry
- Which part of geometry applies to programming
- The reason to start learning geometry today
- What you need to know to get started
- How to make SymPy do the math for you

Geometry is the branch of mathematics that stands behind game engines, computer-aided design applications, 3D printing frameworks, image processing libraries, and geographic information systems. As soon as there are curves, surfaces, or spaces, geometry is involved.

Normally, you don't have to be a geometry expert to use an application with curves and surfaces. Don't worry; this book won't convert you to one. But just as knowing the mechanics behind your car allows you to make the most of driving it, knowing the mathematics behind your tools will allow you to use them in the most efficient way possible.

How much geometry do you have to know to 3D-print a single model, for example? Well, none; you can figure out how to do that with trial and error. But what if you want to automate a 3D printing process for a continuous flow of models? This task is a programmer's job, and it brings in programmer considerations. You want your process to be both fast and robust. Every failed printing attempt means that you lose time and money, so trial and error simply won't cut it.

Normally, to print a model that came from computer-aided design (CAD) software, you need to convert the source model from smooth surfaces to triangles with a special library, ensure that the triangulated model is printable with an analysis algorithm, and turn the model into a list of contours with a slicing function. Only then you can start an actual printing machine. Each part of the process that precedes clicking a switch on a 3D printing machine is geometry wrapped in programming (figure 1.1).

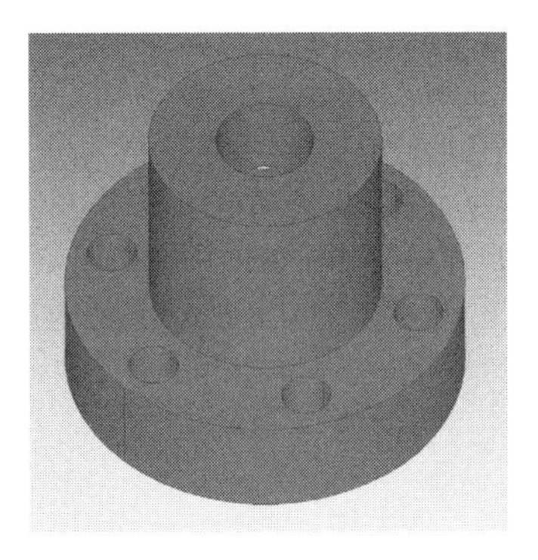
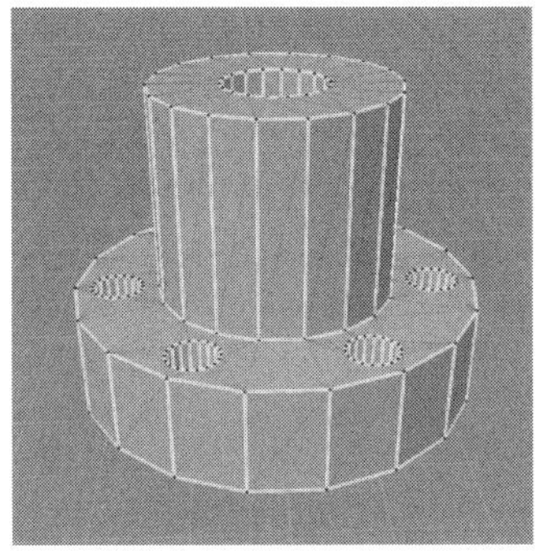
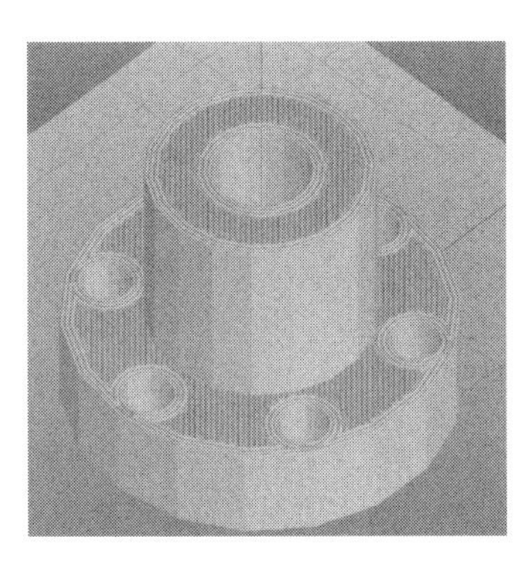

Figure 1.1 Left to right: A model made of smooth surfaces, the same model made of triangles, and the model made of stacked 2D contours

The most efficient surfaces-to-triangles conversion is one that produces something called a manifold mesh with no need for postprocessing correction. You can rely on automatic algorithms to figure out the process, of course, but you often get better results with manual control. The only downside is that to have this control, you have to understand what on Earth a manifold mesh is.

The algorithms that make meshes printable don't necessarily guarantee that the resulting triangles will be high-quality. Certainly, you can increase quality afterward, but you have to know how to assess quality and identify the algorithms that rely on it.

Turning a model into a stack of printable contours is an operation that takes time and can introduce errors. If it takes too long, can you trade some precision for speed? You can, but to do so, you have to know your mesh-reducing algorithms.

So although it's possible to work with modern software without having any considerable knowledge of geometry, every little thing you learn makes you more efficient in your job one way or another. 3D printing is only one example; the same goes for design automation, game programming, image processing, and so on.

1.1 Which parts of programming require geometry?

I've already mentioned a few domains that require geometry—game engines, CAD frameworks, and 3D printing toolkits—but the list doesn't end there. Virtual reality, augmented reality, computer vision systems, computer tomography, 2D graphics and animation, and data visualization all use the same mathematics (although in different quantities). The more you understand the mathematics, the more efficient you'll be in using and writing code for all those tasks. Generally, knowing the math behind your tools makes you a better programmer in any relevant domain.

But that statement is only a truism. Let's go through a couple of domains to see where geometry is and isn't useful:

- *Game engines*—In game engines, geometry is responsible mostly for object positioning, creating smooth transformations, rendering, and modeling physical interaction. Knowing geometry may not make you a better game designer, but programming visual effects in the most efficient way will make your game run smoothly on a wider range of gaming hardware.
- *Virtual/augmented reality*—The situation is more complex in virtual or augmented reality, because now you need to adapt the rendering to each user and consider interactions with real-world objects. Not all these problems are geometric, of course. Rendering calibration for a virtual reality set, for example, is an interdisciplinary problem that involves optics, hardware design, anatomy, and even neurology. Some neurological conditions simply prevent users from following the calibration guides—and in this case, knowing geometry can't possibly help.
- *CAD*—In computer-aided design, the applied geometry is used for precise surface modeling, computing the intersections of surface-bound bodies, and turning these bodies into sets of small chunks for finite element analysis. *Finite element analysis* is a way to simulate the models' physical properties programmatically, thereby saving time and money on real-world crash testing. The process isn't geometrical per se—it has more to do with differential analysis and computation—but converting models from one representation to another and back is a purely geometrical problem.
- *3D printing*—In 3D printing, geometry does everything but the printing itself: preparing the models, placing them efficiently on a printer's platform, building the support structure to ensure that the build will succeed, and then turning all the data into a list of printer instructions. In this domain, material science (such as metallurgy for metal printing or plastics chemistry) is every bit as important as mathematics.
- *Image processing*—In image processing, the applied geometry takes care of turning raw pixel data into smooth curves or surfaces, which is called *segmentation*, and positioning such objects in the image space, which is called *registration*. Transforming the images themselves and positioning them in space are also

geometrical problems. But a lot of discrete mathematics involved aren't related to geometry.

A topologist would say that because every set of objects and a set of relations between them constitute a topological space, relational databases and even source code are geometric objects. This understanding of geometry, however, is too broad to be practical. In this book, we'll stay focused on curves, surfaces, and their transformations.

Geometry supports and enables many programming domains. It may not always be the first violin in an orchestra, but it's often as integral and irreplaceable as a double bass.

1.2 What is geometry for programmers?

We all studied geometry in school. Geometry starts with lines and circles, rulers and compasses, and then moves on to axioms and theorems—lots of theorems. This kind of geometry, called *axiomatic*, is essentially mental calisthenics and unfortunately has little to do with programming (unless you're programming an app that teaches axiomatic geometry to children).

Geometry meets programming only when it becomes *analytic*—when points acquire coordinates, curves and surfaces get their formulas. Even then, some pieces of geometry knowledge remain more relevant to programming than others. Special curves such as cochoid, cissoid, and trisectrix have little significance in modern engineering, for example, because you can model them all with a general thing such as a nonuniform rational basis spline (NURBS for short).

Don't worry if these terms are unfamiliar. This book is designed to explain unfamiliar terms if you need them and omit them if you don't. We'll get back to NURBS in chapter 7, but from now on, you'll never see those three special curves mentioned anywhere in this book or probably in your professional career.

Analytic geometry is tightly connected to linear algebra—so tightly that you can use geometric intuition to learn about linear systems (chapter 3) or vector algebra (chapter 9). To understand curves and surfaces, you also need a small fraction of calculus. Chapter 5 covers all that you have to know; it introduces derivatives as a pragmatic geometric and not some abstract analytical tool. Similarly, chapter 6 introduces polynomials, which represent the digital clay you'll use later to sculpt curves and surfaces.

Speaking of curves and surfaces, chapter 7 introduces NURBS, which is currently the most popular tool for modeling curves and surfaces, and also explains the general approach to splines, allowing you to craft your own spline formulas tuned for the best possible performance.

Chapter 8 explains how to generate and deform surfaces, and chapter 4 shows how to translate, rotate, and scale geometric models with a single operation called matrix multiplication.

The last three chapters are dedicated to three ways to model the real world or even imaginary objects with geometry: signed distance functions, boundary representation, and 3D images. Different models work best in different contexts, so to be most efficient with your algorithms, you have to know all three representations, as well as how to convert a model from one representation to another.

To answer the question "Which part of geometry applies to programming?," everything that made it into the book applies: lines, planes, curves, surfaces, transformations, deformations, and geometric data representations.

1.3 Why not let my tools take care of geometry?

Often enough, modern engines and frameworks try to isolate you from mathematics. Creating in this environment is like driving a car with the hood welded shut. If something goes wrong, you can't fix it, and if you want to tune your car to go faster than the competitors', you'll have to get your hands dirty.

In 2013, I was working on an image processing tool that unbends curved pages from pictures to make them look flat. I created a prototype in C#, and it was slow. I thought about rewriting it in C++, because common wisdom tells us that C++ is faster than C#. But first, I went through the whole workflow with a profiler and found out that the bottleneck was in the initial image projection (figure 1.2).

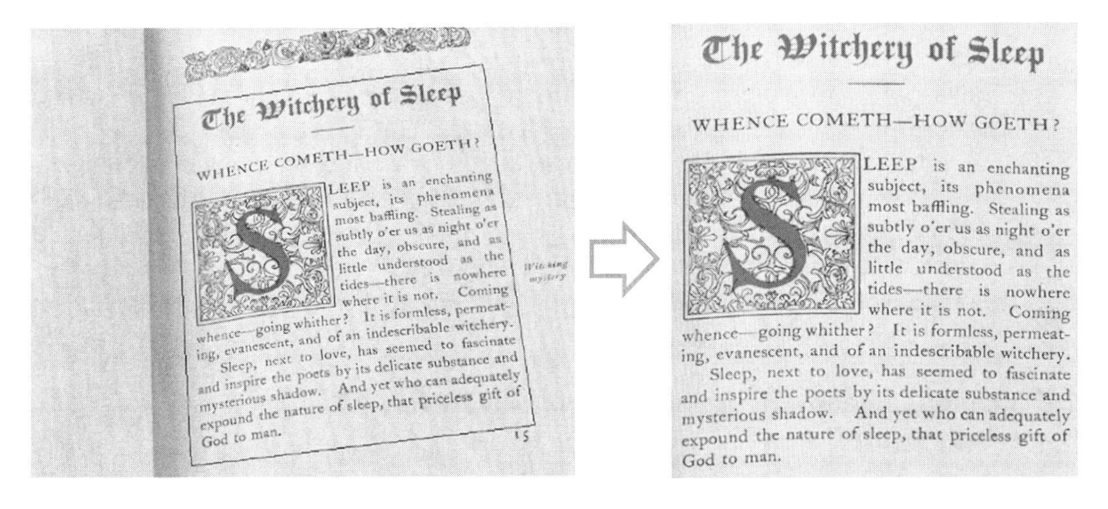

The Witchery of Sleep

WHENCE COMETH—HOW GOETH?

SLEEP is an enchanting subject, its phenomena most baffling. Stealing as subtly o'er us as night o'er the day, obscure, and as little understood as the tides—there is nowhere where it is not. Coming whence—going whither? It is formless, permeating, evanescent, and of an indescribable witchery.

Sleep, next to love, has seemed to fascinate and inspire the poets by its delicate substance and mysterious shadow. And yet who can adequately expound the nature of sleep, that priceless gift of God to man.

Figure 1.2 In 2D, creating a projection is essentially the same as transforming a quadrilateral into some other quadrilateral.

That discovery was an odd one, because this operation is supposed to be fast. All you need to project a single point is perform nine multiplications and six additions. That's it. Usually, you can use matrix multiplication to perform all these operations with a single function call (and you'll see how in chapter 4). Essentially, a geometric transformation is simple number-crunching; it's supposed to be fast.

But I was using a method from a standard library to do the transformation, and under the hood, it made no fewer than four conversions from one representation of a point to another. The process also had an empty constructor call and a method called `CreateInstanceSlow`. The computation itself was so fast that it didn't even show up in a profiler, but the overhead of all the conversions was causing the slowdown.

So I reimplemented the transformation right in the code, and it appeared to be more than a hundred times faster that way. The code was four lines longer, but there were no conversions, no constructors, and no slow instances anymore.

The main takeaway here is that when you know your math, you have the option to bypass inefficient routines you'd have to use otherwise. This option is a competitive advantage on its own. But you could also learn another lesson: distrusting mathematics makes your code slow.

Someone once decided that matrix multiplication is too complex to leave exposed, so they wrapped it with some kind of transformation function. Someone else decided that homogeneous coordinates is not an approachable concept, and they decided to wrap the projective-space point in an affine-space point converter. Then another someone else decided that the words *projective* and *affine* should be banned from the system, and they wrote another wrapper over them. All in all, the weight of the wrappers became a hundred times the weight of the thing wrapped. In the candy-bar industry, this result would have been scandalous.

These terms and concepts may seem scary to an unprepared mind, but the truth is, wrapping them with other terms and concepts doesn't make them go away. In fact, it only introduces more terms and concepts. Wrapping doesn't make things easier; learning does! Besides, these things aren't too complex to begin with.

In this book, you'll learn all you need to know about projective spaces, homogeneous coordinates, and other concepts that are generally considered to be obscure and unapproachable. You'll come to understand and trust the mathematics behind the code of game engines and CAD frameworks. You'll become more proficient as a user of geometry-related code, and you'll also acquire the knowledge to reimplement parts of it by yourself.

1.4 Applied geometry has been around forever; why learn it now?

Two empirical observations explain the technological landscape we live in today. First is Moore's Law, which says that the number of transistors on a chip roughly doubles every two years. Pragmatically, this observation means that year after year, computers get faster, smaller, and cheaper. This effect is limited in time, of course; there's a physical limit to how small a transistor could theoretically be, because there's no such thing as a subatomic transistor. We still observe the effect, though, even if it's not too pronounced. Moore's Law doesn't apply as much to PC central processing units (CPUs) anymore, but mobile devices are still getting faster, and graphics processing units (GPUs) are progressing quite well.

The second observation is Martin's Lawn. In 2014, Robert C. Martin estimated that for the past 40 years, the total number of programmers in the world doubled every five years. This effect is also limited in time, of course; at this rate, we'd run out of nonprogramming people by the middle of the century.

NOTE The *n* in *Lawn* isn't a typo. The observation was published in a blog post titled "My Lawn" at http://blog.cleancoder.com/uncle-bob/2014/06/20/MyLawn.html.

The decline of Moore's Law means that people now seek other ways to increase performance other than waiting until hardware gets faster. New architectures, devices, and business models are available all the time. NVIDIA's server-side GPUs, for example, aren't GPUs at all; they don't produce graphics so much as crunch numbers at an astonishing rate. Since 2016, Google has produced its own computer, but not a general-purpose one; it's a tensor-processing unit specialized for work in artificial intelligence. In the finance industry, where low latency is king, most of the super-quick computations are already done in bare metal.

And of course, cloud technology is a big game-changer—not because of the technology itself, but because of its business model. In the PC world, a user pays for a machine and then buys software that fits this machine. As a result, the software makers should prioritize compatibility over performance. The consumer pays for the run time, but if a product doesn't fit the user's machine, it won't sell.

With the cloud, things are a bit different. Now we in the software industry sell services, not software. We sell access to our software, which runs on someone else's machines. We pay for the machines in the cloud ourselves, so performance is important; the run time is our cost. But we also get to decide what these machines will be. We don't depend on the user's choice much anymore. As a result, we're free to optimize our code for highly specialized devices, and it pays to do so. To optimize algorithms that perform the service you sell, of course, you have to know how they work, and that's why it's important to know the math. You also have to be good at programming, and you have to understand the machine, too. But software frameworks will come and go, and hardware architecture will evolve; the only thing that will stand the test of time is mathematics.

To get back to Martin's Lawn, the demand for competent programmers has been higher than the supply for about half a century. This situation should end at some point. The market will saturate itself, and the competition will become much more serious. To have an advantage in this future competition, you should start learning now.

You can't learn hardware architecture that doesn't exist. You can't get good in a programming language that hasn't been invented. But you *can* learn geometry. It's been around for a few thousand years, and it will stay around for a few thousand more. Unlike with the software or hardware of the future, you can start learning geometry any day, and the sooner, the better. So why not today?

1.5 You don't have to know much to start

This book is called *Geometry for Programmers*, not *Programming for Geometers*, so it doesn't require any specific knowledge of advanced mathematics. The book doesn't teach you how to program either. All the code examples in this book are in Python, but you're more than welcome to use the language of your choice to do exercises or try things yourself. So what should you feel comfortable with to proceed with the book? You should have the following skills:

- *Reading and understanding snippets of Python code*—You don't have to be particularly good in Python programming. In the first half of the book, the Python snippets you'll face are essentially formulas in disguise. We'll use a Python library named SymPy to do the math for us. SymPy is a computer algebra system that, when used correctly, substitutes for years of training in symbolic computation. As we progress toward practical applications, we'll use Matplotlib to draw pictures and NumPy to do numeric computations. You don't have to be familiar with any of these libraries; I'll introduce them briefly later in the book.
- *Picturing things in your head*—The book has pictures, but they're only pictures. To develop your geometrical intuition properly, you need to see how objects interact and how they respond as their parameters change. The book can give you hints and directions, but to see the final "video," you have to use your imagination.

  Also, most of the material in the book is accompanied by links to interactive demos and tutorials. You don't have to rely on reading and programming alone; you can play with the concepts presented in ready-made environments.
- *Coding in whatever language you prefer*—Most chapters have exercises that nudge you to write code in the language of your choice. It doesn't matter which language you use; the geometry is the same in Python, JavaScript, and Rust.
- *Understanding elementary math*—I'll teach you some common mathematical concepts as needed in this book, but you're still expected to understand the basics, such as what equations and coordinate planes are, and what constitutes a function in the mathematical sense.

  This book requires no exposure to higher math. If you're already familiar with linear algebra or calculus, this knowledge will help you in the related chapters, but it isn't mandatory. The nature of programming geometrical entities, however, implies doing a lot of symbolic computations—turning formulas you know into formulas you want. Normally, this task requires considerable training in algebra, but we'll cheat by using a computer algebra system. Nowadays, making a computer do your algebra for you is much simpler than it sounds. We'll get to that task in the following section.

1.6 SymPy will do your math for you

I have one thing to confess: I'm really bad at doing math. I love math, but the feeling isn't mutual.

I love the language of mathematics; I love its explanatory power; I love the feeling of reinvention when I learn new concepts and try them in practice. But when it comes to doing symbolic computations, I can't do a dozen in a row without messing up plus and minus at least four times.

I've always been this way. When I showed my first academic paper to the editor, he took a brief look, pointed at the first formula, and said, "You should have said minus here, not plus." As it turned out, I didn't mess up the sign only once; I did it several times.

This situation complicated my professional career as an engineer, too. Luckily, a friend who is good at math saw me struggling with my equations and asked, "If you hate doing it so much, why do you do it?" "Well, I'm getting paid for it," I said. "Sure," he replied, "but why do you do it by hand?"

In my friend's world, doing math by hand is fine only if you enjoy doing it that way. Professionally, he does math with SymPy, which is a library for symbolic mathematics that aims to become a full-featured computer algebra system.

Both symbolic mathematics and computer algebra may seem too science-y, so let me explain the process the way my friend explained it to me: "You import SymPy; then you feed it your equations and say 'Solve.'" That's it.

Let's try a simple math problem. A train departs from Amsterdam to Paris. Another train, twice as fast, departs from Paris to Amsterdam. The distance between Paris and Amsterdam is 450 km, and the trains meet in an hour. What are the trains' speeds?

When you rewrite this problem in mathematical notation, you get a pair of equations. Let's call the speed of the train departing from Amsterdam V_a and the speed of the Paris train V_p. Also, let's store the speeds in kilometers per hour. Now, the fact that these two trains make 450 km in an hour together turns into this line,

$$V_a + V_p = 450$$

and the fact that the Paris train is two times as fast as the train from Amsterdam turns into this line:

$$V_p = 2V_a$$

These lines give us a two-piece system of equations. I'm sure you can solve the problem in your head easily, but please don't. Let SymPy do it for you. You don't even have to install SymPy on your computer; go to SymPy Live at https://live.sympy.org, and try it all there.

SymPy is an algebra system that does your algebra for you, but it's also a Python module. Being a Python module, it plays by Python rules, which involve a few implications. First, obviously, you have to import it to use it:

```
from sympy import *
```

Next, you have to define symbols. This task wouldn't be necessary if SymPy had its own language, but it reuses Python syntax, and in Python, you have to define symbols before using them. So we have to play by those rules. Let's define our speeds as `Va` for the Amsterdam train and `Vp` for the Paris train:

```
Va, Vp = symbols('Va Vp')
```

Next, let's write the equations one by one, starting with the one for the speed difference. Again, SymPy is a Python module and in Python, `=` is the assignment operator, so we can't write `Vp = 2Va`. Instead, we write all the equations as though they have `0` on the right side. We move everything from the right side to the left while changing its sign. So `Vp = 2Va` becomes

```
Vp - Va * 2,
```

Now the equation for "the trains meet in an hour" looks like this:

```
Va * 1 + Vp * 1 - 450
```

(You don't have to write `* 1`, but let it stay for a while. We'll substitute it with something else in a minute.)

Next, ask SymPy to solve these equations:

```
solution = solve([
    Vp - Va * 2,
    Va * 1 + Vp * 1 - 450
], (Va, Vp))
```

We want to know `Va` and `Vp`, so after the equations, we write these symbols in a separate tuple. This tuple may look like needless repetition, but we'll see why it's necessary when we introduce more symbols.

Finally, we want SymPy to give us the answer. Because it uses Python, we can use a simple `print` command:

```
print(solution)
```

And that's it. The whole program looks like the following listing (ch_01/meet_sympy_numeric.py in the source code for this book).

Listing 1.1 Numeric solution in SymPy

```
from sympy import *

Va, Vp = symbols('Va Vp')

solution = solve([
    Vp - Va * 2,
    Va * 1 + Vp * 1 - 450
], (Va, Vp))

print(solution)
```

Va is the speed of the Amsterdam train, and Vp is the speed of the Paris train, both in kilometers per hour.

When run, the program prints this:

```
{Va: 150, Vp: 300}
```

Nice. But that result isn't the math we should be excited about. These are numbers, and we can get numbers with a calculator. What about letters and formulas? Can SymPy produce a generalized equation that will work for any distance, time, and speed ratio?

Yes, it can. SymPy can do math with letters. We'll give it more symbols and fewer numbers. Let's change all the constants we have in the code, including `1`, to symbols and rerun the program, as shown in listing 1.2 (ch_01/meet_sympy_symbolic.py in the source code for this book).

Listing 1.2 Symbolic solution in SymPy

```
from sympy import *

Va, Vp, Vpx, D, t = symbols('Va Vp Vpx D t')

solution = solve([
    Vp - Va * Vpx,
    Va * t + Vp * t - D
], (Va, Vp))

print(solution)
```

Now Vpx is a ratio, Vp/Va (it used to be 2); D is the distance in between Paris and Amsterdam in kilometers (which was 450); and t is the time before the trains meet in hours (previously, 1 hour).

Note that writing `(Va, Vp)` after the equations makes more sense now. We want to compute speeds from time symbolically, but we can also potentially compute time from speed. We have to be explicit about what we're computing. The symbols in the tuple are the symbols we want to have solved in terms of other symbols. In our case, this solution is

```
{Va: D/(Vpx*t + t), Vp: D*Vpx/(Vpx*t + t)}
```

Yes! We have a generalized solution for the two-trains problem. They can travel from New York to Hong Kong now; we wouldn't care. With symbolic formulas, we have the problem solved for every distance, every time, and every speed proportion!

Now, because we're programmers, let's turn the formulas we've computed into code. I'm sure you can do this by hand easily, but again, let SymPy do it for you. SymPy has the job covered. If you want SymPy to write code for you, all you need to do is ask! If you need some Python code, use the `pycode` function

```
print(pycode(solution))
```

and SymPy will print this:

```
{Va: D/(Vpx*t + t), Vp: D*Vpx/(Vpx*t + t)}
```

Well, yes, this result is exactly the same as before, but only because the formulas we had were already valid Python code. If they contained any language-specific operators and functions, this translation feature would have appeared much more impressive.

We can ask SymPy to produce code in JavaScript (`jscode`), Julia (`julia_code`), FORTRAN (`fcode`), C (`ccode`), Octave and Matlab (`octave_code`), and even Rust (`rust_code`).

SEE ALSO SymPy's online documentation is great. If you want to learn more about code generation, please visit http://mng.bz/ElmO.

But wait—there's more! Let's say you want to publish your equations. Mathematicians love to publish things in LaTeX, so why don't we try it as well? With SymPy, you don't have to translate your solution into LaTeX by hand. Use the `latex` function instead of `pycode`:

```
print(latex(solution))
```

The result will be perfectly printable LaTeX:

```
\left\{ Va : \frac{D}{Vpx t + t}, \  Vp : \frac{D Vpx}{Vpx t + t}\right\}
```

And in print, the formulas will look like this:

$$\left\{ Va : \frac{D}{Vpx t + t}, \ Vp : \frac{D Vpx}{Vpx t + t}\right\}$$

Nice! In this book, however, we'll use mathematical notation sparingly. Remember, it's *Geometry for Programmers*, not *Programming for Geometers*.

A small recap: SymPy is a Python library that does your math for you. To make it do so, you have to import it, declare the symbols for your equations, combine your equations into a system, and run `solve`. If possible, SymPy will solve equations numerically; if necessary, it will solve them symbolically, writing a generalized solution with symbols in it. Then you can turn this solution into code in a language of your choice or even into a publishable LaTeX formula.

Now you know most of what you'll ever need to be prolific with SymPy. You'll get the rest on the go. Congratulations! You spent maybe 15 minutes reading this introduction, and you can already add a computer algebra system to your résumé. Isn't that a good start?

Summary

- Applied geometry is geometry that has applications in computer graphics, animation, CAD, 3D and 2D printing, augmented and virtual reality, computer vision, and image processing in general.

- You can start learning geometry at any time, but now is probably the best time. Your learning will pay off immediately, if you already work as a programmer of CAD applications or games, or in the future, as a long-term, nonperishable investment in your education.
- SymPy is a Python library that does your math for you. It can do some number-crunching, but its true calling is symbolical computations. It turns formulas you have into formulas you want. Moreover, it can turn formulas into source code in the programming language of your choice and prepare them for publication as a bonus.
- Learning applied geometry doesn't require rare skills or knowledge. We'll use SymPy to do our math for us as a shortcut.

2 Terminology and jargon

This chapter covers

- The similarities and differences between numbers, vectors, and points
- The informal terminology of triangles and functions
- Equations of lines and planes
- The function types you will often meet in practice
- The shortest possible introduction to matrix algebra

Every profession has its own language, a family of terms and concepts bound together by a web of relations. Applied geometry has its own language, too; it consists of rigid definitions from mathematics and more flexible jargon coined to simplify communications. For an outsider, however—even a mathematician or programmer—this combination of formal and intuitive naming creates a barrier.

This chapter gives you an introduction to the basic language of applied geometry. It covers a lot of new words, but don't worry; at this point, you don't need to develop a deep understanding of every one of them. This level of understanding comes from practice, and we'll have plenty of that in the following chapters.

By the end of the chapter, you'll be comfortable with terms such as near-degenerate triangle, nonmanifold mesh, and continuous deformation field. This familiarity should help you not only go farther through this book, but also effectively seek more knowledge in libraries' documentation, code comments, or online communities.

2.1 Numbers, points, and vectors

Numbers, points, and vectors are three separate concepts, not generally interchangeable. They do have a lot of similarities, however. Recognizing these similarities not only makes learning these concepts easier, but also boosts mathematical intuition, eventually enabling you to write highly efficient code.

2.1.1 Numbers

Historically, every time people were unhappy with the numbers they knew, they extended them by inventing even more numbers of a different kind.

Numbers in mathematics

It all started with *natural numbers*, numbers such as 1, 2, and 3. Now they have their names and even a special letter to denote them: $\mathbb{N}$. But 5,000 years ago, these numbers were the only ones discovered.

Sumerians had an advanced numeric system, which was positional, pretty much like the one we use today. In a positional system, the value of a digit depends on its position. In 101, for example, the digit 1 means both a hundred and one. A zero in the second position doesn't add anything, but it occupies the position so that the number doesn't collapse to 11.

Positional systems need a zero—not necessarily a number in its own right, but a digit. So the Sumerians extended the natural numbers with a zero forming a new entity: $\mathbb{N}_0$. Today, $\mathbb{N}_0$ is standardized as natural numbers by ISO 80000-2. Heads up: the numbers that start from one $\mathbb{N}_1$ are sometimes also called natural numbers.

At some point, people invented bookkeeping and started counting debt. So people in ancient China and then India enriched their numerical systems with negative numbers such as -1, -2, and -3. Combined with negative numbers, natural numbers with zero form a new type of numbers called *integers*: $\mathbb{Z}$.

Integers are fine for counting rocks and sheep—things you either can't or shouldn't cut in pieces—but for measuring sacks of grain and barrels of oil, ratios such as $\frac{1}{2}$ and $\frac{3}{4}$ are too useful to be neglected. Ancient Greeks even thought that every number can be represented as a ratio of integers if the integers are long enough. Today, these representable-as-a-ratio numbers are called *rational*: $\mathbb{Q}$.

Although there were other entities that weren't rational, such as π, they were denied citizenship in the realm of numbers until the Age of Enlightenment. European mathematicians found out that π isn't an outcast: a whole class of entities simply doesn't fit into the set of rational numbers. You can't even write the square root of 2 as a ratio, for example. These entities are still numbers, though, and you can still treat them as such. Today, these numbers are called *irrational*.

Eventually, mathematicians combined rational numbers and irrational newcomers such as π, e, and the square root of 2 in a new set. They called these numbers the *real numbers*: ℝ.

But this development isn't the end of the story. While trying to solve polynomial equations in radicals, using only +, –, ×, /, and the root function, mathematicians presumed that although you can't take a square root of a negative number in real numbers, perhaps in some other numbers, you can. Once again, the whole numeric system needed an extension.

The extension came from introducing so-called *imaginary numbers*, which are essentially a quantity of square roots of –1. This concept is a genius idea, really. You can't compute a square root of –2 in real numbers so . . . don't. Make a "square root of –1" an accountable entity on its own, and carry on.

This leads us to complex numbers. A *complex number* consists of two parts: a real part, which is a real number, and the imaginary part which is, well, imaginary.

Then mathematicians invented *quaternions*, numbers that consist of 4 parts; *octonions*, numbers consisting of 8 parts; and *sedenions*, the ones that consist of 16 parts. All these numbers are called *hypercomplex*. As you can see, every time mathematicians faced a new practical problem that couldn't be solved with the numbers they had, they didn't give up. Instead, they extended the numbers by introducing a generalization of the then-current numeric system. Integer numbers generalize natural numbers, meaning that natural numbers become a special case of integers; real numbers generalize integers; complex numbers generalize real numbers, and so on.

SEE ALSO This chapter provides a brief exposition of concepts and ideas that we'll rely on throughout the book. Its main goal is merely an introduction of new terms, not necessarily a comprehensive explanation of every one. Ideally, though, every idea or concept deserves a chapter of its own. At the same time, beautiful explanations are available from other sources. Case in point is "A Visual, Intuitive Guide to Imaginary Numbers" from Better Explained, at http://mng.bz/Opzw. Sources like these will be referenced in the "See also" callouts, so please don't hesitate to visit them if you feel that the material in this book isn't enough.

Numbers in programming

Programming is slightly different from mathematics. Every piece of computer memory is a finite amount of elements hosting only ones and zeroes. In computing, we don't really have infinite sets like natural numbers or continuous entities such as real numbers. We can't afford true generalization in programming, so instead, we must resort to modeling.

A 32-bit signed integer type, for example, is an array of 32 bits we use to model some pragmatic subset of genuine integer numbers. This model comes with a limitation: as you can write only 100 different numbers with 2 decimal digits, you can write only 4,294,967.296 different numbers with 32 binary digits. You simply can't fit all the infinite amount of integer numbers into a finite amount of bits.

You may object that some languages—Python and Haskell, among others—have "unlimited" integers from the box, meaning that you can add numbers as long as you want and never get an integer overflow. That's true, but the larger a number becomes, the more CPU power it takes to operate on it. Large numbers take more RAM, use more cache space, take more time to go through the network, and so on, which limits the use of "unlimited" integers in practice. So quite often, even in Python, we willfully resort to limited integers: integers that occupy only a predefined chunk of memory, usually 32 or 64 bits.

The real problem is that no single standard regulates all the aspects of how computers implement integer numbers. How do you write negative numbers in memory? How do you store the operation result? Do you interrupt on overflows? Which byte comes first: the lowest one or the highest one?

There are different standards, conventions, and practical considerations for each of these questions, but there's rarely consensus. These unresolved ambiguities bring out the most unpleasant consequences. In MSVC C++, for example, this expression is true:

```
uint16_t(50000) * uint16_t(50000) == -1794967296
```

For historical reasons, the default integer type of an expression in C++ is the signed integer type unless at least one of the operands can be represented only in the unsigned one. In MSVC, the default integer type is 32-bit, and `uint16_t` is the unsigned 16-bit integer—which can, of course, be represented in the 32-bit signed form. But when you multiply two 32-bit signed integers together, the result may overflow, and in our case, it does.

The lack of a universal specification for integer numbers makes them dangerous to work with—less so in Python, as you lose only performance, and more so in C++, because you gradually lose your mind. If your program passes unit tests and integration tests but fails manual testing, integer overflow is the usual suspect. At some point, we were thinking about making a "_ days since the last integer overflow" board at the office; that's how often integers cause trouble.

> **SEE ALSO** If you're interested in learning more about integer-related defects in C and C++ code, please take a look at *Secure Coding in C and C++*, 2nd ed., by Robert C. Seacord (O'Reilly, 2013).

Integer numbers are cumbersome for programming. But what about real numbers? Well, there are no real numbers in programming. They can't possibly exist, and they don't.

Think of it this way. Every piece of computer memory has a limited number of bits, and any continuous interval hosts an unlimited amount of real numbers. You can't dedicate sequences of bits to represent every possible real number. So once again, you have to compromise. You have to emulate real numbers with some good-enough model.

The closest thing to real numbers we can afford is the rational numbers made of "unlimited" integers. These rational numbers don't cover all the real numbers—you can't write π in this model, for example—but they're good enough for most of the computations that require precision over performance. Like "unlimited" integers, "unlimited" rational numbers get slower and slower as the computation progresses, because every operation makes them larger.

There's another way to emulate a real number, and it's fast and simple. Instead of writing a real number, you write an integer number of some preagreed real values. As with integer numbers or cents, you can emulate realish-looking numbers such as $9.99 (999 cents) and $100.00 (10,000 cents). This is the same as putting a decimal point in some specific position of an integer number and keeping it there through all the operations. So this model of real numbers is called *fixed-point numbers*.

The downside of fixing the point is that if you suddenly need more digits to the right of the decimal point than you agreed on beforehand, you can't have them in this model. If you need a percentage of a cent to compute interest with more precision, for example, $9.99 becomes 99900 cent percent. To operate on this representation, you have to introduce another model with more digits and agree on conversion rules.

This process gets cumbersome with every new model, and at some point, this "change model when needed" thing starts asking for automation. So the most common model that emulates real numbers today is the so-called *floating-point numbers*. In this model, the position of the decimal point is written down along with the integer value. The decimal point changes its position automatically as the computation goes on, which makes the decimal point "floating."

This floating-point model is convenient to use but, as you might imagine, rather complex to govern. Consider this fact: because we don't have infinite precision, we still truncate to some kind of integer, so the result of any operation depends on the floating-point positions of operands. Operations shift the floating point, so the precision of operands depends on the previous operations the operand had to go through, which brings up one of the most unpleasant implications of floating-point numbers:

```
(a+b)+c ≠ a+(b+c)
```

In plain words, unlike in real numbers, in floating-point numbers, the exact result of the sum depends on the order of operations. Pragmatically, this means that the optimizations compilers do for you, such as turning `a+b+c+d` into `(a+b)+(c+d)`, may or may not affect the precision of your computation.

In some languages, you have some control of the optimizations the compiler does for you. In Fortran, when you write

```
a/b/c
```

you let your compiler turn it into a slightly faster thing:

```
a/(b*c)
```

But this operation is, strictly speaking, not the same. It is the same in math, but in floating-point numbers, this optimization may or may not introduce a small error. If you want the compiler to retain the expression intact, you write it like this:

```
(a/b)/c
```

So technically, in Fortran's floating-point numbers, even `(a/b)/c` ≠ `a/b/c`. This isn't the oddest thing yet. Floating-point numbers also have two values that are interpreted as zero. One value is positive, and the other is negative. The rationale is simple. Because floating-point numbers are inherently imprecise, the zero may not actually be a zero. It might be an extremely small number. Like the half-penny in the "integer numbers of cents" model, a small number, when it can't be written in our model due to the lack of precision, becomes a pseudozero instead. But even though this small number has no nonzero integer digits, it's still entitled to its sign.

NOTE Having both +0 and -0 in one system is rarely helpful in practice, but you still have to be aware of their existence. I once had a problem when the test system reported a bug and I couldn't quite reproduce it with a unit test in C++. It took me way too long to realize that the test system was comparing strings, not numbers. It was complaining about `"-0"` being not the same as `"0"`, and of course, in the text representation, they aren't. In C++, however, `0.` and `-0.` are treated as the same value, so my unit test was always green.

Floating-point numbers also have two special values for the positive infinity and the negative infinity, and as with zeroes, they're not real infinities; they're used to denote the numbers that are too large to fit into the model.

What's most peculiar, the floating-point model may host values that aren't numbers at all. These values are called *not-a-numbers*—NaNs for short—and can be used to carry error codes or any other data that should conceptually be embedded in your computation.

So yes, floating-point numbers are tricky, but at least they're well standardized. An IEEE 754 Standard for Floating-Point Arithmetic does a great job of covering many of the computational aspects, so that even if this model looks insane in some of its aspects, it looks consistently insane in all possible languages and platforms.

Keep in mind that as a programmer, you have the duty and the privilege of selecting an appropriate model for the computation. There are "unlimited" integers; there are rational numbers to substitute for real ones; there are arbitrary length floating-point numbers; and there are fixed-point numbers if you prefer speed over flexibility. Pick what you need, not what your language of choice gives you as the default.

SEE ALSO If you want a brief visual explanation of floating-point numbers, see Fabien Sanglard's "Floating Point Visually Explained" post at http://mng.bz/DZyw. And if you want a comprehensive paper, the classic "What Every

Computer Scientist Should Know About Floating-Point Arithmetic," by David Goldberg, is what you're looking for. You can find it at http://mng.bz/lJ08. Throughout his book, we mostly meet the limitations of the floating-point model in practical examples, such as in chapter 3 when intersecting a ray with a triangle and in chapter 8 when making a bitmap into a continuous and smooth function.

Let's switch from decimal points to geometric points. A point in geometry doesn't have width or weight, or any property other than its location. In 2D, we denote a point with a pair of real numbers, called *Cartesian coordinates.* This isn't the only possible way to write points with numbers, but it's the simplest and most pragmatic one.

We usually place the first axis horizontally from left to right and the second vertically from bottom to top. The conventional names for these axes are x and y. Figure 2.1 shows a point a with coordinates (0.5, 1.5).

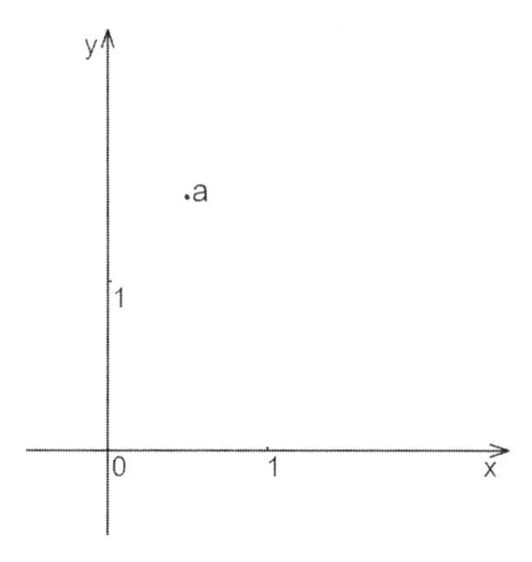

Figure 2.1 A point in Cartesian coordinates: *a* = (0.5, 1.5)

This convention isn't written in stone, however; it's something that people agreed on a long time ago. Sometimes, it's practical to follow another convention. Coordinates on a screen, for example, usually start from the top-left corner, not the bottom-left corner. Moreover, when we address a screen, we address not real points with coordinates in real numbers, but pixels, which are essentially color-containing cells in a grid that represents a picture—picture cell, pic-cell, pixel. Unlike geometric points, pixels have widths and heights, and they're addressed by integer coordinates. On a 1-to-1 scale, our point a (0.5, 1.5) should correspond to a pixel [0, 1], as shown in figure 2.2.

Sometimes, even for bitmaps, we use x for the horizontal axis and y for the vertical one. But another convention comes from matrices: the i-j row-column convention. The i is the row index, and j is the column index. Note that this notation is inverse of x-y. The j index corresponds to the horizontal axis, which is usually x, and the i index is vertical, which is usually y.

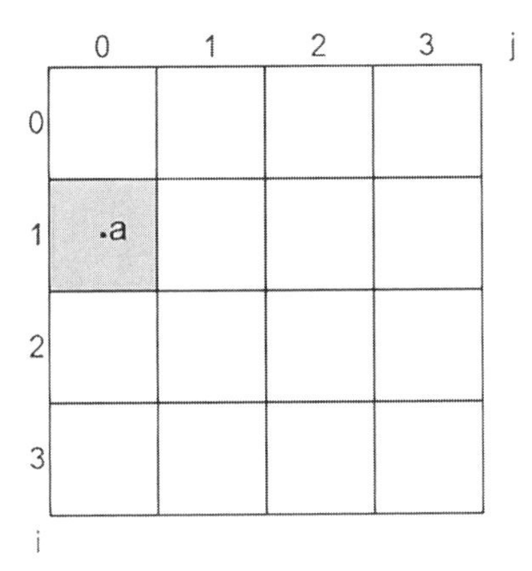

Figure 2.2 A pixel isn't a point, but a pixel contains point *a* and an infinite amount of other points.

Please keep in mind that people mistake *x* for *y* and *i* for *j* all the time. But please know and respect the conventions yourself, which will save you and your colleagues a lot of effort.

> **SEE ALSO** The difference between mathematical points and pixels is pronounced in computer graphics. Two interactive tutorials from Red Blob Games do a good job of showcasing the differences: "Line Drawing on a Grid," at http://mng.bz/BlQ8, and "Circle Fill on a Grid," at http://mng.bz/dJaO.

A point is a basic concept in geometry, so it's applicable in every chapter of this book. Pixels are featured explicitly in chapter 8 when we discuss image interpolation and in chapter 12 when we learn about representing geometric data as 3D images.

2.1.2 Vectors

Somewhat similar to a point, a *vector* is a direction with length. Sometimes, vectors are even written down as points. The vector that points toward point *a* is written as (0.5, 1.5), as shown in figure 2.3.

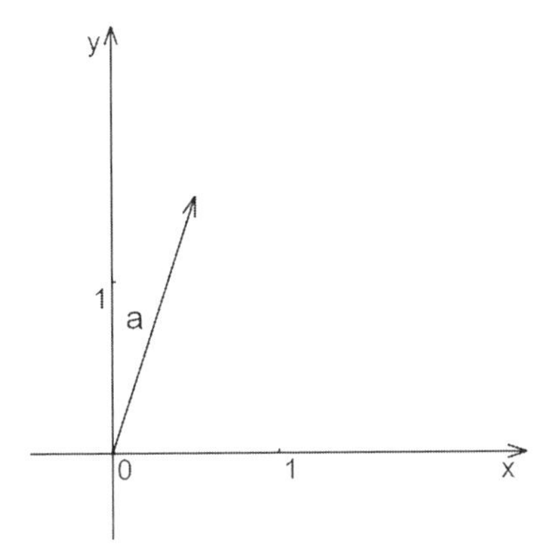

Figure 2.3 The vector a pointing at point *a* = (0.5, 1.5)

But the terms *vector* and *point* are not synonymous! A *point* is something that geometric objects are made from, and a *vector* is something that shows where things are. A vector is about showing a thing, not being a thing.

There are vectors that denote points, however, and they're sometimes called *point vectors*. We draw them as though they're arrows starting at (0, 0) and ending at the points they represent. It's only a convention, though. A vector doesn't have origin—only length and direction.

The vector **a** towards point *a* is 2D. But in "vector as a point" notation, adding a dimension is as simple as adding another number to the tuple. (0.5, 1.5) is a 2D vector example, and adding one more number to the tuple makes the vector 3D: (0.5, 1.5, 0.0).

NOTE We can have vectors of any dimension, but in practice, we rarely have to go beyond 3D.

Unlike points, vectors can be treated as numbers. You can add them (figure 2.4):

$$(a_1, a_2) + (b_1, b_2) = (a_1 + b_1, a_2 + b_2)$$

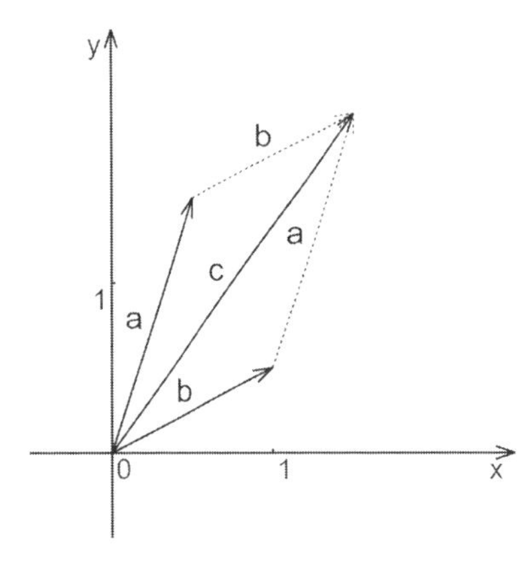

Figure 2.4 Vector addition: a = (0.5, 1.5), b = (1, 0.5), c = a + b = b + a

And you can multiply vectors by numbers (figure 2.5). This operation is essentially scaling vectors. That's why real numbers in the context of vector operations are called *scalars*:

$$(a_1, a_2) \times d = (a_1 \times d, a_2 \times d)$$

Because we can add and scale vectors, we can make vectors out of other vectors. Let's introduce a pair of vectors, **i** and **j**:

$$i = (1, 0)$$

$$j = (0, 1)$$

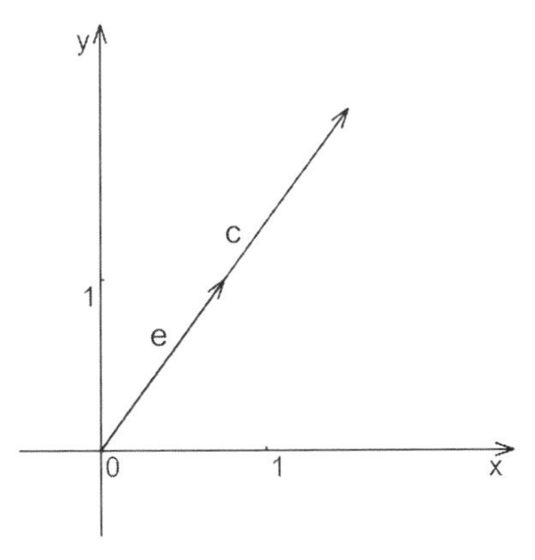

Figure 2.5 A vector by scalar multiplication: c = (1.5, 2), e = 0.5 * c

These vectors are called *basis vectors*, and they form a so-called *orthonormal basis*. We'll see what this term means in a moment. In this basis, the vector **a**, which is (0.5, 1.5), can be written as the sum of scaled **i** and **j** (figure 2.6):

$$a = (0.5, 1.5) = 0.5\mathbf{i} + 1.5\mathbf{j}$$

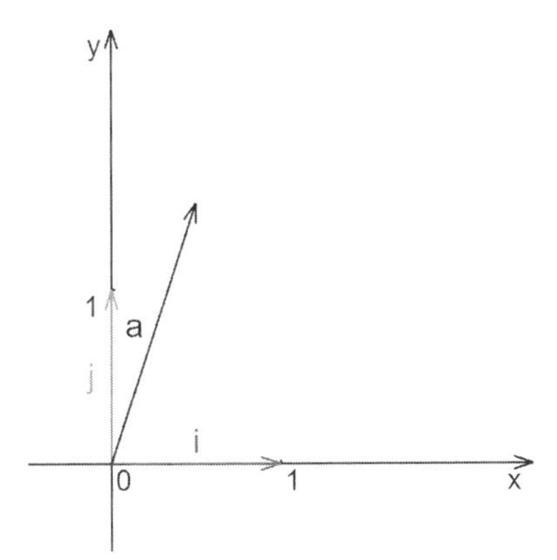

Figure 2.6 A point vector a written in orthonormal basis as 0.5i + 1.5j

NOTE There are other well-respected notations for basis vectors, such as $\mathbf{e}_1$=(1, 0) and $\mathbf{e}_2$=(0, 1). SymPy, however, uses **i**, **j**, and (for 3D) **k**, so we'll stick to that notation.

You might think that writing vectors in this way is unimportant because the vector as a sum of basis vectors is no different from a tuple of coordinates. If anything, the tuple is more compact and easier to program. With tuple notation, in programming terms, a vector is an array of numbers.

And you're right. Most of the time, we're fine with tuples. But this "vector as a sum of basis vectors" notion opens new possibilities. You see, basis vectors don't necessarily have to be (1, 0) and (0, 1).

The basis vectors we introduced as (1, 0) and (0, 1) are convenient, though. First, they form a right angle with each other. They're perpendicular or, to use another word, *orthogonal* to each other. Also, they both have a length of 1, and this property is called *normalization*. The vectors are *normalized*. Both properties together make the basis *orthonormal*. But there could also be nonorthogonal and non-normalized bases.

Don't worry if the words won't stick in your head; the words aren't too important. The most important idea is that the basis may change from context to context. And if a context demands basis vectors to be of different lengths or arbitrary directions, we can live with that.

In chapter 3, we'll write a ray-cast check to see whether a 3D ray intersects a 3D triangle. To make this check quick, we'll go from the usual orthonormal basis of the 3D space to the space created by the triangle edges and the ray. The space will also be 3D space, but its basis won't be orthonormal. We'll use this new basis to simplify the check and gain some performance.

SEE ALSO This book is about geometry, so its introduction to adjacent topics such as vector algebra is minimal. But a comprehensive, free, fully interactive linear algebra book is available online for free, and its chapter on vectors is fabulous! It starts with a brick game and goes through all the important concepts with figures you can play with. How cool is that? You can find this book at http://mng.bz/rdED.

Like the point, the vector is a basic concept that I'll refer to in every chapter. Moreover, chapter 9 is dedicated to the geometry of vector operations.

2.1.3 Section 2.1 summary

Numbers in mathematics can be integer, real, or complex, and in programming, we don't have any of them. We have only models that usually work well enough. Choosing the right model for the computation is part of the programmer's job.

Points are what curves and surfaces are made of. Points on a coordinate plane, or a coordinate system in general, could be written as tuples of real numbers. There are different conventions about how to write them in different contexts; please know and respect these conventions.

A vector consists of a direction and a length. Point vectors show where points are. Like points themselves, they could be written as tuples of real numbers, but like numbers, they could be used in mathematical expressions. A vector could be written as a sum of scaled basis vectors.

2.2 Vertices and triangles

I'm certain that you know what a triangle is. But less-known definitions and jargon are used sporadically in code comments and documentation, so let's concentrate on them.

2.2.1 Being pedantic about triangles

Do you know the difference between a triangle point and a triangle vertex?

In figure 2.7, A, B, and C are the triangle's *vertices;* they define the triangle. A triangle may have only three vertices, and you can't build a triangle unless you have all three. A, B, C, D, and infinitely more points from inside the triangle are the triangle's points. Any point that belongs to a triangle, including the triangle's interior surface, is a triangle's point.

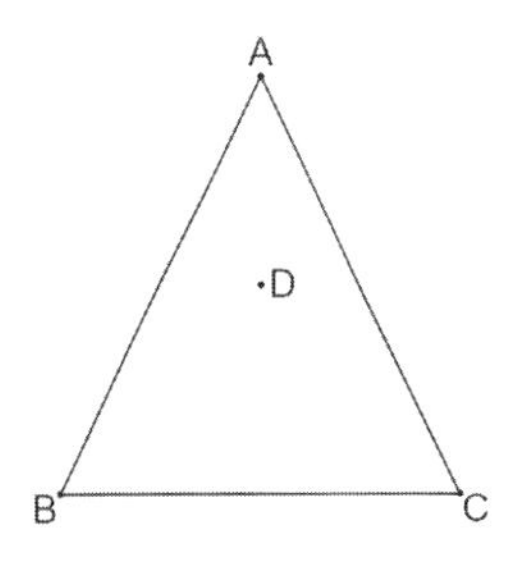

Figure 2.7 A, B, C, and D are all points but only A, B, and C are called vertices.

I mention this difference because in practice, these terms are often used interchangeably, but this practice only causes needless confusion. A vertex is definitely a point, but a point isn't necessarily a vertex.

While we're being pedantic, do you know the difference between a point *on* a triangle and a point *in* a triangle? In figure 2.8, point D is *on* the triangle, but it's not *in*. The edge AC doesn't belong to the triangle's interior surface, but it's part of the triangle. The difference between a vertex and a point, or the difference between in and on, may not seem to be dramatic. Still, following conventions like these consistently makes your code and documentation easier to read and understand. And because it doesn't cost you any additional effort to follow these conventions, there's no good reason not to follow them.

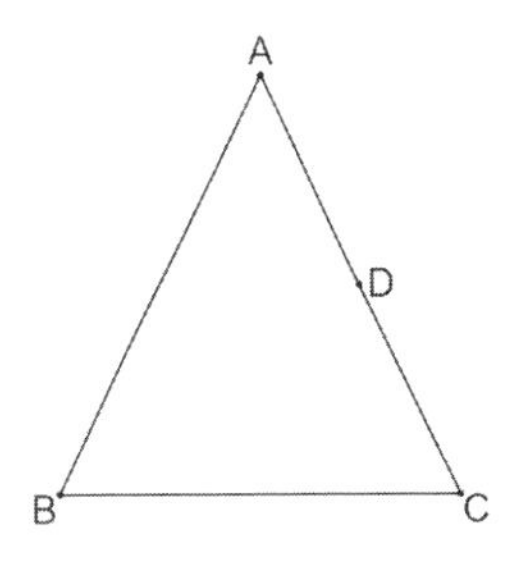

Figure 2.8 *On a triangle* and *in a triangle* have different meanings. Here, point D is on the triangle but not in it.

SEE ALSO If you need a refresher on triangle properties in general, please see Mathigon's "Triangles and Trigonometry" course at http://mng.bz/VpD0.

2.2.2 Triangle quality

Trick question: which one of the objects in figure 2.9 is not a triangle?

Technically, the third object isn't a triangle. But in programming, it represents the triangle type that causes the most trouble: a degenerate triangle, a special case in which

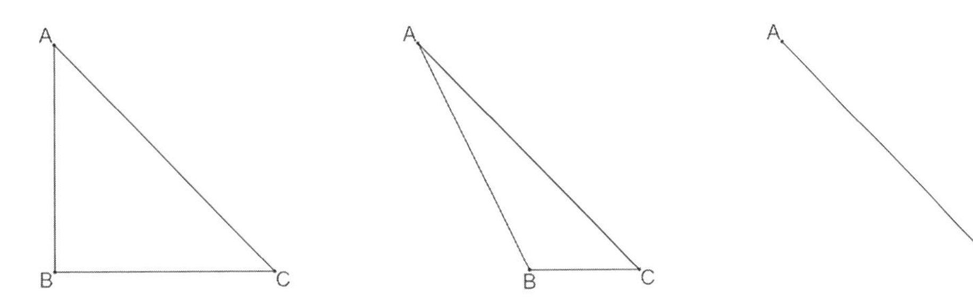

Figure 2.9 Triangles or not?

a triangle has zero area. A triangle essentially degenerates into a segment that still disguises itself as a triangle in the sense that it still has three vertices.

Degenerate triangles can be either of two types: needles and caps (figure 2.10). Can you see which is which?

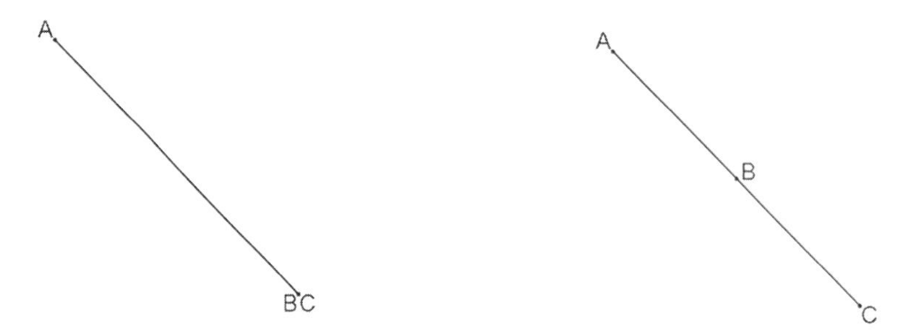

Figure 2.10 Not all degenerate triangles are the same. Some are needles, and others are caps.

When two points coincide, that's a needle. It has a "sharp" end, which is A in our case, and an "eye," where B meets C.

When three points lie on the same line, but none of them coincide, it's a cap. I guess that when the triangle isn't completely degenerate, it *does* look a bit like a cap (figure 2.11).

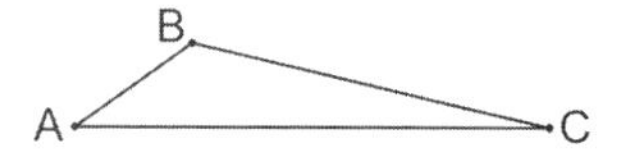

Figure 2.11 Not a completely degenerate cap

Now, exactly how do caps and needles cause trouble? First of all, if we model something with a set of triangles, degenerate triangles mess up the model's topology. Topology isn't only a mathematical discipline, but also a slang word for topological properties, which are properties that don't change under continuous deformations. It gets tricky with real surfaces, but for a set of triangles, the topology is basically "who touches whom."

Normally, when you model some shape with a bunch of triangles, your model implies that a triangle may have up to three neighbors. In figure 2.12, for example, triangle ABC is a neighbor to triangles ADB, BEC, and CFA.

Segments AB, BC, and CA are called the *edges* of the ABC triangle. Normally, each edge has at most two adjacent triangles. Edge AB has two neighboring triangles: ADB and ABC. But the edge AD has only one: ADB. If you organize your geometric model this way, the triangles that represent the model are called a *triangle mesh*—more specifically, a *manifold triangle mesh*.

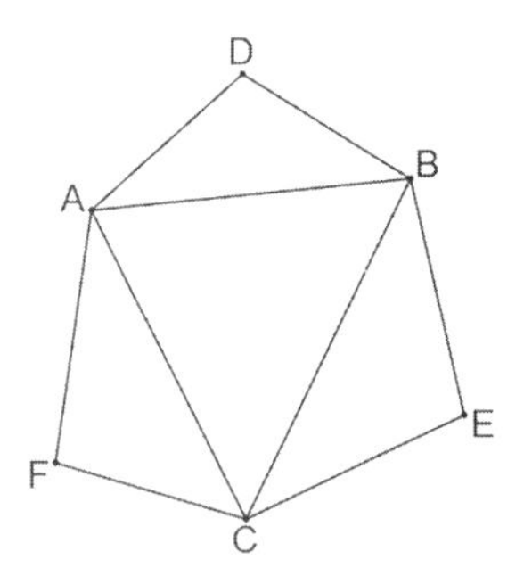

Figure 2.12 Triangle ABC and its three neighbors: FAC, CBE, and ADB

We'll get back to manifolds later in the book. This word has a specific meaning, concerning not only surfaces, but also curves and solid bodies. For now, let's say that in 3D, one manifold mesh models a single surface. Most of the models you see in games and 3D printing software are made of manifold meshes.

You can lift the "two triangles per edge" restriction, of course, and allow your edges to have more neighbors, making your mesh *nonmanifold.* It's possible, but normally, you wouldn't want to do that. Nonmanifoldness makes a lot of algorithms that process meshes as surfaces way more complicated. Besides, nonmanifoldness complicates the data model itself and makes the code slower to run and harder to debug. If an edge has at most two triangles attached, you can assign it a pair of indices to denote the edge's neighbors there. If you don't know how many triangles a single edge can belong to, you have to assign each edge a dynamic array of neighbors, and because there are usually only one or two inhabitants, the data structure is wastefully extended.

So what happens when a needle creeps into your perfect manifold triangle mesh? In figure 2.13, the BCD and ABE triangles look like neighbors, but they aren't. There's a proxy between them: ABC. It has no area, but it's still there, wedged between BCD and ABE.

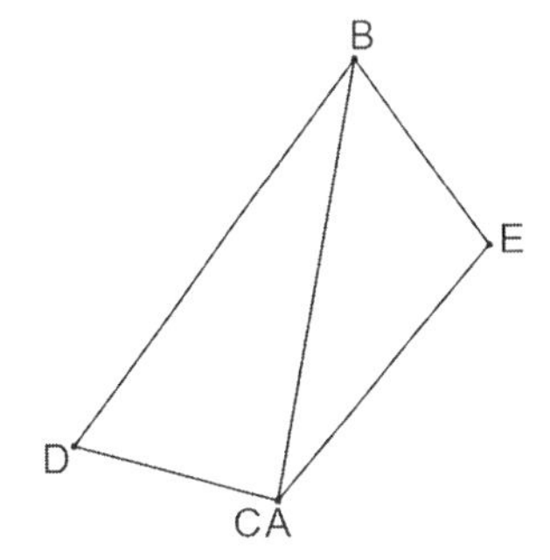

Figure 2.13 A degenerate triangle ABC in an otherwise manifold mesh

Let's say you want to pick all the triangles on the same plane as CBD. This task is common in computer-aided design (CAD) applications when you want to pick a specific face of a model.

The algorithm for plane picking works like this: you pick the CBD triangle and start traversing its neighbors; then you make a list of triangles that share the same plane and start to traverse their neighbors, followed by the neighbors of their neighbors, and so on until you reach the triangles that don't share the plane anymore. Now, you can't say that ABC triangle shares a plane with BCD because, as a degenerate, ABC doesn't belong to any specific plane at all. So if you traverse your neighbors in the width-first fashion described here, you don't even try to probe the AEB.

This algorithm doesn't know about degenerate triangles; it doesn't take them into account. But this ignorance makes the algorithm simple and fast.

Degenerate triangles make mesh processing needlessly difficult. When you're using a compound mesh operation, it's usually advisable to clean any degenerate triangles out of your model before doing anything else. But these triangles are easy to filter out, so in reality, they're not the biggest troublemakers. Near-degenerate triangles are. These triangles do have area, but because they're extremely thin, all operations on near-degenerate triangles are error-prone.

What's a *near-degenerate triangle*? I'm afraid that there's no single formal definition. There's no single way to tell what makes a triangle bad for computation. There are multiple ways to measure a triangle's quality, though. One way, perhaps the most popular, is to measure the proportion of its inscribed circle radius to its circumscribed circle radius. This proportion is also known as the *Rin/Rout metric*.

The *incircle* is the largest possible circle contained in the triangle, and the *circumcircle* is the smallest circle that contains the whole triangle itself. The incircle touches every edge of a containing triangle, and the circumcircle passes through all its vertices. So the Rin/Rout is the radius of the inner circle divided by the radius of the outer.

The equilateral triangles have the best proportion of the incircle's radius to the circumcircle's. Then, as the triangle gets thinner, the proportion declines toward zero (figure 2.14).

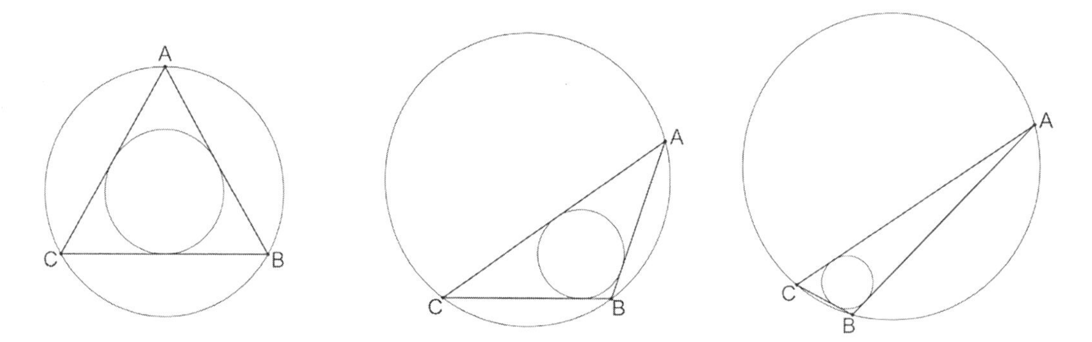

Figure 2.14 The Rin/Rout metric gets smaller as a triangle heads to the degenerate state.

For strictly degenerate triangles, the Rin/Rout is considered to be 0. This means that you can filter both types of degenerate and near-degenerate triangles by filtering out all the triangles with the Rin/Rout less than some predefined small-enough number.

> **SEE ALSO** Assessing triangle quality is an advanced topic, and no entry-level tutorials are available, I'm afraid. But an academic paper by Jonathan Richard Shewchuk contains a survey of quality measures for triangles. You can find it at http://mng.bz/xdPe.

Chapter 11 is dedicated to modeling surfaces with triangle meshes. Although the triangle is a basic concept, it will reappear throughout the book in seemingly unrelated places, such as chapter 3, which is about the geometry of linear equations. A practical example in chapter 3 is intersecting a ray with a triangle.

2.2.3 Section 2.2 summary

Triangles are important when you want to represent a surface with a set of simple elements. Triangle quality isn't completely formalized but is essential in practice. If you want your algorithms to work perfectly, you have to supply them a mesh of high-enough quality.

2.3 Lines, planes, and their equations

As with triangles, I expect you to have some intuition about lines and planes in 2D and 3D. This section shows how to formalize this intuition with specific formulas.

2.3.1 Lines and planes

In 2D space, this equation is a line equation:

$$y = ax + b$$

This way of describing lines is called an *explicit equation*. The equation consists of two variables, y and x, and two constant numbers, a and b. You can pick any a and b, plot a function graph, and see for yourself that the result is a line. For $a = 0.5$ and $b = 1$, the line looks like the one in figure 2.15.

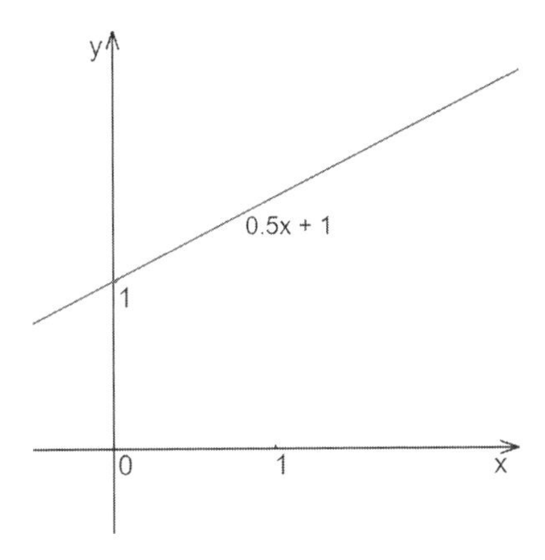

Figure 2.15 An explicit line equation: $y = 0.5x + 1$

This $y = ax + b$ equation, however, has a problem: it doesn't cover all the possible lines. If a line is strictly vertical, as in figure 2.16, we need a different type of equation: $x = c$.

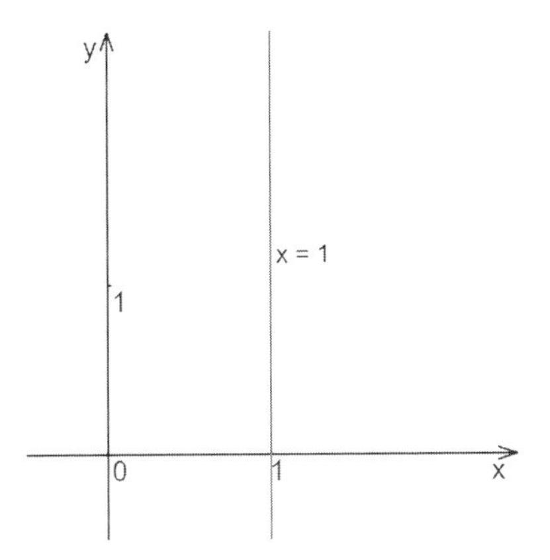

Figure 2.16 An equation that can't be put explicitly in y = something(x) form

Having two formulas for the same primitive is inconvenient, both in mathematics and in programming, so we often use a *general form*, in which we can describe every possible line:

$$a_1x + a_2y + a_3 = 0$$

Here, a_1, a_2, and a_3 are the coefficients that describe the line, and x and y are variables. This way of describing geometric objects is also called *implicit* because unlike in the explicit formula, we don't have a constructive way to build the object one point at a time.

Another, slightly more complicated line description is called a *canonical equation*:

$$\frac{x - x_0}{a_x} = \frac{y - y_0}{a_y}$$

Now, not only a_x and a_y, but also x_0 and y_0 are coefficients, so with little algebra, we can get back to the general form unless either a_x or a_y equals zero.

This new representation doesn't seem pragmatic for now, but we'll need it in 3D because in 3D, the general form of this example isn't a line equation anymore:

$$a_1x + a_2y + a_3z + a_4 = 0$$

It's not apparent in this form, but if you make the equation explicit, its planar nature becomes visible (figure 2.17):

$$z = -\frac{a_1}{a_3}x - \frac{a_2}{a_3}y - \frac{a_4}{a_3}$$

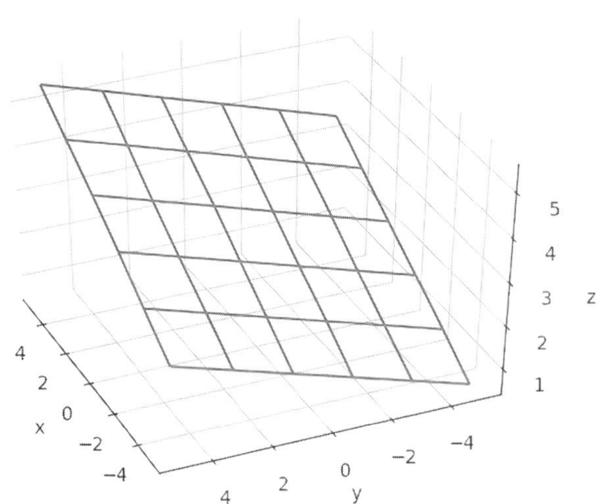

Figure 2.17 This plane is z = 0.3x + 0.2x + 3 or 0.3x + 0.2y - z + 3 = 0.

The formula produces z for every combination of x and y. We cover our x-y plane with a z-function. It can't be a mere line; it's a whole plane on its own. So the general line equation doesn't generalize to 3D when we simply add a new scaled variable, because it becomes a plane equation there. What is the general equation for a 3D line, then?

Well, there is none. A line in 3D isn't set by a single equation, but by a pair of equations.

You know that on a plane, two nonparallel lines intersect at a point. The situation is almost the same in 3D, but for planes. Two planes intersect at a line—unless, of course, they're parallel or coinciding. Pragmatically, you can describe your line as an intersection of planes (figure 2.18).

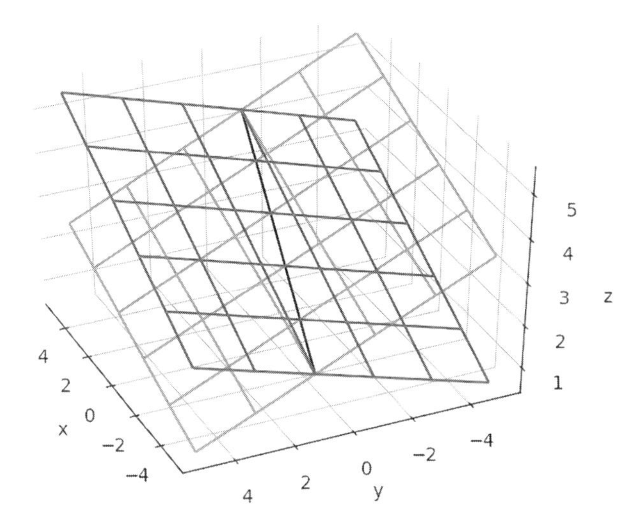

Figure 2.18 An intersection of two nonparallel planes is a straight line.

This isn't the one and only way to describe a line in 3D, but the other way deserves its own subsection.

> **SEE ALSO** For a more elaborate explanation of how to build a plane, please see "Equations of Planes" on this University of Texas web page: http://mng.bz/Al4K.

Understanding lines, planes, and their equations is essential for understanding linear algebra as well. Chapter 3 capitalizes on geometric intuition to explain methods of solving systems of linear equations as though they are looking for lines' and planes' intersections.

2.3.2 The parametric form for 3D lines

As a reminder, in 2D, we can use an explicit equation for a line:

$$y = ax + b$$

This approach is convenient, but as we've seen, it doesn't cover all the possible lines. We can't program a vertical line with this equation.

But what if instead of bundling both variables in a single equation, we make them independent? Can we set both x and y as functions of some other independent variable? Well, we can try. Let's call this new variable t:

$$x = f_x(t)$$

$$y = f_y(t)$$

This way of denoting lines is called a *parametric form*. The idea is that in this parametric form, we can compute a pair of (x, y) for every possible t. These pairs, when put together, form a line of a shape that is governed entirely by the functions. This approach allows us to model all kinds of curves, not only straight lines. But to model a straight line specifically, we need to define our functions like this:

$$x = x_d t + x_o$$

$$y = y_d t + y_o$$

Here, (x_o, y_o) is a point on a line—any point will do—and (x_d, y_d) is the directional vector showing the way the line goes. The t is a *parameter*, a variable put into the system that produces a point for every value. There's a point for $t = 0$, and there are other points for $t = 1$, $t = 5.43$, $t = -123.45$, and so on.

The equations in this form resemble the explicit line equation, but this representation is completely different, although it can represent explicit equations, among others. A line from figure 2.15, ($y = 0.5x + 1$), could be written in parametric form like this:

$$x = 1t + 0 \text{ (or simply } x = t\text{)}$$

$$y = 0.5t + 1$$

The first equation says that x and t are equal. Therefore, the second equation, $y = 0.5t + 1$, is the same as the explicit equation $y = 0.5x + 1$.

More useful still, by using a pair of equations, we can describe a vertical line, such as ($x = 1$) from figure 2.16, with the same parametric notation:

$$x = 1 \text{ (short for } x = 0t + 1\text{)}$$

$$y = t \text{ (the same as } y = 1t + 0\text{)}$$

Now the x is always 1, and the y could be any number. This line is a vertical line. So by using a pair of independent equations, we can denote both nonvertical lines that could have been described with an explicit formula and vertical lines that couldn't have been described that way. What's more, this notation translates nicely to 3D:

$$x = tx_d + x_o$$
$$y = ty_d + y_o$$
$$z = tz_d + z_o$$

Computationally, this representation is a little better than writing each line as an intersection of planes, because in this notation, you have only six coefficients to carry around, not eight. Also, we can rewrite the same equation in its vector form (figure 2.19) to save some screen space:

$$\mathbf{x} = \mathbf{x}_o + t\mathbf{x}_d$$

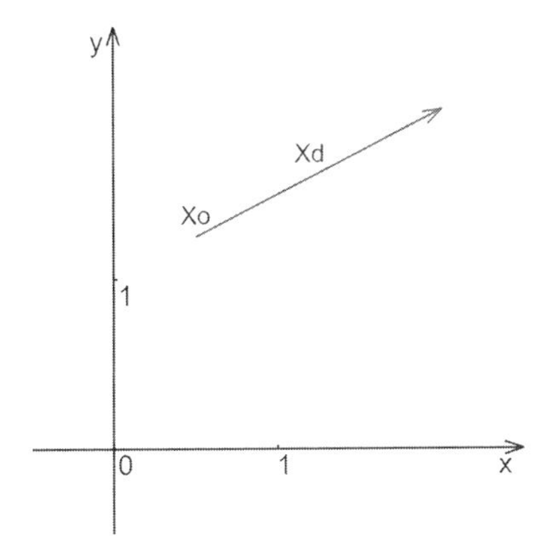

Figure 2.19 A point vector x_o and a vector (not a point vector) x_d together define a line. In our case, that line is 2D, because it's a plot on a planar book page, but the same principle works in 3D as well.

This form is handy in programming. If we want to make a line out of sparkling particles in some 3D engine, for example, we can do that by adding a particle at $(x(t), y(t), z(t))$ for different values of t.

Let's rewrite the equations to get rid of the parameter. First, let's extract it:

$$t = \frac{x - x_o}{x_d}$$
$$t = \frac{y - y_o}{y_d}$$
$$t = \frac{z - z_o}{z_d}$$

Then let's reform the equations without the t:

$$\frac{x - x_o}{x_d} = \frac{y - y_o}{y_d}$$
$$\frac{y - y_o}{y_d} = \frac{z - z_o}{z_d}$$

Don't these equations look familiar? Yes, they do! You saw the canonical line equation in the preceding section. But now there are two of these equations, and they describe a line in 3D. That's why we didn't dismiss the canonical form right away. It may not be pragmatic in 2D, but it comes back in 3D.

You don't have to remember formulas for every form; you can always search for them online. But building some intuition about how lines and planes work together in both 2D and 3D allows you to see the parallels between these concepts. As a significant bonus, the same intuition will help you see how the linear systems work and understand them better. We'll get back to linear systems in chapter 3.

> **SEE ALSO** An online calculus tutorial from Harvey Mudd College explains this topic with more examples and motivations. You can find it at http://mng.bz/Zo5O.

The parametric form may not be the most important instrument for drawing lines per se, but we'll use the same parameterization principle to build curves in chapter 7 and surfaces in chapter 8.

2.3.3 Section 2.3 summary

There are several ways to describe a line in 2D with formulas, of which a general or implicit form is the most universal. A line in 3D, however, can't be described as an implicit formula. Instead, we have to use a system of equations, a parametric vector formula, or a canonical representation.

2.4 Functions and geometric transformations

There are no functions in classical geometry—only rulers and compasses. In analytic geometry, however, there are no rulers and compasses. But there are coordinate spaces, curves, surfaces, and their formulas. A function, as a mathematical concept, becomes an important geometric tool now, and we need to know our tools well.

2.4.1 What is a function?

We all have good intuition about what a function is. In programming, a *function* is something that accepts some input and returns some output. If it doesn't do anything else, such as writing to a file or printing on the console, it's called *pure*. Pure functions are pretty close to functions in the mathematical sense.

We can take a peek at the formal definition of mathematical functions, but fair warning, it may not be helpful on its own. Here it goes: a function is a special kind of binary relation that connects every element of its domain to one and only one element of its codomain.

This definition isn't helpful unless you already know about relations and domains. On the other hand, you probably have good intuition about functions, so let's try a little trick. We'll reverse the definition, and learn about relations and domains from a function we know and love: the sine (figure 2.20).

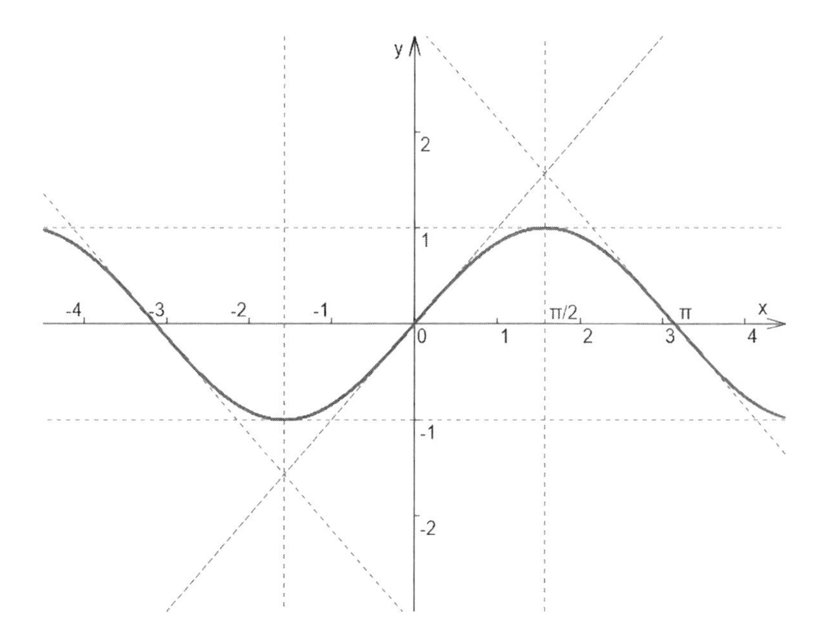

Figure 2.20 The sine function: y = sin(x)

The sine defines a y for every x in $\mathbb{R}$, and all the *y*s lie in the range [-1, 1]. The sine is both a relation and a function.

We can invert the relation and say that now for every y, there's a set of *x*s. This relation will no longer be a function, though. Although for every y in $(-\infty, -1)$ and $(1, \infty)$, the corresponding set will be empty, for every y in $[-1, 1]$, it will be infinite.

This is still a relation but not a function. What's the difference, then? A *relation* is a rule that links input values with output values. It can link any set of input values to any set of output values. A *function* is a special case of a relation. It's also a rule, and it links one or many input values with the output values, but it's not allowed to link many output values to a single input. A function can't have multiple outputs. It's exactly this many-to-one linkage that defines the term *function;* it separates functions from all other types of relations.

To avoid possible confusion, you should understand a special terminology related to functions. The set to which all the *x*s belong is called the *domain.* For the sine, that set is the whole set of real numbers. We have a letter for this set: $\mathbb{R}$.

The set to which all the *y*s belong is called the *codomain*. It's also ℝ for the sine. But the sine function also has a distinct *range*, which is not all the numbers, but specifically [-1, 1]. The range is a separate term, not the same as the codomain. Speaking in programming terms, the *codomain* is the output type of the function, whereas the *range* is all the possible values the function can produce.

The sin(x) and tan(x) share the codomain ℝ, but tan(x) covers all the ℝ with its range, and sin(x) does not (figure 2.21).

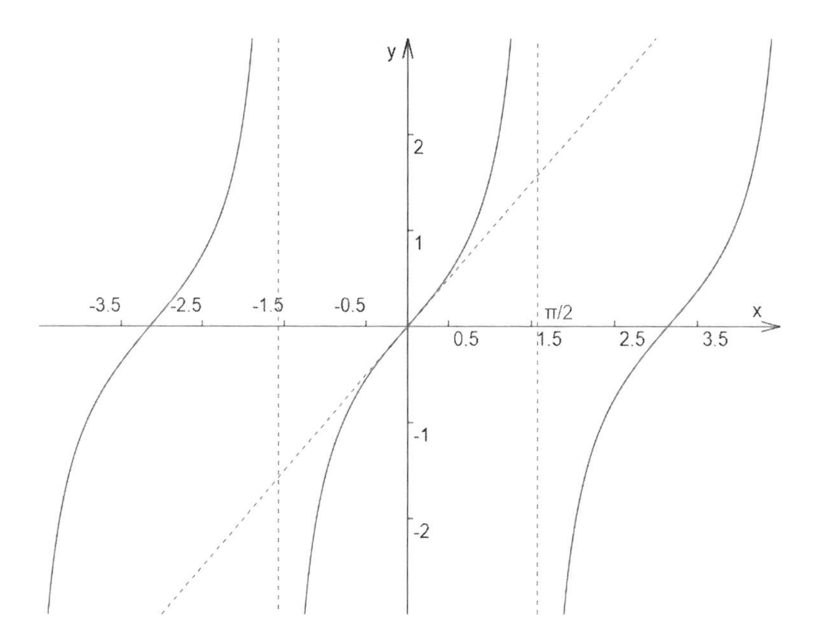

Figure 2.21 The tangent function: y = tan(x)

The domain, confusingly, isn't the type, as in a codomain, but all the possible values, as in a range. This time, it's all the possible values for the input. The domain of tan(x) is ℝ except for $\pi/2+i\pi$, where i is any integer number.

These terminological nuances here are thin and often neglected in jargon. People often mistake one thing for another, so don't be surprised if someone says that sin(x) has the range of ℝ or tan(x) has the domain of ℝ. If you're unsure, ask whether they're talking about the type of the possible values or the possible values themselves. The answer usually clears things up.

Here's another important difference between the sine and the tangent: the former is *continuous*, and the latter is not. *Continuity* is important as a property, especially in geometry, for the reasons explained in the next section.

I won't get into a formal definition of continuity, though, I'm sure that you have an impression of what it is. Let's synchronize our intuition. First, a continuous function should not have points excluded from its domain, as the tan has. Second, even if its domain is the whole $\mathbb{R}$, it can't have abrupt changes in value. A `floor(x)` function, for example, accepts a real number and returns its truncated integer value. It's defined for all the numbers, but it's still discontinuous (figure 2.22).

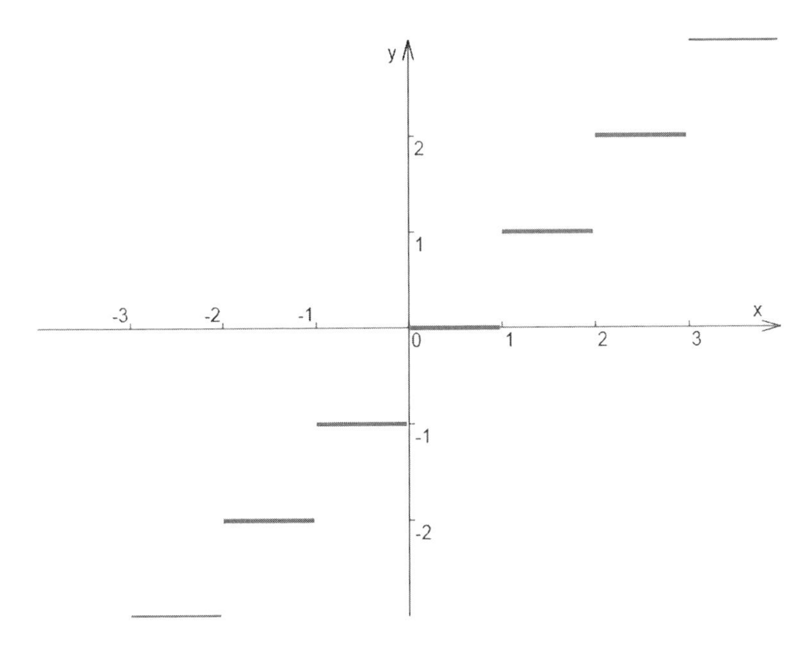

Figure 2.22 The truncation function (also known as the *floor*) is defined everywhere but still discontinuous.

As with lines and planes, we have a few ways to describe a function. When we describe how exactly it turns input values into its output, the function is *explicit*. The following equation is a cubic polynomial function in explicit form:

$$f(x) = ax^3 + bx^2 + cx + d$$

The letter f here names the function, x names the function's argument, and the equation on the right of the equation sign is the way to compute the function f for any x. But if we want to explain what types it operates on, we use slightly different notation. If f is a polynomial over real numbers, we can state that it turns a real number into a real number:

$$f : \mathbb{R} \rightarrow \mathbb{R}$$

Conceptually, a function is a special case of relation. It's a rule that binds output values to input values. Even in programming, it doesn't necessarily require computation. A few decades ago, it was common in programming to substitute trigonometric functions for lookup tables. Mathematically, a lookup table is still a function. Exactly how a function binds a pair of values is irrelevant as soon as there's a single output value in the codomain for each input value from the domain.

Functions aren't geometrical objects, but a geometric concept that's closely associated with a function is the function's graph. A *graph* is a curve or a surface that represents a function on a plot. Unlike the function itself, the graph is a geometrical thing.

In 2D, a graph is a set of points: $(x, f(x))$. It's a curve. In 3D space, a graph is a set of: $(x, y, f(x, y))$. It's a surface.

The terms *function* and *plot* are usually interchangeable in jargon. When we show a plot, we don't say "Look at the graph"; we say "Look at the function." It's also common to speak about a function's inclination when, strictly speaking, the inclination is the property of its graph. As long as these terminological inaccuracies don't hurt understanding, we can accept them as part of applied-geometry lingo.

SEE ALSO The Math Is Fun website has a good explanation of functions and related terms at https://www.mathsisfun.com/sets/function.html.

Functions are the most important instruments of analytic geometry. In chapter 6, we'll learn about polynomials, which appear to be a fast and memory-efficient, economical way to model arbitrary functions with simple numeric operations. In chapter 7, we'll learn to turn polynomials into continuous splines to model curves, and in chapter 8, we'll learn how to build surfaces with functions. Chapters 10 to 12 feature functions when discussing data conversion from one representation to another.

2.4.2 Function types most commonly used in geometric modeling

You should meet a few types of functions. A type of function is usually specified by its domain and codomain. In programming, this is almost the same as the input type and the output type.

The most important function, discussed in two chapters later in the book, is called a *geometric transformation*. It's a function that turns a point into another point, normally, in the same space.

We used $\mathbb{R}$ to denote a set of real numbers, and this is essentially the same as 1D space. So let's use $\mathbb{R}^2$ to denote 2D space, $\mathbb{R}^3$ to denote 3D space, and $\mathbb{R}^N$ to denote dimensionality of any positive integer value.

The domain of a geometric transformation is $\mathbb{R}^N$, and its codomain is also $\mathbb{R}^N$. The simplest transformation is called *translation*; it shifts all the points the same distance and in the same direction (figure 2.23).

Note that even if we apply the transformation to specific points, such as points of a square as in figure 2.23, the transformation itself is defined for all the $\mathbb{R}^2$. We define a rule that applies to all the points of the space, so essentially, it's a transformation of the whole space, not only the points of a particular model.

A similar kind of function is called a *vector field*. It defines an M-dimensional vector for every point of N-dimensional space. Its domain is $\mathbb{R}^N$, and its codomain is $\mathbb{R}^M$. M and N can be the same, of course, but this isn't always the case.

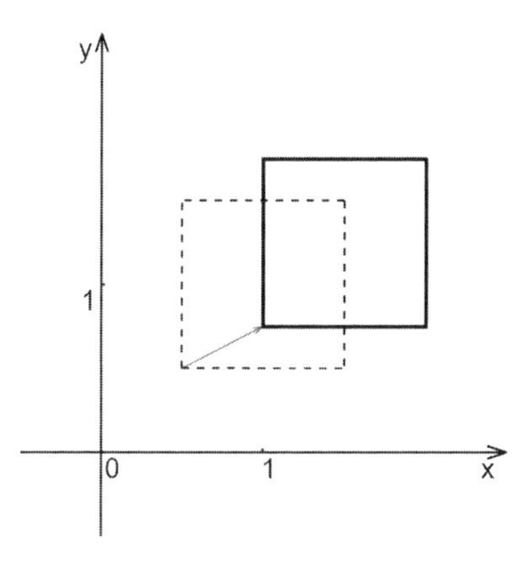

Figure 2.23 The simplest geometric transformation: translation

A vector field is conceptually different from a geometric transformation, as vectors are different from points. A geometric deformation moves points around, whereas a vector field can show only where to move the points. It can do much more, of course. It can carry a coloring scheme, for example, in which each vector is defined not in conventional Euclidean space, but completely unrelated color space.

Like a single variable function, a vector field can be continuous. But even a continuous vector field is usually displayed as a bunch of vectors sticking out of a plane with some regular intervals between them (figure 2.24). This illustration is only a convention, though. With a continuous vector field, you have to imagine the vectors you don't see between the ones you do.

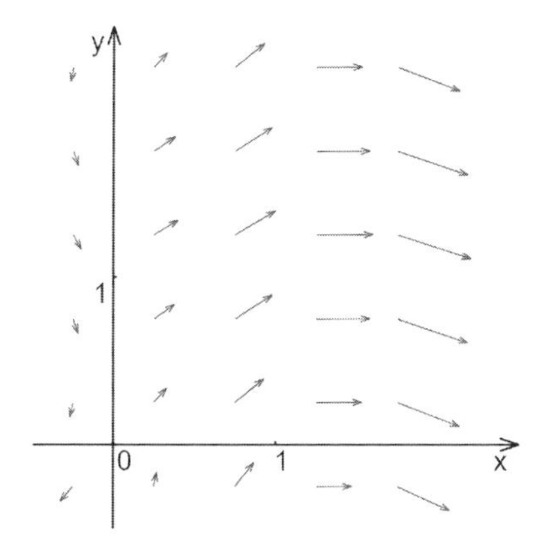

Figure 2.24 An illustration of a vector field. You have to imagine that it's continuous.

A vector field that stands behind a geometric transformation specifically is called a *deformation field* (figure 2.25). It's a rule that describes how you should translate or move points of some space, depending on where these points are.

The field shows that some points may be moved rather far away, and some may not be moved at all. Once again, the field itself may be continuous, even if we don't see it in a picture.

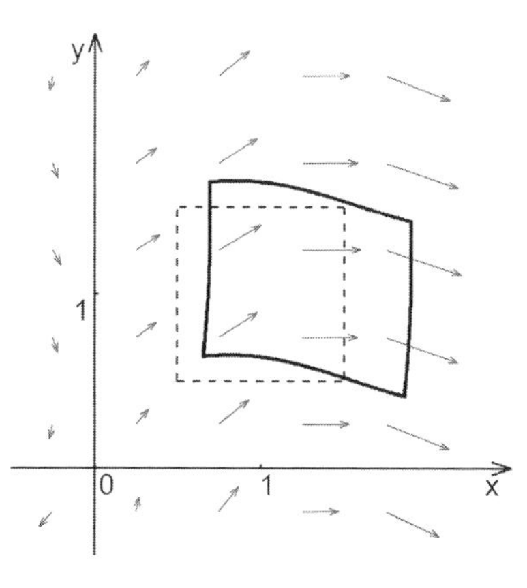

Figure 2.25 The vector field from figure 2.24 used as a deformation field. It moves, bends, and stretches the square.

Deformation fields are useful in 3D modeling. In chapter 8, we'll build a deformation field to introduce randomization to the code that produces 3D models.

SEE ALSO Khan Academy has nice extensive tutorials on both vector fields and transformations at http://mng.bz/Y6lz and http://mng.bz/GRMD, respectively.

Another special case of a function is a *scalar field*, which is like a vector field but assigns a scalar, also known as a real number, for every point. Its domain is $\mathbb{R}^N$, and its codomain is $\mathbb{R}$.

One example of these fields is a *signed distance function*, or SDF for short (figure 2.26). An SDF is a distance to a curve or a surface in 3D defined for all the points in space. The distance function is called *signed* because we set it positive for points outside the body and negative for the points inside the body.

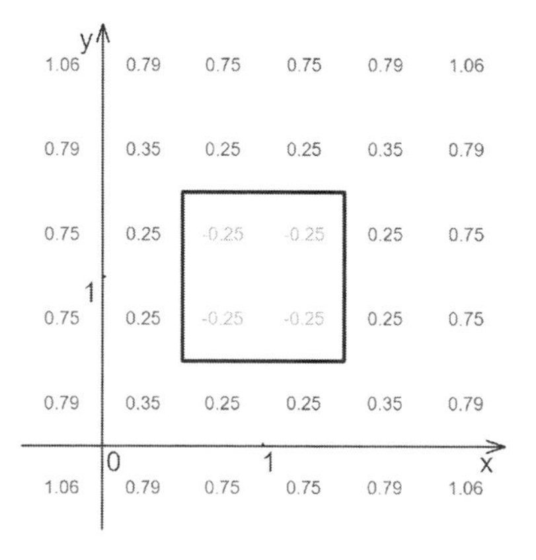

Figure 2.26 An SDF illustration

These kinds of functions are widely used in modern CAD, for good reason. A lot of operations on 3D models represented as SDFs are simpler and faster than those on equivalent models represented by surfaces or triangle meshes. In chapter 10, we'll see how to use SDFs to make a solid 3D model into a hollow shell with a few lines of code.

You don't have to know all the function types by their names. You can always specify them by their type signatures: input and output types. But it helps to know at least the most common terms to make communication with other geometers easier.

2.4.3 Section 2.4 summary

A *function* is a relation that binds a given codomain value to each of the domain values. The function may have properties, of which continuity is the most important for modeling curves and surfaces. Special types of functions are determined by their domain and codomain types, which you should know because they're widely used in geometry (table 2.1).

Table 2.1 Function types used in geometry

Name	Domain	Codomain
Geometric transformation	The whole space	The same space
Vector field	The whole space	Another space representing vectors of some dimensionality
Deformation field	The whole space	Vectors in the same space
Scalar field	The whole space	Real numbers
Signed-distance function	The whole space	Real numbers

2.5 The shortest possible introduction to matrix algebra

Before moving to chapter 3, we have one more topic to cover. In school, you could do geometry with rulers and compasses. In analytic geometry, we use different tools. One of these tools is matrix algebra. We use matrices for stacking up transformations, for building curves that go through a given set of points, for finding planes' intersections, and for performing many other tasks.

It's a little awkward to write about algebra in a geometry book, but at some point, doing so becomes unavoidable. That point is here.

2.5.1 What is algebra?

Algebra is a mathematical discipline that studies algebraic systems up to their isomorphisms. Sure, this definition is awfully unhelpful, but it's also rather precise. Once again, because you already have a notion about what algebra is from school, we'll invert the definition to learn about algebraic systems and isomorphisms.

An *algebraic system* is a bit like a type in programming; it's a set of values, also known as the *domain*, and a set of operations that do something to the values from this domain.

A set of integers and an integer addition, for example, form an algebraic system. This system is characterized by its properties:

- An integer plus an integer is still an integer. This property is called *closure.* The integers are closed under addition.
- Integers could be summed up in any order: $(a + b) + c = a + (b + c)$. This is called *associativity.* You don't have to memorize all the properties by name, by the way. But be aware of their existence.
- Integers could be swapped places in the addition: $a + b = b + \mathrm{a}$. This is called *commutativity.*
- There is an *identity element* that does nothing to the addition: $a + 0 = 0 + a = a$.
- For every integer in the set there is an *inverse element* for every element that turns it into the identity when added: $a + (-a) = 0$.

All these properties characterize the algebraic system. All the algebraic systems that have the same properties have the same name. Integers and addition, for example, form an algebraic system called a *group.*

There are other algebraic systems, such as groupoids, semigroups, fields, and rings. At this point, fields and rings are of no use to us. The group that matrices and the matrix multiplication form together is useful, however.

NOTE Technically, matrices under multiplication don't form the same group as the integers under addition, because matrix multiplication isn't commutative, but the rest of the properties are shared.

To complete our algebraic education, let's get back to isomorphism. Two algebraic systems are *isomorphic* if, and only if, they operate on sets that have one-to-one correspondence that allows all the algebraic properties from one set to work on another. It's not important what these sets are, and it's not important whether we call their operations additions or multiplications. We only want these systems to share properties.

The main idea of abstract algebra is that isomorphism is enough to transfer knowledge from one system to another. If you wrote a book about matrix algebra, and it didn't sell well, you could autoreplace *matrice* with *integer* and *multiplication* with *addition.* All the theorems would transfer from matrices to integers automatically, and you could republish the book under a new name, such as *The Mysterious Adventures of Noncommutative Integers.*

SEE ALSO The concept of isomorphism is so potent that it's already being adopted in nonmathematical areas of software engineering, as you can find in this article about software integration: http://mng.bz/RlW0.

That's why we started discussing matrix algebra to begin with. We wouldn't care about matrices in a geometry book at all unless we're planning to reuse their algebraic system for something geometrical. And we are.

In chapter 4, we'll see how reusing algebra for geometric transformations lets us stack a whole series of rotations, translations, and scalings into a single matrix—one example of how algebra saves us memory and wins performance.

2.5.2 How does matrix algebra work?

Let's start from the beginning. This is a number or, in terms of vector algebra, a scalar:

$$1$$

This is a vector:

$$(1, 2, 3)$$

A vector looks like an array of numbers, and in several programming languages, most notably C++ and Scheme, arrays are also named vectors. What distinguishes a vector from some arbitrary array is its algebraic system, or algebra for short—a set of rules and operations that determine how vectors interact. You can add (1, 2, 3) and (4, 5, 6) when they're vectors, and according to their algebra, you'll get (5, 7, 9). But adding two arrays is ambiguous. The result of addition could be an array of sums, as with vectors, or a concatenation (1, 2, 3, 4, 5, 6), as with strings. So a vector is like an array plus the set of vector-specific operations.

Here's a matrix:

$$\begin{pmatrix} 1 & 2 & 3 \\ 4 & 5 & 6 \\ 7 & 8 & 9 \end{pmatrix}$$

It's a 2D array. Like vectors, matrices have their own algebraic system. You can write a matrix in a more programming-friendly way by using nested brackets:

```
[ [1, 2, 3],
  [4, 5, 6],
  [7, 8, 9] ]
```

As soon as you respect the algebra, the exact notation is irrelevant.

A term that generalizes both vectors and matrices is *tensor*. In geometry, we rarely have to work with tensors other than vectors and matrices, but it's still nice to know the general term for them. An inertia tensor, which in 3D is a 3x3 matrix, is called a tensor for historical reasons, and this fact alone makes modeling physical interactions sound way more complicated than it should be.

Essentially, any 2D tensor with a name is a matrix with a specific purpose, and matrices aren't hard to use. We need to know only a few operations to use them properly, starting with matrix multiplication. Here's how it's done:

$$\begin{pmatrix} a_{11} & a_{12} & a_{13} \\ a_{21} & a_{22} & a_{23} \end{pmatrix} \times \begin{pmatrix} b_{11} & b_{12} \\ b_{21} & b_{22} \\ b_{31} & b_{32} \end{pmatrix} = \begin{pmatrix} a_{11}b_{11} + a_{12}b_{21} + a_{13}b_{31} & a_{11}b_{12} + a_{12}b_{22} + a_{13}b_{32} \\ a_{21}b_{11} + a_{22}b_{21} + a_{23}b_{31} & a_{21}b_{12} + a_{22}b_{22} + a_{23}b_{32} \end{pmatrix}$$

Every cell with indices [i, j] in the product matrix is a product of the i-row of the first matrix and the j-column of the second. You take a row and a column, multiply the corresponding elements, and sum them up so that they make a number.

Consequently, you can multiply a matrix with width N only by a matrix with height N. If this width-height equivalency isn't met, no multiplication is possible.

You can think of a vector-by-matrix multiplication as a special kind of matrix multiplication with a thin matrix:

$$\begin{pmatrix} a_1 & a_2 & a_3 \end{pmatrix} \times \begin{pmatrix} b_{11} & b_{12} & b_{13} \\ b_{21} & b_{22} & b_{23} \\ b_{31} & b_{32} & b_{33} \end{pmatrix}$$

$$= \begin{pmatrix} a_1b_{11} + a_2b_{21} + a_3b_{31} & a_1b_{12} + a_2b_{22} + a_3b_{32} & a_1b_{13} + a_2b_{23} + a_3b_{33} \end{pmatrix}$$

This kind of multiplication is especially useful in geometry, because this is how we transform points. If applied to a point indicated by a point vector, a matrix multiplication turns it into another point, and the way it does so is applicable in real-world applications, as we'll see in chapter 4.

Now, you may probably have noticed that this matrix multiplication isn't commutative. In other words, A × B isn't the same as B × A. As with division in real numbers, 1/2 is obviously not the same as 2/1. This fact gets even more obvious with rectangular matrices. Indeed, we can multiply a 2 × 3 matrix by a 3 × 2 matrix, but not vice versa. Or wait—shouldn't it be 3 × 2 by 2 × 3?

That's the problem. The concept of rectangular matrices is crystal-clear in textbooks, but in practice, people mess up the implementation way too often. There's no single convention about how you should store your matrices and vectors in memory. If you store your vectors as columns instead of rows, you should apply your matrix multiplication to a vector from the left side or keep the matrix transposed beforehand.

It gets messy. The worst thing is that even if you don't understand matrix multiplication, you can still make it work with trial and error, because it's all about keeping multiplications in the right order. People do that all the time. They try things until they suddenly work without understanding how they should have worked properly. Then they leave the code as is, leading to even more mess. The only way to combat this problem is to develop solid intuition about matrix operations.

SEE ALSO Better Explained offers "A Programmer's Intuition for Matrix Multiplication" at http://mng.bz/Nm01.

The next important matrix operation helps us clean up this mess a little bit. *Transposition* is basically flipping a matrix over the top-left to bottom-right diagonal so that the rows become columns and columns become rows:

$$\begin{pmatrix} a_{11} & a_{12} & a_{13} \\ a_{21} & a_{22} & a_{23} \\ a_{31} & a_{32} & a_{33} \end{pmatrix}^T = \begin{pmatrix} a_{11} & a_{21} & a_{31} \\ a_{12} & a_{22} & a_{32} \\ a_{13} & a_{23} & a_{33} \end{pmatrix}$$

One observation: If you come to a new project, and the matrices don't work, they're probably transposed from what you're used to. Another observation: If a routine has more than one matrix transposition in it, there's a chance that the code was written by trial and error and probably can be rewritten to work more efficiently.

Another operation is called *inversion*. We'll look into inversion in chapter 4, using this property to undo a geometrical transformation. For now, remember that a matrix may or may not have an inverse element and that when you multiply these matrices, this multiplication results in an identity matrix.

Here's an example. The second matrix is the inverse of the first one:

$$\begin{pmatrix} 2 & 3 & 2 \\ 4 & 2 & 3 \\ 9 & 6 & 7 \end{pmatrix} \times \begin{pmatrix} -4 & -9 & 5 \\ -1 & -4 & 2 \\ 6 & 15 & -8 \end{pmatrix} = \begin{pmatrix} 1 & 0 & 0 \\ 0 & 1 & 0 \\ 0 & 0 & 1 \end{pmatrix}$$

The first matrix, of course, is also the inverse of the second, so when you swap them, the equation holds:

$$\begin{pmatrix} -4 & -9 & 5 \\ -1 & -4 & 2 \\ 6 & 15 & -8 \end{pmatrix} \times \begin{pmatrix} 2 & 3 & 2 \\ 4 & 2 & 3 \\ 9 & 6 & 7 \end{pmatrix} = \begin{pmatrix} 1 & 0 & 0 \\ 0 & 1 & 0 \\ 0 & 0 & 1 \end{pmatrix}$$

SEE ALSO You don't have to know how to inverse matrices by hand. SymPy can do these things for you (chapter 4). But if you still want to know how it's done, please see "Inverse of Matrix" at http://mng.bz/1MxV.

Elements of matrix algebra are discussed in chapter 4 along with the geometric transformations. We'll use matrices to build functions with desirable properties in chapter 6, and in chapter 12, we'll discuss how the field of computer tomography is essentially solving huge matrix equations.

2.5.3 Section 2.5 summary

Matrices are like 2D vectors: tables of numbers with specific operations that define matrix algebra. A matrix operation may have some geometrical meaning. A matrix-on-vector multiplication, for example, is a projective transformation.

Tensor is the term that generalizes vectors and matrices. Tensors are like multidimensional arrays, but with their own algebras (sets of rules and operations).

2.6 Exercises

Exercise 2.1 Which one of these mathematical concepts hasn't had its algebra mentioned in this chapter?

1. A vector
2. A point
3. A matrix

Exercise 2.2 Which of these function types were introduced in this chapter?

1. A vector field
2. A manifold mesh
3. A scalar field
4. A geometric transformation

Exercise 2.3 In which numbers is a function $f(x) = x^2$ continuous?

1. Natural numbers without zero
2. Integer numbers
3. Real numbers
4. Natural numbers with zero

Exercise 2.4 Invent a relevant metric to assess triangle quality.

2.7 Solutions to exercises

Exercise 2.1 A point. A set of things and a set of operations and rules under which they apply to things constitute an algebraic system (algebra for short). Vectors were shown with addition and scalar multiplication, and matrix algebra was presented as a noncommutative group.

Exercise 2.2 All except manifold mesh. It was introduced in this chapter, but it's not a function.

Exercise 2.3 It's continuous in real numbers only. Continuity doesn't exist in integers.

Exercise 2.4 This question is an open one, so there's no single right answer here. Tip: if your metric shows "horrible" for a degenerate triangle, and "excellent" for an equilateral one, it's probably relevant.

Summary

- In programming, we don't have true real numbers, but we can choose among several models, all of which have pros and cons.
- A point is a building block for curves and surfaces, but nothing more. There's no algebra for points.
- Vectors are like geometric pointers. They have their own algebraic system, so they behave both like points and numbers simultaneously.

- Triangle quality is essential for computation, but it doesn't have a single well-established definition.
- A general line equation in 2D becomes a general plane equation in 3D, and the line in 3D can be described only parametrically, canonically, or by a system of equations.
- A function is a rule that links a single codomain value to each and every domain value.
- Domain and codomain types establish a function type. The most important function types for us are a geometric transformation, a vector field, a scalar field, and a distance function.
- An algebraic system, or algebra for short, is a toolbox of operations and rules applicable to some set of mathematical entities such as numbers, vectors, and matrices.
- Tensors are multidimensional arrays with algebras. Matrices and vectors are tensors, too.

3 The geometry of linear equations

This chapter covers

- Learning the geometrical sense of systems of linear equations
- Telling which systems could possibly be solved
- Understanding iterative solvers, including convergence, stability, and exit condition
- Understanding direct solvers and algorithmic complexity
- Picking the best solver for any particular system

Systems of linear equations are everywhere. In fact, we solved one in the first chapter. Remember the two-trains problem?

```
solution = solve([
    Vp - Va * 2,
    Va * 1 + Vp * 1 - 450
], (Va, Vp))
```

Yes, this is a small system of linear equations. It has two equations, which makes it a system, and each equation is a sum of scaled variables supplied optionally with a number, which makes equations linear.

You've probably heard that half of the whole domain of machine learning is linear systems in disguise, and this saying isn't far from the truth. These systems are more than welcome in traditional engineering and economics, too. And of course, they are used vastly in applied geometry, from small systems that occur naturally in problems like ray and triangle intersection to large systems that turn thousands of x-ray images into a single 3D computer tomography scan (chapter 12). These systems are popular in computational mathematics for two reasons. First, a lot of real-world processes are inherently linear and can be modeled with simple linear equations. Second, because computational mathematics isn't good at solving nonlinear systems, we tend to linearize our equations whenever possible, even if we lose a little modeling accuracy.

Geometry stands privileged among other mathematical disciplines: it not only relies on linear equations to solve its own problems, but also gives these equations a geometrical meaning, allowing us to understand them through plots and graphics. The most obscure algorithms become comprehendible when you see them in action. This understanding, obtained by watching them work, also helps us competently choose the best algorithms to solve our problems.

3.1 Linear equations as lines and planes

A *linear equation* is a sum of variables multiplied by some coefficients that equals some number. The variables may occur only in power 1. Having an x^2 in an equation, for example, disqualifies it as linear, making it quadratic instead. Grouped together, linear equations form a *system of linear equations.*

A *solution* to a linear system is a set of variables' values for which all the equations hold. Depending on the system, there could be a single combination, a whole class of them, or none at all. Sometimes, a solution exists, but we have trouble finding it numerically. Sometimes, there are no solutions mathematically, but the one we compute despite its theoretical impossibility is good enough for our needs. Everything depends on the system.

This dependency, however, isn't immediately apparent when all we see are equations made of numbers and letters. But if we rewrite them as geometrical entities, we literally see which systems are easy to solve and which are not.

3.1.1 Introducing a hyperplane

This is a linear equation of two variables, x and y:

$$ax + by = c$$

If we move the coefficient from the right side to the left side, we get

$$ax + by - c = 0$$

Never mind the minus sign; this equation is a line equation in 2D. So a linear system of two equations is essentially a system of two lines. The solution to this system is the point where the lines intersect.

Awesome! Now we can take our two-trains problem from before and draw it (figure 3.1):

$$V_p - 2V_a = 0$$

$$V_a + V_p - 450 = 0$$

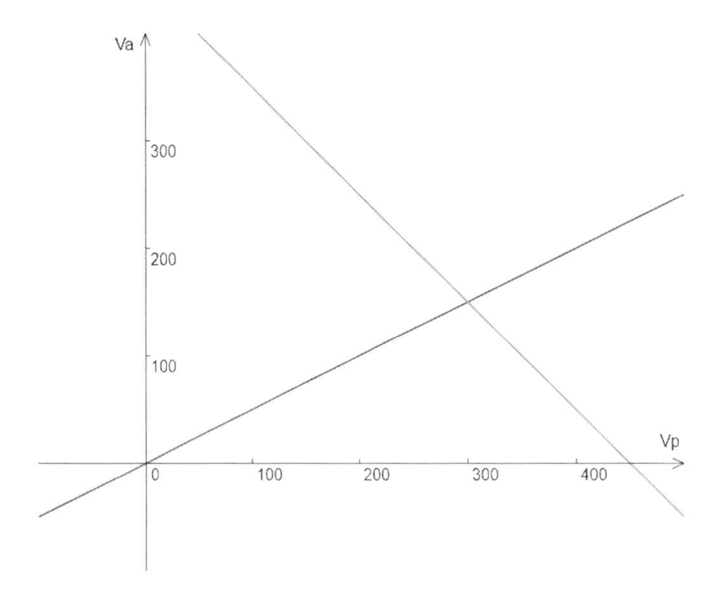

Figure 3.1 The two-trains problem as an intersection of lines

The solution is immediately apparent: It's where the lines intersect. In figure 3.1, the solution is the point (300, 150).

We can do the same thing in 3D. A linear equation like this

$$ax + by + cz = d$$

turns into a plane equation:

$$ax + by + cz - d = 0$$

A system of three variables, therefore, is a system of three 3D planes. The point of the solution is where all the planes intersect.

We don't want to stop at equations with only two or three variables, of course. But to keep the geometrical sense as we add more variables, we need a new term, a new geometrical concept that generalizes lines and planes, and that corresponds to a linear equation of N variables:

$$a_1x_1 + a_2x_2 + \dots + a_Nx_N = b$$

There is such a concept: a *hyperplane*. It's a subspace of N-dimensional space with its own dimensionality N-1. A line in 2D is a 2-dimensional hyperplane. A plane in 3D is a 3-dimensional hyperplane. The formula for an N-dimensional hyperplane follows the pattern of lines and planes but allows as many variables as you want:

$$a_1x_1 + a_2x_2 + \ldots + a_Nx_N + a_{N+1} = 0$$

With a small variable substitution, we can turn any linear equation into a hyperplane equation:

$$b = -a_{N+1}$$

$$a_1x_1 + a_2x_2 + \ldots + a_Nx_N + a_{N+1} = 0 \quad \Leftrightarrow \quad a_1x_1 + a_2x_2 + \ldots + a_Nx_N = b$$

So any linear system is a system of N-dimensional hyperplanes.

3.1.2 A solution is where hyperplanes intersect

In 2D, a line intersection is a solution of the system of two linear equations. With hyperplanes, the same principle scales nicely to higher-dimensional spaces.

In 3D, three intersecting planes intersect at a single point. In 4D, four intersecting hyperplanes intersect at one point. In 5D . . . well, I hope you see the pattern here. As a general rule, if N different N-hyperplanes intersect in N-dimensional space, they all intersect at exactly the same point.

But what if, for whatever reason, some of the hyperplanes don't intersect at all? What if—and let's get back to 2D for a while—two lines that constitute a system are parallel? They'll never intersect, and there won't be a solution (figure 3.2).

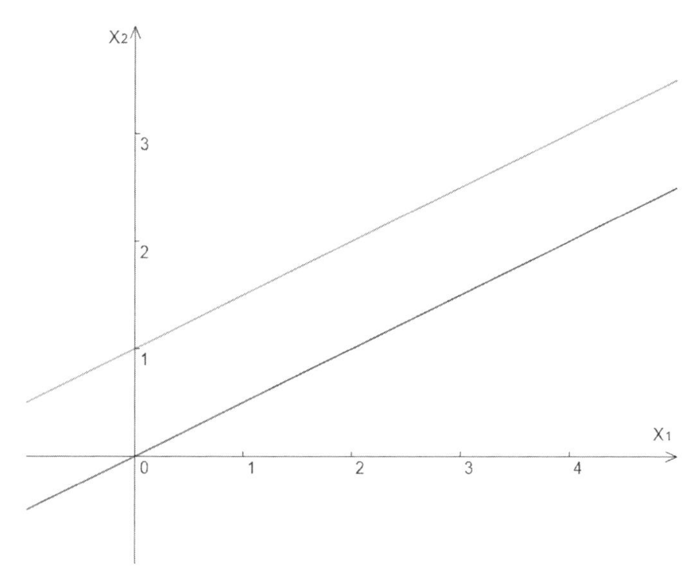

Figure 3.2 No solutions exist if lines, or hyperplanes in general, are parallel.

The same thing may happen in 3D, of course. In a cubic room, for example, a ceiling doesn't intersect the floor. A back wall is exactly parallel to the front wall, too. A back wall, a flat roof, and a front wall will never intersect at a single point.

Well, some systems don't have a solution. Luckily, we can detect parallel hyperplanes from the equations. The equations for the case in figure 3.2 are

$$x_1 - 2x_2 = 0$$

$$0.5x_1 - x_2 = -1$$

We can rewrite the second one to look more like a familiar line equation. Let's move the *-1* from right to left, and by doing that make the right side *0*:

$$x_1 - 2x_2 = 0$$

$$0.5x_1 - x_2 + 1 = 0$$

Now let's multiply the second equation by 2, which we can do without changing the essence of the equation. Multiplying everything by the same nonzero number doesn't change the fact that the left side of an equation equals the right side:

$$x_1 - 2x_2 = 0$$

$$x_1 - 2x_2 + 2 = 0$$

That's what parallel lines' equations look like. They're exactly the same except for the free coefficient—a number that doesn't belong to any of the variables.

Now, if we want to intersect two lines, and the lines are exactly the same, every point of any line will work as a point of intersection. In terms of equations, if a system has two equations, and both equations describe the same line, the system will have an infinite continuum of solutions. Such systems are called *indeterminate.* As an example, these lines coincide:

$$x_1 - 2x_2 + 2 = 0$$

$$0.5x_1 - x_2 + 1 = 0$$

You can take any pair of x_1 and x_2, that satisfy the first equation, such as (1, 1.5) or (2, 2), and the pair will immediately satisfy the second because, geometrically speaking, both equations describe the same line. All the points of the first line are the points of the second line, and each point is also a system's solution. Equations that represent the same hyperplane, even with different coefficients, are called *linearly dependent.*

Systems that contain linearly dependent equations are often considered to be unsolvable, as you often see in the documentation for numerical solvers. That's not because these systems have no solutions, but because they don't have a single solution, and numerical solvers are often tailored to get one and only solution for you. A continuum of solutions may overwhelm a numeric algorithm and make it crash and burn.

Also, if the hyperplanes' equations are the same except for the free coefficient—the one that has no variable assigned—the hyperplanes aren't the same, but parallel, and the system of linear equations that they represent doesn't have a solution. Furthermore, systems that include almost-parallel hyperplanes may be technically solvable, but they don't play well with numeric solvers. Solving these systems numerically may take a lot of time, and the result may come with a large error, too.

But what if a system of three linearly independent equations has four variables? Or what if there are two variables but four equations? We'll look at these systems in the following section.

> **SEE ALSO** An interactive applet on the Math Warehouse website is dedicated to linear equations. You can find it at http://mng.bz/Px99. Interestingly, the applet doesn't recognize coinciding lines, considering them only to be parallel, so it reports no solutions where infinite solutions exist.

3.1.3 Section 3.1 summary

A system of N linear equations of N variables is geometrically the same as a set of N N-hyperplanes. In 3D, they're regular 3D planes, and in 2D, they're lines. We're exploiting this property to showcase the systems and their solutions in this chapter.

Where all the hyperplanes intersect, there is a point of the solution. If some of the hyperplanes are parallel, which happens when only the left sides of their equations are linearly dependent, there is no solution. If some of the hyperplanes coincide, which happens when their equations are fully linearly dependent, the solution is no longer a single point, but a continuum of points. This situation sometimes upsets numerical solvers, causing them to refuse to give any answer.

3.2 Overspecified and underspecified systems

If there are as many variables as there are linearly independent equations, the system is called *well-specified* or *well-defined*. This means systems with two equations and two variables, three equations and three variables, four equations and four variables, and so on. These systems probably have single points as solutions.

But if the number of equations and the number of variables don't match, things get complicated. Let's take a look at such systems and see what we can get from them.

3.2.1 Overspecified systems

When there are more equations than variables, the system is called *overspecified, overconstrained,* or *overdetermined.* A system of three equations with only two variables, for example, is overspecified. As you might imagine, in 2D, such a system looks like figure 3.3.

Normally, such a system doesn't have a solution. A solution is a point where all the lines, or hyperplanes in general, intersect, and what are the chances of that? Well,

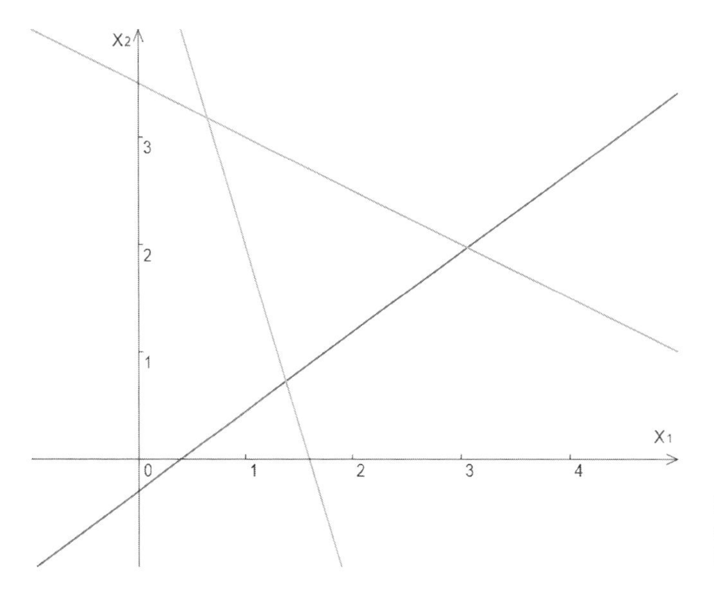

Figure 3.3 An overspecified system with three equations but two variables

technically, we can't exclude the possibility of such an intersection, so even overspecified systems may have a solution after all (figure 3.4).

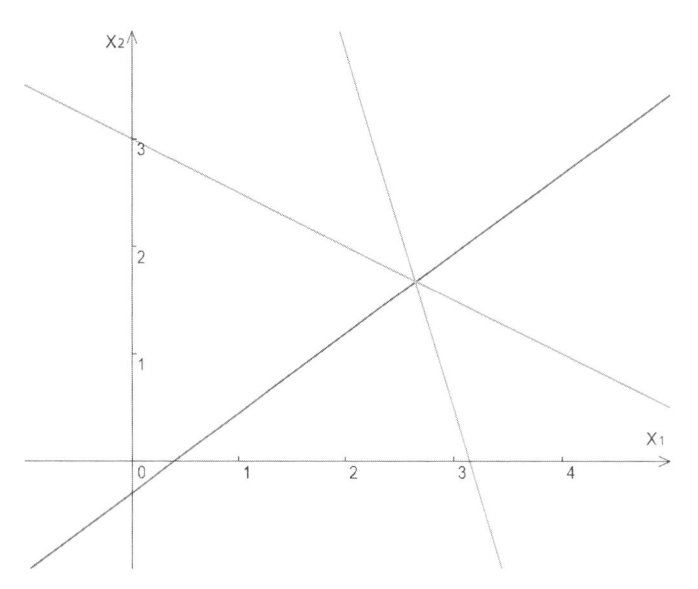

Figure 3.4 An overspecified system that oddly enough has a solution

If some of the hyperplanes coincide, or their equations are linearly dependent, a system might have a solution, too—or even, as we'll see in the following section, an infinite number of solutions. Also, some numeric methods can solve overspecified systems, assuming that all the data confirms the same solution. Even if geometrically, the hyperplanes of equations don't have a single point of intersection, there's some point to which most of them are close enough. A solution in this sense isn't geometrically correct but still pragmatic enough to be valuable.

Case in point: in computed tomography, the final 3D image is computed from a huge array of x-ray measurements that come with some inevitable error. We'll see how the computation goes in chapter 12. Although it's technically possible to compute a 3D image from a well-specified system, overspecification there helps minimize the error.

3.2.2 Underspecified systems

If a system has more variables than equations, it's called *underspecified, underconstrained,* or *underdetermined.* A system with three variables but only two equations would be such an example.

In 2D, the underspecified system consists of a single equation, the equation denotes a single line, and every point of this line automatically becomes a solution. That's a bit boring.

In higher dimensions, underspecified systems usually have a continuum of solutions. When two 3D planes intersect, they intersect at a straight line, and every point of this line is a solution to the system. But when two 3D planes are parallel, or their equations are linearly dependent except for the free coefficient, there's no intersection at all, and no point in space constitutes a solution to such a system.

Underdetermined systems are rarely solved numerically; no single set of numbers describes the solution. But these systems still could be, and often are, solved symbolically. Who says that only a point can be a solution? A line, a plane, and a hyperplane are valid mathematical objects, and we can find good use for them all.

Symbolic computation is relatively slow, however. We can solve systems of a million equations numerically, but symbolically, it takes forever to solve even an 8×8 system.

Sometimes, though, when the equations allow, we can try to split the system into two. First, we solve only part of the system—symbolically, of course—and then the solution is another set of equations. We can use these equations to solve the rest of the system.

We'll see an example of this trick in chapter 4. For now, be aware that having a hyperplane as a solution isn't such a bad thing.

3.2.3 Section 3.2 summary

Even if there's no single point where all the hyperplanes meet, so there's no single tuple of coordinates as a solution, the system may still be solvable. Overspecified systems are often solved numerically and inaccurately. Underspecified systems are solved symbolically; their solutions are hyperplanes, including lines in 2D and planes in 3D, not specific points.

3.3 A visual example of an interactive linear solver

I've already mentioned inaccurate but pragmatic numeric solvers that can solve even systems that are mathematically unsolvable. They sound like magic, but they're only more math in disguise.

Now we'll craft our own algorithm to solve a system of linear equations for us. This algorithm isn't practical, and much better ones have already been invented, but it's supposed to illustrate the general concept behind the whole class of numeric methods.

3.3.1 The basic principle of iteration

A whole branch of computational mathematics called *numerical methods* handles solving things with tolerable error; it's vast and complex, so we won't go there. But to use these methods effectively, we need to understand them. Well, this book is a geometry book, so we'll use geometry to build this understanding.

Let's solve a system of two equations iteratively. First, we need to pick a point. Any point will do. Then we project the point onto one of the lines that the equations make (figure 3.5). A *projection* here is simply a point on a line that's the closest to the original one. The vector from a point to its projection forms the right angle with the line.

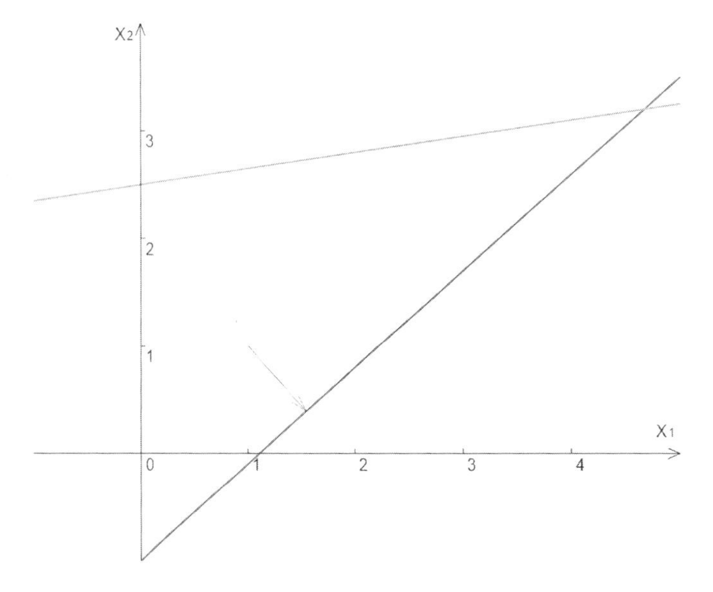

Figure 3.5 A point is being projected on the bottom line.

The point of projection is closer to any point of the bottom line than the original point, the point of the solution included. Now let's project the point from the bottom line to the top one (figure 3.6).

The point is getting closer to all the points of the top line, including the point of the solution. Because the point of solution belongs to both lines by definition, simply projecting the initial point back and forth moves it closer to the point of solution (figure 3.7).

So wherever we start initially, every projection brings the point closer to the point where lines intersect, which is the solution to our system. The step that brings us closer to the solution is called *iteration*. The iterations we use in real-world algorithms are usually way less graphic than a simple projection, but they do bring us to the solution faster.

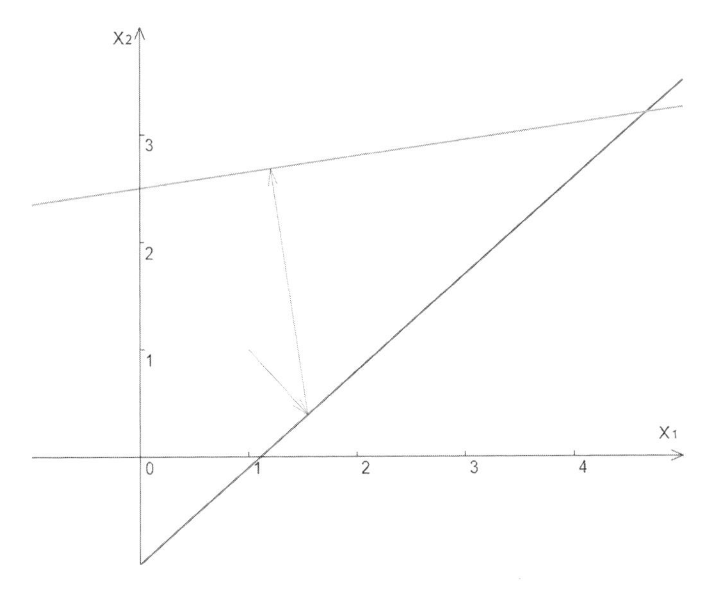

Figure 3.6 A point is projected to the bottom and then to the top line.

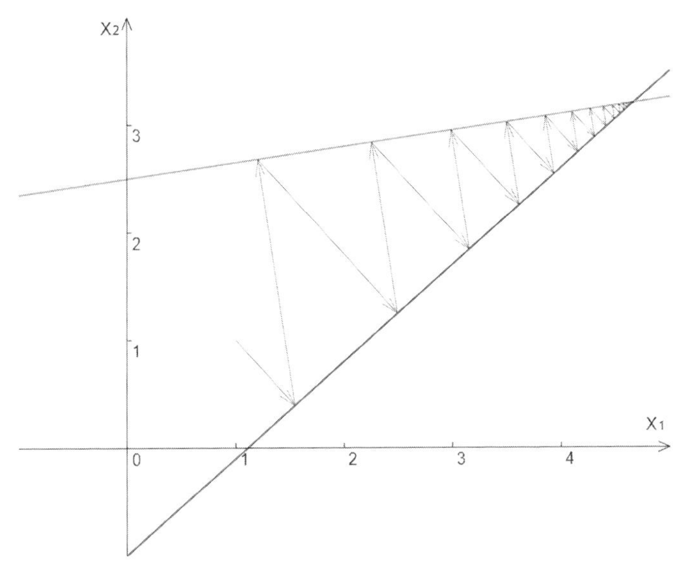

Figure 3.7 The point of projection gets closer to the point of solution with every projection.

3.3.2 Starting point and exit conditions

Now we know how to get closer to the point of the solution, but that knowledge isn't enough to craft a workable iterative algorithm. We also need to think about where we start iterating from and—most important—how we stop.

Is the starting position relevant to our task? Not really. Ideally, of course, we want it to be as close to the solution as possible, but we don't know where that might be. That's the whole point. When there are no criteria for choosing the best point, any point will do.

Are the exit conditions relevant? Definitely! Look, although we're getting closer to the solution with every iteration, strictly mathematically, we'll never get there. In the general case, the step with which we're getting closer gets smaller and smaller but never equals zero.

At some point, literally and figuratively, we have to stop and say we're close enough to a pragmatic solution. One possible exit condition for our problem is the length of the last step. When it gets small enough, we should be close enough. In figure 3.8, we allow our algorithm to stop when the step size becomes smaller than 0.3.

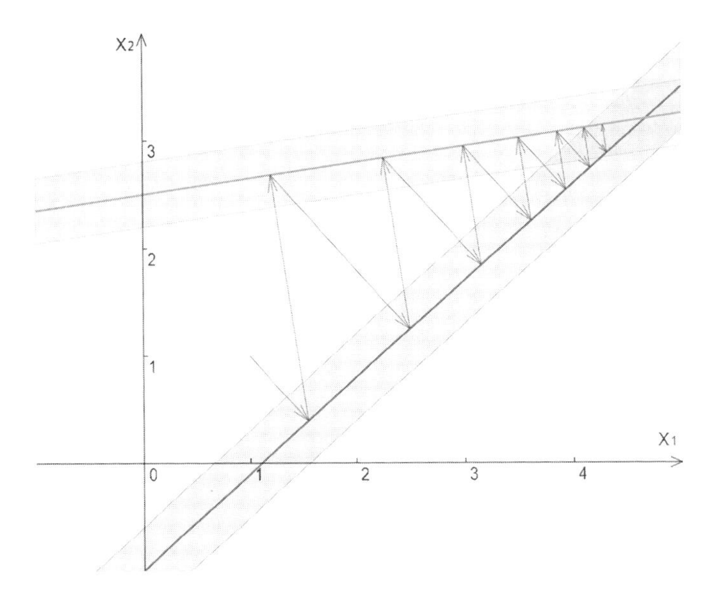

Figure 3.8 Iterative algorithm with an exit condition: when the step is shorter than 0.3, the algorithm stops.

This approach has a fault, however: the number 0.3 doesn't reflect how near we got to the actual solution. In our case, the solution is (4⅔, 3.2), and the solution found by the algorithm is (~4.273, ~3.141). The actual error is more than 0.3; it's closer to 0.4. The situation gets worse as the angle between the lines gets more acute (figure 3.9).

Sometimes, the number that characterizes the exit condition—in our case, 0.3—is called *tolerance*. Please note that the algorithm's tolerance doesn't necessarily reflect its actual error, although some correlation is usually implied.

In our example, the correlation depends on the problem itself. The more parallelish the lines are, the more error we get for the same tolerance. When the lines are truly parallel, there's no solution, and our algorithm even fails to fail; it goes from one line to another, alternating infinitely between the same two points. On the other hand, when the lines are orthogonal (figure 3.10), we get a precise solution, and in only two iterations!

The majority of real-world problems, as you might imagine, lie between these two cases: the infinite search, and the immediate solution. A special term characterizes where the iterative process falls on this spectrum, so let's look into the related terminology.

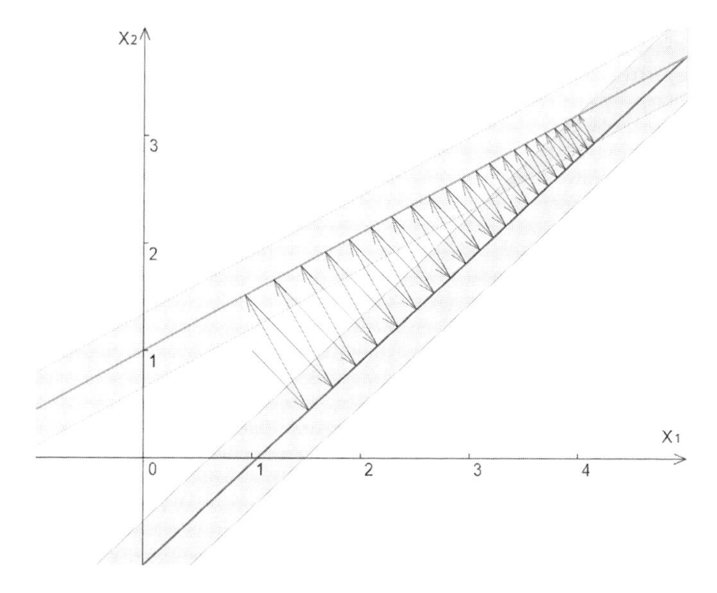

Figure 3.9 The difference between the last step length and the actual error gets larger as the angle gets smaller.

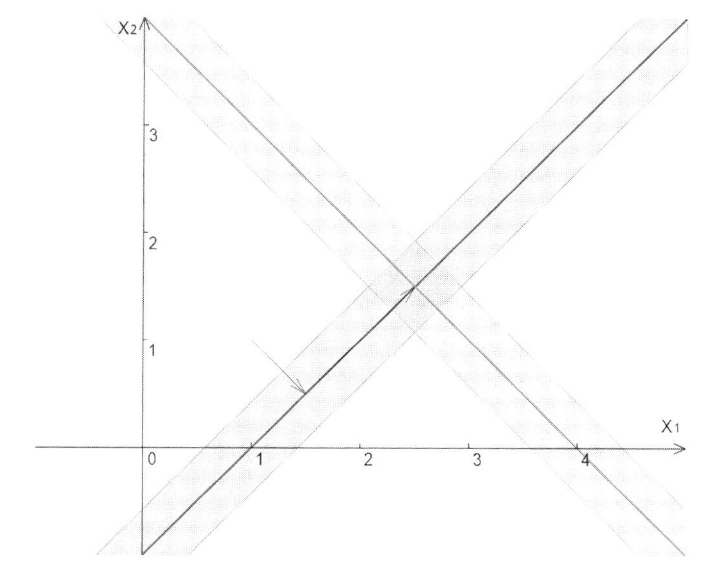

Figure 3.10 Ideal problem: no error, and two iterations

3.3.3 Convergence and stability

The term *convergence* reflects how fast the algorithm progresses, if at all. The algorithm in the preceding section either converges or doesn't; there's no other option. But some other iterative algorithms may *diverge*. In some specific cases, they get farther from the solution with every iteration.

Convergence isn't always a measurable characteristic. In programming, some algorithms can be described helpfully while using this term vaguely, which is fine because for most usable algorithms, making a proper convergence analysis may be harder than creating a new algorithm from scratch.

Another term that's closely associated with convergence is *stability*. Stability is the ability of the algorithm not to accumulate an error as it goes, which is extremely important for iterative algorithms.

Let's say that for some other iterative algorithm, each step introduces 0.1 percent of error. We reuse the step's output as the input for the following step, and the error accumulates. After the second step, the error becomes more than 0.2 percent. This progression doesn't seem too bad yet. But if the error accumulates consistently, in 100 iterations, it becomes more than 10 percent, and after 694 iterations, the error exceeds 100 percent, rendering the algorithm by itself useless.

Our particular algorithm is stable, meaning that even if each step, being a projection, introduces some computational error, this error is random and doesn't accumulate.

Convergence and stability are concepts that come from numerical methods, and as such, they're not geometrical. To solve a geometric problem with a linear system solver, however, we need to pick the right algorithm, and to do so, we need to understand its capabilities and limitations.

3.3.4 Section 3.3 summary

To build an iterative algorithm, we need to answer three questions: how do we select where to start, how do we proceed, and how do we know we're close enough? The speed at which the algorithm proceeds toward the solution is characterized by the convergence, and the computational error it might accumulate while going there is characterized by its stability.

3.4 Direct solver

Not all linear solvers are inherently iterative. Some of them are *direct,* meaning that instead of converging toward the solution for nobody knows how long, they compute the solution directly in a predetermined number of operations.

3.4.1 Turning an iterative solver into a direct one

We've seen that when the lines are orthogonal, our iterative algorithm finds the solutions in two iterations. Because when lines are orthogonal to each other, the point "travels" toward the line in parallel to the next line. The fact that it moves orthogonally to the line we project the point onto doesn't matter anymore. So can't we make our steps always go in parallel with the other line so that our algorithm always works in two iterations? Yes, we can (figure 3.11).

Awesome! We have no more convergence and no more exit conditions. We move the point twice, and there we go!

So why did we bother learning about iterative algorithms? Well, direct algorithms are all fun and games when we have only two lines. When the dimensionality of the

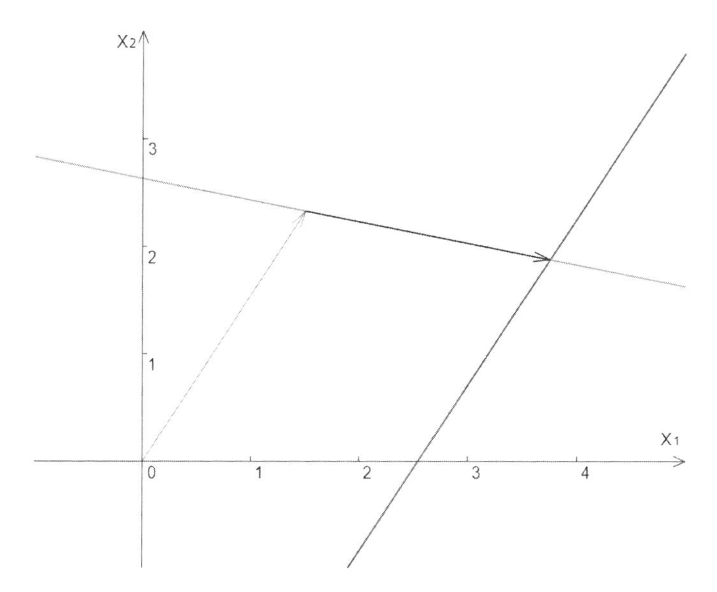

Figure 3.11 An iterative solver turned into a direct one by changing the way we do an iteration

problem rises, however, so rises the complexity of the direct algorithms, and for some algorithms, it rises faster than for others.

This "go along the other line" approach is simple in 2D, but even in 3D, you have to move the point in parallel to both other planes to make a single iteration, and that task isn't trivial. You have to find a line where these planes intersect and use it as a guiding vector. To do that you'll have to solve a system of two linear equations. So to solve a system of three equations, you have to solve a system of two equations three times, because you need three steps to get to the point of the solution.

In 4D, there are three 4D hyperplanes. They also intersect at a straight line, but now you have to solve a system of three linear equations four times. Every one of these equations implies solving three systems of two equations, and every system of two equations boils down to solving a pair of linear equations. So the total number of steps for solving a four-equation system is $4 \times 3 \times 2 = 24$.

Generally, to solve an N-dimensional hyperplane intersection, we have to solve N (N–1)-dimensional hyperplane intersections first. And each (N–1)-dimensional problem implies solving N–1 (N–2)-dimensional ones until we get down to solving a lot of 1-dimensional problems, which are trivial.

All in all, the solution of *N* equations by the naive implementation of our direct algorithm requires $N \times (N-1) \times (N-2) \ldots 2 \times 1$ trivial solutions. This number is also called *N-factorial* and has an exclamation sign as its mathematical notation:

$$N \times (N-1) \times (N-2) \ldots 2 \times 1 = N!$$

This notation seems unintimidating in little numbers, but to get an impression of how fast this number grows, see table 3.1, which lists the first 20 values of the N-factorial.

Table 3.1 First 20 values of N!

N	N!
1	1
2	2
3	6
4	24
5	120
6	720
7	5,040
8	40,320
9	362,880
10	3,628,800
11	39,916,800
12	479,001,600
13	6,227,020,800
14	87,178,291,200
15	1,307,674,368,000
16	20,922,789,888,000
17	355,687,428,096,000
18	6,402,373,705,728,000
19	121,645,100,408,832,000
20	2,432,902,008,176,640,000

NOTE I chose the number 20 because it's the last factorial value that fits into a 64-bit integer. The factorial function is a common interview question for junior developers. It's funny that the most effective implementation for limited integers in C++ or Java is not a real function, but a short lookup table.

So the direct algorithm, although it always works in N iterations, becomes complex when N grows large. At some point, it becomes impractically complex.

This brings us to the question of algorithm complexity. How do we assess that?

3.4.2 Algorithm complexity

We usually don't estimate algorithm complexity by counting actual processor cycles. This kind of estimate is possible, and we have tools that can do the work, but then we'd have to reestimate complexity every time a new processor came out.

We don't estimate complexity by measuring actual computational operations, either. This kind of estimate is also possible and sometimes even reasonable for small algorithms or for comparisons of micro-optimizations. But the number of computational operations in the machine code depends heavily on how the compiler optimizes the code, and we would have to redo this estimate with every new compiler version.

In fact, we don't even estimate complexity by measuring anything in a particular scenario; doing that would make the estimation dependent on the scenario itself. As we saw in the "parallelish lines" case, the speed of the algorithm may rely heavily on the problem. Instead, we use a theoretical estimate as a limiting function of the problem instance size.

In our naive implementation of the direct solver, for example, we see that to solve a system of N equations, it needs N! solutions of trivial systems. No matter how long it takes to gather these solutions in practice, even if we can "cheat" and cache some of the solutions to avoid recomputation, the fact that the whole computation requires no more than N! steps already gives us a good-enough characteristic of the algorithm. We can write this fact down by using so-called asymptotic notation, specifically big-O:

$$O(N!)$$

Big-O was designed to explain how fast functions grow long before computers were invented. It knows nothing about caches and latencies. All it does is abbreviate a function to the part that grows fastest, neglecting all the uninteresting details.

Let's say you have an algorithm that computes the sum of all the matrix elements. You need to go through all the rows and all the columns, so the complexity function of your algorithm will be quadratic:

$$\text{complexity}(N) = N^2$$

In big-O notation, this complexity function remains the same, but you write it differently:

$$O(N^2)$$

Now if you expand your algorithm to also compute the sum of the free coefficients, you make N more additions:

$$\text{complexity}(N) = N^2 + N$$

The function gets more complicated. But as N^2 becomes larger and larger, N becomes less and less relevant in comparison. In big-O, we simply let it go. The big-O estimation for our "more than quadratic" function is still quadratic:

$$O(N^2)$$

If instead of simply summing the elements, we want to sum something more complex, such as the maximal element in a row and in a column for every position in a matrix, the complexity function could be something like this:

$$\text{complexity}(N) = N^2(2(N-1) + 1)$$

Let's expand it:

$$\text{complexity}(N) = 2N^3 - N^2$$

In big-O, as we remove every member except the one that contributes most to complexity, we also remove the constant factors. It doesn't help us to describe an algorithm; in complex numbers, for example, the same algorithm will run twice as long anyway. We want to estimate the algorithm as a general idea, not a specific implementation, so the constant has to go. Losing constants and lesser members turns a slightly complicated formula into a simple and elegant $O(N^3)$.

In practice, of course, losing our constants may be misleading. The big-O for a hash map lookup, for example, is $O(1)$, and the big-O for a binary tree lookup is $O(\log(N))$. Does this mean that a hash map always outperforms the tree? No, it doesn't. When a tree is short enough, it can easily outperform the map. Their big-Os start to make sense only as the data grows larger and larger. At some point, at some N, the hash map will start outperforming the binary tree; the exact value of that N depends not on the complexity estimates, but on both tree and map implementations.

Complexity estimation in big-O doesn't seem too specific, but paradoxically, it's theoretical enough to be practical. Most of the time, we don't want an exact prediction of how long will it take for the algorithm to finish its work anyway. The time relies heavily on multiple factors, mostly hardware, compiler, and optimizations. We want a complexity estimation to compare algorithms when we're picking the right one for the job. And for that purpose, the more the metric is disjoined from reality, the better.

Suppose that we want to solve a system of four equations, and we have two algorithms. One algorithm has its complexity estimated as $O(N^3)$, and the other's complexity estimate is $O(N!)$:

$$4^3 = 64, \text{ and } 4! = 24$$

This doesn't necessarily mean that the latter algorithm will be faster. Remember that the estimation is theoretical; it doesn't tell us which is better. But it shows that the algorithms are comparable at speed and that both could be used for this problem.

Now suppose that we have a system of 1,000 equations. $1000^3 = 1{,}000{,}000{,}000$, and $1000! = \ldots$ I probably shouldn't write this number, as it would be a whole page full of digits. Yes, it's that large. Clearly, the first algorithm wins, because it's capable of finishing its work before the Sun runs out of fuel, and the second one isn't.

There are other notations for algorithm complexity estimations, but big-O is used most. All the algorithms in the C++ standard library, for example, have their complexity specified in big-O. So when we pick an algorithm for a job, we can look briefly at what the standard says about its complexity and make a well-reasoned choice.

3.4.3 Section 3.4 summary

Some algorithms for solving linear equation systems aren't iterative, but direct. They operate on different principles that allow them to solve systems in a predictable amount of time.

Although the efficiency of iterative algorithms is characterized mostly by their convergence, direct algorithms should be assessed by using complexity estimation instead. This complexity estimation usually comes in big-O notation, which is a function showing how the amount of abstract operations grows with the dimensionality of the problem.

3.5 Linear equations system as matrix multiplication

We've done something similar before. Matrices multiply in row-on-column order, so if you present your N-dimensional vector as a 1 × N matrix, you can multiply it by an N × N matrix and get another 1 × N matrix as a result.

Now let's turn things around. Let's multiply a matrix by an N × 1 vector and see what happens.

3.5.1 Matrix equations

Let this be our matrix:

$$\begin{pmatrix} a_{11} & a_{12} & a_{13} \\ a_{21} & a_{22} & a_{23} \\ a_{31} & a_{32} & a_{33} \end{pmatrix}$$

We want to multiply it by this 3 × 1 matrix, which is essentially a vector (x_1, x_2, x_3) in disguise:

$$\begin{pmatrix} x_1 \\ x_2 \\ x_3 \end{pmatrix}$$

This kind of vector is sometimes called a *column vector*, as opposed to the row vector that we saw in chapter 2. Let's write the result of the multiplication as another column vector:

$$\begin{pmatrix} b_1 \\ b_2 \\ b_3 \end{pmatrix}$$

When put together, these matrices form a matrix equation:

$$\begin{pmatrix} a_{11} & a_{12} & a_{13} \\ a_{21} & a_{22} & a_{23} \\ a_{31} & a_{32} & a_{33} \end{pmatrix} \times \begin{pmatrix} x_1 \\ x_2 \\ x_3 \end{pmatrix} = \begin{pmatrix} b_1 \\ b_2 \\ b_3 \end{pmatrix}$$

Let's say that we know the *a*s and the *b*s, and we want to know the *x*s. So far, we're not fluent with matrices, so the reasonable thing to do would be to revert to good old scalar formulas, which we can do by making both sides of the equation into column vectors. The column vectors are equal when their elements are equal and their elements are scalars. So let's do the matrix multiplication for the left side, row-on-column, business as usual:

$$\begin{pmatrix} a_{11} & a_{12} & a_{13} \\ a_{21} & a_{22} & a_{23} \\ a_{31} & a_{32} & a_{33} \end{pmatrix} \times \begin{pmatrix} x_1 \\ x_2 \\ x_3 \end{pmatrix} = \begin{pmatrix} a_{11}x_1 + a_{12}x_2 + a_{13}x_3 \\ a_{21}x_1 + a_{22}x_2 + a_{23}x_3 \\ a_{31}x_1 + a_{32}x_3 + a_{33}x_3 \end{pmatrix}$$

Now we can make each pair of corresponding values in the column vectors into equations:

$$a_{11}x_1 + a_{12}x_2 + a_{13}x_3 = b_1$$

$$a_{21}x_1 + a_{22}x_2 + a_{23}x_3 = b_2$$

$$a_{31}x_1 + a_{32}x_2 + a_{33}x_3 = b_3$$

This system is well-defined because there are three variables—x_1, x_2, x_3—and three equations. We already know that systems of this type usually have a single solution, and even when they don't, we know why.

This exercise shows that we can solve matrix equations as linear systems. More important, we can solve systems of linear equations solely by playing with matrices. But first, we have to learn a little more about matrices themselves.

3.5.2 What types of matrices we should know about

We've already seen an identity matrix, which is the best matrix we can meet:

$$\begin{pmatrix} 1 & 0 & 0 \\ 0 & 1 & 0 \\ 0 & 0 & 1 \end{pmatrix} \times \begin{pmatrix} x_1 \\ x_2 \\ x_3 \end{pmatrix} = \begin{pmatrix} b_1 \\ b_2 \\ b_3 \end{pmatrix}$$

Essentially, this isn't an equation but already a solution, because what the equation says is

$$x_1 = b_1$$

$$x_2 = b_2$$

$$x_3 = b_3$$

The next-best thing is called a *diagonal matrix*, a matrix in which all the nonzero elements reside on its top-left to bottom-right diagonal:

$$\begin{pmatrix} a_1 & 0 & 0 \\ 0 & a_2 & 0 \\ 0 & 0 & a_3 \end{pmatrix} \times \begin{pmatrix} x_1 \\ x_2 \\ x_3 \end{pmatrix} = \begin{pmatrix} b_1 \\ b_2 \\ b_3 \end{pmatrix}$$

This is also a nice matrix to meet in an equation because the solution boils down to this:

$$x_1 = \frac{b_1}{a_1}$$

$$x_2 = \frac{b_2}{a_2}$$

$$x_3 = \frac{b_3}{a_3}$$

Next, a *triangular matrix* is one that has all the nonzero elements below the diagonal, diagonal included, or above it. The former type is called a *lower triangular* matrix:

$$\begin{pmatrix} a_{11} & 0 & 0 \\ a_{21} & a_{22} & 0 \\ a_{31} & a_{32} & a_{33} \end{pmatrix} \times \begin{pmatrix} x_1 \\ x_2 \\ x_3 \end{pmatrix} = \begin{pmatrix} b_1 \\ b_2 \\ b_3 \end{pmatrix}$$

The latter type is called an *upper triangular* matrix:

$$\begin{pmatrix} a_{11} & a_{12} & a_{13} \\ 0 & a_{22} & a_{23} \\ 0 & 0 & a_{33} \end{pmatrix} \times \begin{pmatrix} x_1 \\ x_2 \\ x_3 \end{pmatrix} = \begin{pmatrix} b_1 \\ b_2 \\ b_3 \end{pmatrix}$$

These systems may seem to be nontrivial to solve at first, but with a little patience, you can still solve these types of matrix equations or linear systems one row at a time. For the lower triangular matrix, the computation goes like this:

$$x_1 = \frac{b_1}{a_{11}}$$

Then, when we know the x_1:

$$x_2 = \frac{b_2 - a_{21}x_1}{a_{22}}$$

Finally, with a known x_1 and x_2:

$$x_3 = \frac{b_3 - a_{31}x_1 - a_{32}x_2}{a_{33}}$$

These three types of matrices form equations that are easy to solve—no iterations and no factorial-like algorithm complexity, but only subtraction, multiplication, and division. The best thing is that by doing a series of operations, we can turn any arbitrary matrix into a triangular one!

3.5.3 *Things we're allowed to do with equations*

We've seen that multiplying all the coefficients of a line by the same nonzero number doesn't change the line. The same thing goes for planes and hyperplanes. Certainly, because scaling the coefficients equally doesn't change a hyperplane, it doesn't change the point at which this particular hyperplane intersects all the other hyperplanes in a system. In other words, we can take any equation from a system and multiply it by a nonzero number. The system will retain the same solution because geometrically, none of the lines, planes, or hyperplanes that constitute the system have been changed.

Another operation that doesn't affect the system's solution is adding one equation to another. In 2D, for example, if we add one equation to another, we definitely change the latter equation. A line becomes a completely new line everywhere except at the point where two source lines used to intersect. At that point, and only at that point, both equations in a sum become equivalent, so adding one to another becomes the same as multiplying the equation by 2.

By adding equations together, we change the equations themselves, but the point of the solution remains intact. And because we can also multiply equations by nonzero, we can safely multiply the second equation by -1 before addition, effectively turning addition into subtraction.

The third thing we're allowed to do is swap equations between themselves. This should be fairly obvious geometrically. If line *a* intersects line *b* in point *c*, line *b* intersects line *a* in point *c*, too. They share the point of intersection by definition, so there's no difference between line *a* intersecting line *b* or line *b* intersecting line *a*.

To sum things up, we can safely multiply any equation by a nonzero number; we can add any equation to another equation, which also implies subtraction; and we can make any two equations swap their places in a system. Although each of these operations changes some of the equation's coefficients, the system as a whole remains solvable at the same point as before. We'll see how we can combine these operations to solve a system in section 3.6.

3.5.4 Section 3.5 summary

A linear equations system may be written as a matrix by column vector multiplication resulting in another column vector. Three operations allow us to bring any solvable linear system to a form where the coefficients matrix becomes the identity and the system itself becomes a solution. The first two operations are scalar multiplication and addition over equations, and the third operation is a swap.

3.6 Solving linear systems with Gaussian elimination and LU-decomposition

The terms in the section title look intimidating, but they become transparent as soon as you learn what the algorithms behind them do. So before discussing the terms, let's go back to our two-trains problem from the beginning of the chapter one last time.

3.6.1 An example of Gaussian elimination

To remind you of the two-trains problem from the beginning of the chapter (and also chapter 1), the equations are

$$V_p - 2V_a = 0$$

$$V_p + V_a = 450$$

The system graph looks like figure 3.12.

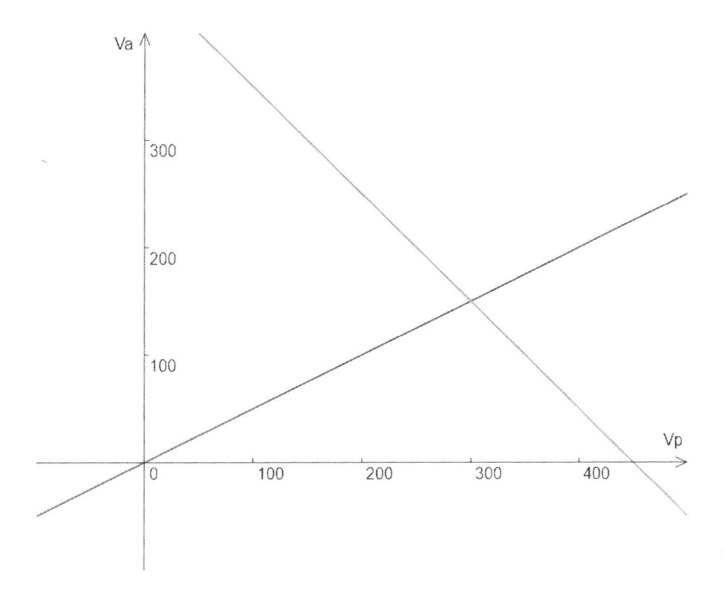

Figure 3.12
The two-trains problem

Let's solve this system once more, this time by eliminating the equation's coefficients one by one until the solution appears by itself. First, let's multiply the second equation by the coefficient before V_a from the first equation, which would be -2:

$$V_p - 2V_a = 0$$

$$-2V_p - 2V_a = -900$$

This operation doesn't change the lines.

Next, let's subtract the second equation from the first one. We're allowed to subtract, because subtraction is the same as multiplying the second argument by -1 and adding that to the first one. Then the equations become

$$3V_p = 900$$

$$-2V_p - 2V_a = -900$$

and the lines change (figure 3.13).

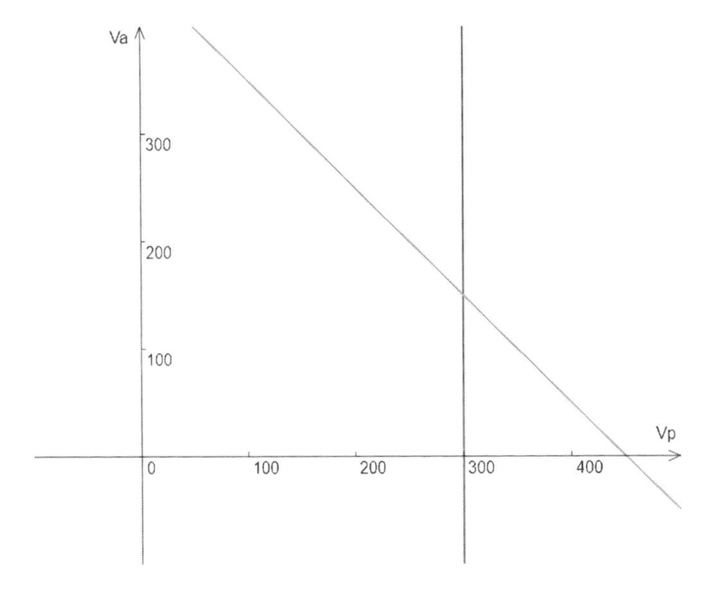

Figure 3.13 Lines after the equations subtraction

Let's divide the first equation by the V_p coefficient and multiply it by the V_p coefficient from the second equation instead. Once again, this operation doesn't change the lines, only the coefficients:

$$-2V_p = -600$$

$$-2V_p - 2V_a = -900$$

But now we can subtract the first equation from the second, and this operation will change the second line the way that the earlier subtraction changed the first line. The equations become

$$-2V_p = -600$$

$$-2V_a = -300$$

and the lines are like those in figure 3.14.

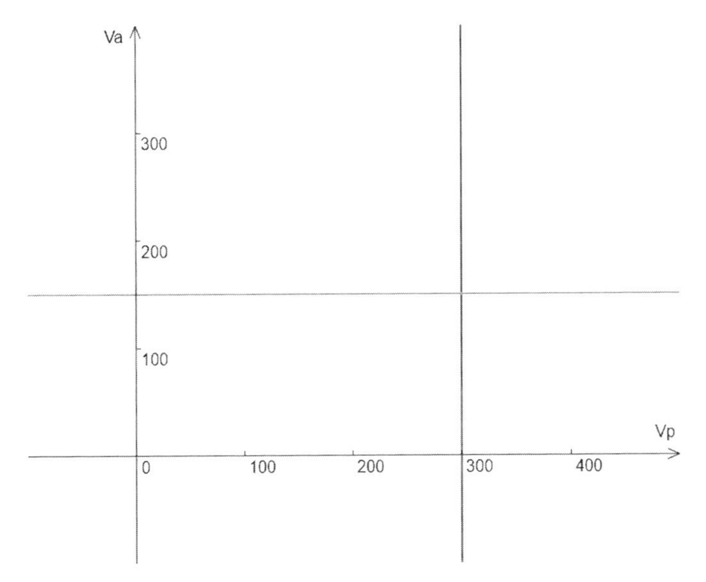

Figure 3.14 Lines after the second subtraction

The last step is to divide both equations by the coefficients before V_p and V_a, respectively:

$$V_p = 300$$

$$V_a = 150$$

That's how a system becomes a solution. If we put these numbers in the original equations instead of the variables, the equations hold:

$$300 - 2 \times 150 = 0$$

$$300 + 150 = 450$$

The process of solving a system in matrix form involves the initial matrix, the vector of variables, and the vector of free coefficients:

$$\begin{pmatrix} 1 & -2 \\ 1 & 1 \end{pmatrix} \begin{pmatrix} V_p \\ V_a \end{pmatrix} = \begin{pmatrix} 0 \\ 450 \end{pmatrix}$$

When we multiply the second row and a corresponding element from a vector of free coefficients by -2, the equation becomes

$$\begin{pmatrix} 1 & -2 \\ -2 & -2 \end{pmatrix} \begin{pmatrix} V_p \\ V_a \end{pmatrix} = \begin{pmatrix} 0 \\ -900 \end{pmatrix}$$

Then the first subtraction makes the 2 × 2 matrix lower-triangular:

$$\begin{pmatrix} 3 & 0 \\ -2 & -2 \end{pmatrix} \begin{pmatrix} V_p \\ V_a \end{pmatrix} = \begin{pmatrix} 900 \\ -900 \end{pmatrix}$$

We're halfway there. We know that that triangular matrices are easily solvable. But wait—it gets better. Let's multiply the first row and the first element of a vector on the right by -2/3. This operation doesn't change the lines, and it doesn't change the matrix form either:

$$\begin{pmatrix} -2 & 0 \\ -2 & -2 \end{pmatrix} \begin{pmatrix} V_p \\ V_a \end{pmatrix} = \begin{pmatrix} -600 \\ -900 \end{pmatrix}$$

But then the second subtraction converts it to a diagonal one:

$$\begin{pmatrix} -2 & 0 \\ 0 & -2 \end{pmatrix} \begin{pmatrix} V_p \\ V_a \end{pmatrix} = \begin{pmatrix} -600 \\ -300 \end{pmatrix}$$

Now all we have to do is divide each row and a corresponding coefficient by –2. This operation turns the diagonal matrix into the identity matrix:

$$\begin{pmatrix} 1 & 0 \\ 0 & 1 \end{pmatrix} \begin{pmatrix} V_p \\ V_a \end{pmatrix} = \begin{pmatrix} 300 \\ 150 \end{pmatrix}$$

That's our solution to the two-trains problem.

3.6.2 *What do "elimination" and "decomposition" mean?*

The principle of changing a matrix equation until it becomes a solution scales nicely to higher dimensions. No matter how many dimensions you have, you can always convert your coefficients matrix to an identity matrix by using only three operations:

- You can scale a row and a free coefficient by some nonzero number.
- You can add one row and a free coefficient to another row and a free coefficient.
- You can swap two rows and the respective free coefficients.

We didn't have to do the last operation in our two-trains example. But consider this: Our first move in the example was multiplying the second equation by the coefficient before V_a from the first one. But what if the coefficient is 0? We can't scale an equation by zero.

No problem! In this case, we swap the equations and scale by the other coefficient before V_a. If that coefficient is also 0, we're in trouble indeed, but not because we don't have enough operations—because there's simply no V_a in the system! The system is overspecified.

If a system is properly specified, and there are no linearly dependent equations, applying the tree operations in a specific order, as in the example, turns the matrix into the identity matrix and the vector of free coefficients into the vector of the solution. This method of solving a linear system is called *Gaussian elimination,* because we eliminate coefficients one by one and also because Carl Friedrich Gauss came up with this method first.

NOTE Except, of course, he didn't. According to Stigler's Law of Eponymy, no scientific discovery is named after its original discoverer. The idea of elimination has been known to the Chinese for at least 1,000 years and rediscovered repeatedly.

Another method close in heart to Gaussian elimination is *LU-decomposition. LU* stands for *lower-upper,* and it indeed decomposes a matrix into a product of lower-triangular and upper-triangular forms. Sometimes, it's also called *factorization* because these matrices work as factors or multipliers of the source one.

LU-decomposition may be used not only to solve linear equations, but also to find inverse matrices. It has several more applications, but at this point, they're not particularly helpful to us. What's important to know is that all three methods—Gaussian elimination, LU-decomposition, and matrix factorization—are essentially the same approach to solving linear equations.

Let's do one last thing before moving on to practical advice: estimate the complexity of this approach. To turn every coefficient from some arbitrary number into 0 or 1, we need no more than N^2 operations. Each operation, however, consists of multiplying an equation by a number, and that's N plus 1 operations per every coefficient. All in all, this approach results in $N^2(N+1)$ or N^3+N^2 operations, and in big-O notation, we ignore everything but the member that grows fastest, so the estimation would be $O(N^3)$. That's way better than the N! we had before.

Most of the direct methods in use today rely on some form of decomposition (Gaussian elimination or another). In practice, the algorithms we have are heavily optimized and have even better complexity estimation than $O(N^3)$. One algorithm has $O(N^{2.372})$.

Even though $O(N^{2.372})$ is rather fast for a direct algorithm, people still use other algorithms, including iterative solvers, and sometimes, the best algorithm for solving a linear system is no algorithm at all. In the following section, we'll see why.

3.6.3 Section 3.6 summary

Gaussian elimination, LU-decomposition, and matrix factorization are all based on the same principle: using scalar multiplication, addition, and row swapping to convert a matrix we have into a matrix we want. Both Gaussian elimination and LU-decomposition have $O(N^3)$ theoretical complexity in their naive implementation, but in practice, they work even faster.

3.7 Which solver fits my problem best?

Being a programmer, you're unlikely to write your own solver. Quite a few solvers have already been written, they've been improved through the decades, and most of them are in the public domain anyway. Competing with them isn't an easy task; neither is it economically rewarding.

Using solvers properly is important, however. And to use something properly, it helps to know how the thing works. That's why we've looked at a primitive iterative algorithm, a direct solver, and an example of Gaussian elimination. Now it's time to summarize what we've learned, sprinkle it with a few German surnames, and make it into a practical guide to choosing the best solver for a job.

3.7.1 When to use an elimination-based one

As we saw earlier, all solvers based on Gauss elimination, LU-decomposition, or factorization are based on essentially the same idea. They all have their complexity roughly estimated as $O(N^3)$, which makes them effective for small to midsize systems. If you have only 5 to 100 equations, don't look any further; any of these algorithms will do. They can deal with much larger systems, sometimes even those with millions of equations, but the matrices in which these systems are represented should be sparse. A *sparse matrix*, as opposed to a *dense matrix*, is a matrix in which most of the elements are zeros. This term isn't mathematical; no definite criteria separate sparse matrices from dense ones.

> **NOTE** There could be a rule of thumb, though. NVIDIA documentation on Compute Unified Device Architecture (CUDA)-accelerated libraries defines sparse matrices this way: *The cuSPARSE library contains a set of basic linear algebra subroutines used for handling sparse matrices. The library targets matrices with a number of (structural) zero elements which represent > 95% of the total entries.*

Remember that we discussed matrices being two-dimensional arrays? This representation is a good mental model, but it's not always best for representing data in actual computation. If a matrix is sparse, we can store only the nonzero elements along with their indices. We don't have to store zeroes; we know that they're there.

This data model works especially well with elimination-based algorithms, because their whole point is the elimination of nonzero elements. The data starts small and gets smaller as the algorithm progresses.

Most modern libraries provide data structures and routines that support sparse matrices. One time, a colleague of mine rewrote a project to use one library instead of another—namely, Eigen instead of LAPACK—and some tests became several times faster because of the change. I was initially skeptical about the change because both LAPACK and Eigen are well-established and well-polished libraries. Eigen is a C++ library, and LAPACK, although originally written in Fortran, is universally available via its C interface. Changing one for another shouldn't make much difference performancewise, but somehow, it did.

As it turned out, my colleague not only changed the library, but also switched some of the solvers to work on sparse structures. LAPACK doesn't have that functionality out of the box. So the real change was not about adopting a C++ library instead of a Fortran one, but adopting a whole new set of algorithms with it.

True, a modern compiler can make your code run a little faster. But to make it dramatically faster, you need to revise your algorithms.

SEE ALSO You can learn more about sparse matrices and how to deal with them in Eigen's documentation at http://mng.bz/JlO0.

3.7.2 When to use an iterative one

Although even direct algorithms bring in some computational error, iterative ones are all about changing computational error for speed. They make sense if you have a lot of equations and need to solve them fast—not necessarily accurately, but fast.

Also, unlike algorithms based on elimination or decomposition that work only with well-defined problems, iterative algorithms can work with overspecified systems as well. Even if there's no single point of intersection for all the hyperplanes, there's a point (or even a continuum of points) that lies as close to all the hyperplanes as possible, which pragmatically can work as a solution.

Popular iterative methods such as Jacobi and Gauss-Seidel are tailored to work with diagonally dominant matrices. Unlike sparse and dense, this term is a formal mathematical term. *A diagonally dominant matrix* is a matrix in which the absolute values of each diagonal element are greater than or equal to the sum of absolute values of other elements in a row.

With diagonally dominated matrices, these methods converge quickly and don't accumulate much error. They might work if a matrix isn't strictly diagonally dominant, but their convergence isn't mathematically guaranteed.

You can use methods called *preconditioners* to prepare your matrix for an iterative solution, so having a noncompliant matrix isn't a show-stopper. Sometimes, you can rearrange your matrix simply by swapping the rows, or you can use something like partial LU-decomposition to make your matrix diagonally dominant.

These methods are usually supplied along with a solving library, so you don't have to rearrange the rows by yourself. Be aware, however, that convergence isn't a given and that more often than not, simply running a solver on your problem isn't the best approach.

3.7.3 When to use neither

In this day and age, the programming community is pretty much librarycentric. It's considered to be bad manners to reimplement a piece of functionality that someone else coined for you. When you're working with small linear systems, however, reinventing the wheel may be well worth your time, especially if you have the right tools.

Solving small problems in general, and solving small systems of linear equations in particular, may suffer dramatically from the overhead of data structure conversion, passing the data, and even making a function call. If you can solve your problem with a handful of additions and multiplications, even extra memory access will show itself in a profiler.

You may think that because the problem is small, nobody cares if the solution is inefficient. But small problems often form a groundwork on which larger things are built. A raytracing rendering detects whether a ray intersects a triangle, which is a small problem, but it does this for every ray it emits, and there could be millions of rays per frame. Making your intersection check a little more effective helps you make your raytracer a little bit faster, so you can make better visual effects with it and win a larger market share when your game comes out.

So instead of calling some numeric solver for a 2×2, 3×3, or 4×4 system, solve it symbolically, so that the solution isn't specific numbers, but formulas that work for any input, and use the solution instead of the function call. I don't suggest that you solve the system manually; you can use SymPy to do that. SymPy is the library that does your math for you. Specify the equations you want to solve symbolically, and run `solve`. SymPy will turn your equations into a solution, a bit like Gaussian elimination does, and will even translate the symbolic solution into the language of your choice. How cool is that!

The symbolic solution grows in both length and time by the factorial. Yes, symbolic solvers also have $O(N!)$ complexity. Also, turned into code, the solution accumulates more computational error with every new dimension. These two things make this SymPy-based approach impractical for any system larger than 4×4.

But for small problems, a plain solution obtained by solving a system symbolically will still work faster than generic Gaussian elimination or any iterative algorithm, not because you used an efficient solution but solely because you cut out the overhead.

In the following section, we'll use SymPy not only to generate a symbolic solution, but also to improve its readability and performance.

3.7.4 Section 3.7 summary

For midsize problems, elimination-based methods work best. If your problem implies a sparse matrix, you can benefit greatly by choosing an appropriate data representation and specialized algorithms. Iterative methods help with large or overspecified problems that direct methods, elimination included, can't solve conceptually or because of time constraints. For small problems (up to four equations), using a symbolic solution in-place instead of a third-party function is often the best way to go.

3.8 Practical example: Does a ray hit a triangle?

Linear systems occur in a lot of small geometrical problems, such as determining a segment intersection, finding the distance from a point to a triangle, or putting an edge offset on an arbitrary triangle. These small problems form the core of any geometric library.

When I first went to work on a 3D game engine, I was overwhelmed by all the routines I had to learn: intersections, projections, distances, point-to-plane, edge-to-sphere, triangle-to-triangle . . . a whole space of functions. But after a month or so, I started noticing that most of the code in these routines does the same thing: solves a small linear system. Apparently, most of these problems are linear by nature. If you know how to form a problem into a system, you're halfway done with the code. And if you know how to run SymPy . . . well, that concludes the other half.

3.8.1 The ray-triangle intersection problem

In 3D space, a ray comes from point *P* by the direction set by a vector **d**. A triangle in this space is formed by a triplet of points: *A*, *B*, and *C*. We want to know whether a ray hits a triangle (figure 3.15). And while we're at it, if a ray does hit the triangle, let's determine the point of intersection, too.

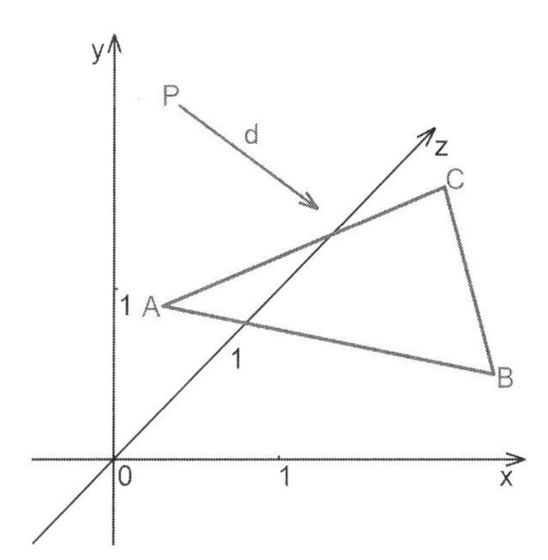

Figure 3.15 A ray may or may not intersect a triangle in 3D.

This problem is typical for raytracers. *Raytracing* is a rendering technique. For each pixel of the screen, we emit a ray, and by finding out which triangles the ray hits, we choose the pixel color. Raytracing is more than that, however. We use a similar technique to detect whether a point lies inside or outside a triangle mesh, which is important for collision detection in 3D models and for related operations such as gluing two bodies together. Finding the ray-to-triangle intersection is a simple but practical problem.

3.8.2 Forming a system

First, let's write our ray as a parametric equation. We saw this equation as a 3D line in chapter 2. Now, because we want to model a ray and not the whole line, our parameter will be restricted:

$$\mathbf{R} = \mathbf{P} + t\mathbf{d},\ t >= 0$$

Here, **R** is a point vector for any point on a ray; **P** is the point vector for point P; and t is a parameter, a number going from 0 to infinity.

Now let's write an equation that specifies points on a triangle. To do that, let's rely on the fact that a point on a plane could be put in a nonorthonormal linear basis. Our usual orthonormal basis consists of three basis vectors that follow the coordinate axes: (1, 0, 0), (0, 1, 0), and (0, 0, 1). All these vectors are of length 1—hence, the "normal" part of the name—and they're orthogonal to one another, meaning that the angle between any of them is always right.

The triangle, being set as a triplet of points, forms another basis. In the triangle's basis, the edges of a triangle form the basis vectors. They don't have to be at the right angle, and they don't have to be of length 1, so this basis is nonorthonormal. Any point S that lies in the same plane as a triangle ABC could be positioned by a point vector **S** (figure 3.16).

$$\mathbf{S} = \mathbf{A} + u\mathbf{AB} + v\mathbf{AC}$$

- **S** is a point vector for any point on a triangle.
- **A** is the point vector for A.
- **AB** is the vector pointing from A to B.
- **AC** is the vector from A to C.
- u and v are the coordinates of S in the triangle's basis.

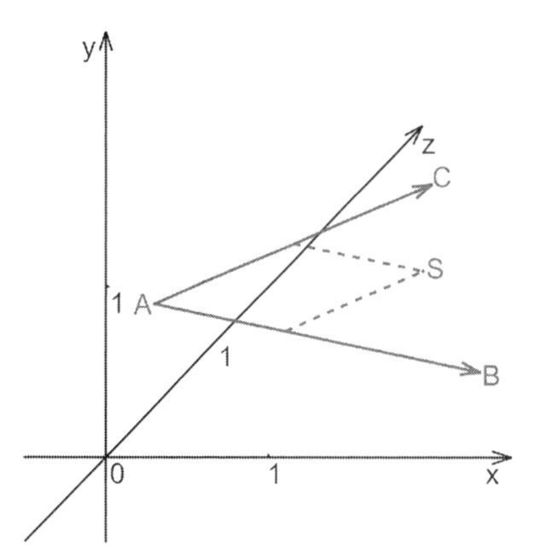

Figure 3.16 Point S in the basis made by AC and AB

In this basis, every point that lies on the triangle's plane has a pair of coordinates (u, v). If u = 0 and v = 1, the point S is the same as C. If u = 0.5 and v = 0.5, the point lies in the middle of the *BC* segment. A 2D basis covers a whole plane, of course. If $u = 100$ and $v = 200$, the point is still in the triangle's plane, but it's too far from the triangle to be of any use to us. Because we're hunting for the ray-triangle intersection, we want to filter out the points that don't belong to the triangle. Luckily, in a triangle's own basis, this task is easy (figure 3.17).

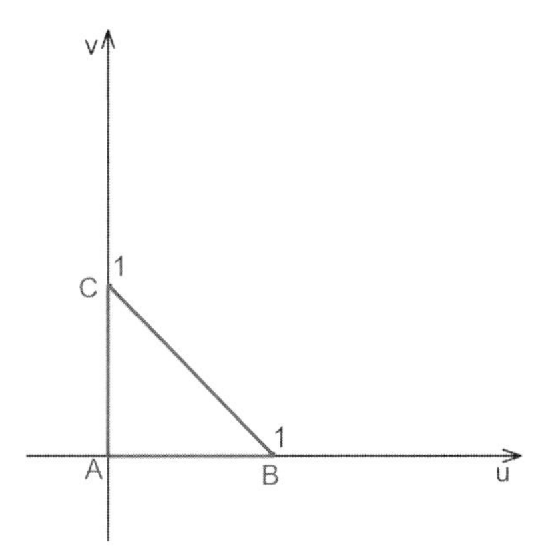

Figure 3.17 The triangle in its own basis

In figure 3.17, the triangle is parametrized in u and v coordinates, where u goes along the **AB** vector and v goes along the **AC** vector. Then the equation and constraints that specify all the points on a triangle are

$$\mathbf{S} = \mathbf{A} + \mathbf{AB}u + \mathbf{AC}v,$$

$$u \geq 0,$$

$$v \geq 0,$$

$$u + v \leq 1$$

The constraints on u and v guarantee that the point lies on the triangle. The equation for the right border is $u = 0$, and $v = 0$ constitutes the bottom border; $u + v = 1$ is the same as $v = 1 - u$; and this is the equation for the line that goes diagonally from (0, 1) to (1, 0).

Now let's form the system for a ray-triangle intersection. We use R to denote a point on a ray and S to denote a point on a triangle, so a ray hits a triangle if and only if $\mathbf{R} = \mathbf{S}$, respecting all the parametric constraints. So our final equation is

$$\mathbf{P} + t\mathbf{d} = \mathbf{A} + \mathbf{AB}u + \mathbf{AC}v$$

These are our constraints:

$$t >= 0,$$

$$u >= 0,$$

$$v >= 0,$$

$$u + v <= 1$$

Never mind the constraints; it looks as though we have one equation and three variables. Doesn't this make a system underspecified?

Not in 3D space. In 3D, every vector equation expands into three scalar equations: one for x, one for y, and one for z. In these coordinates, our system appears to be well-specified after all:

$$P_x + td_x = A_x + AB_x u + AC_x v$$

$$P_y + td_y = A_y + AB_y u + AC_y v$$

$$P_z + td_z = A_z + AB_z u + AC_z v$$

Now let's give the system to SymPy and see how it does.

3.8.3 Making the equations into code

After we rewrite the equations in Python, we get the code in the following listing (ch_03/ray_triangle_intersection.py in this book's source code).

Listing 3.1 Solving the ray-triangle intersection symbolically

```
from sympy import *

Px, Py, Pz = symbols('Px Py Pz')
dx, dy, dz = symbols('dx dy dz')
Ax, Ay, Az = symbols('Ax Ay Az')
ABx, ABy, ABz = symbols('ABx ABy ABz')
ACx, ACy, ACz = symbols('ACx ACy ACz')
t, u, v = symbols('t u v')

solution = solve([
    Px +t*dx - (Ax + ABx*u + ACx*v),
    Py +t*dy - (Ay + ABy*u + ACy*v),
    Pz +t*dz - (Az + ABz*u + ACz*v)
], (t, u, v))

print (pycode(solution))
```

The point from which the ray starts

The ray's direction vector

Point A of the ABC triangle

Vector AB

Vector AC

Parameters that specify the point on a ray and the point on a triangle's plane

Then the solution will be

```
{t: (ABx*ACy*Az - ABx*ACy*Pz - ABx*ACz*Ay + ABx*ACz*Py -
➥ ABy*ACx*Az + ABy*ACx*Pz + ABy*ACz*Ax - ABy*ACz*Px +
➥ ABz*ACx*Ay - ABz*ACx*Py - ABz*ACy*Ax + ABz*ACy*Px)/
➥ (ABx*ACy*dz - ABx*ACz*dy - ABy*ACx*dz +
➥ ABy*ACz*dx + ABz*ACx*dy - ABz*ACy*dx),
```

```
u: (ACx*Ay*dz - ACx*Az*dy - ACx*Py*dz + ACx*Pz*dy -
➥ ACy*Ax*dz + ACy*Az*dx + ACy*Px*dz - ACy*Pz*dx +
➥ ACz*Ax*dy - ACz*Ay*dx - ACz*Px*dy + ACz*Py*dx)/
➥ (ABx*ACy*dz - ABx*ACz*dy - ABy*ACx*dz +
➥ ABy*ACz*dx + ABz*ACx*dy - ABz*ACy*dx),
v: (-ABx*Ay*dz + ABx*Az*dy + ABx*Py*dz - ABx*Pz*dy +
➥ ABy*Ax*dz - ABy*Az*dx - ABy*Px*dz + ABy*Pz*dx -
➥ ABz*Ax*dy + ABz*Ay*dx + ABz*Px*dy - ABz*Py*dx)/
➥ (ABx*ACy*dz - ABx*ACz*dy - ABy*ACx*dz +
➥ ABy*ACz*dx + ABz*ACx*dy - ABz*ACy*dx)}
```

It's probably not a good idea to paste it in the code base right away. Let's groom the formulas a bit to make them more readable. First, let's separate the divisor, which is common to all three equations:

```
div = ABx*ACy*dz - ABx*ACz*dy - ABy*ACx*dz +
➥ ABy*ACz*dx + ABz*ACx*dy - ABz*ACy*dx
```

Then let's shorten the expression by collecting the coefficients for repetitive variables. A `collect` function in SymPy does that work for us, collecting coefficients for the variables we specify:

```
collect(ABx*ACy*dz - ABx*ACz*dy - ABy*ACx*dz +
➥ ABy*ACz*dx + ABz*ACx*dy - ABz*ACy*dx, (dx, dy, dz))
```

In this example, the function collects the coefficients for *dx*, *dy*, and *dz*, and puts them in the corresponding brackets. This collection not only shortens the expression for readability, but also makes the computation three multiplications shorter for better performance:

```
div = dx*(ABy*ACz-ABz*ACy) + dy*(-ABx*ACz+ABz*ACx) + dz*(ABx*ACy-ABy*ACx)
```

We can do the same with the parameters, although we should select different collecting variables for the *t*:

```
t = ACx*(-ABy*Az + ABy*Pz + ABz*Ay - ABz*Py) +
➥ ACy*(ABx*Az - ABx*Pz - ABz*Ax + ABz*Px) +
➥ ACz*(-ABx*Ay + ABx*Py + ABy*Ax - ABy*Px) / div

u = dx*(ACy*Az - ACy*Pz - ACz*Ay + ACz*Py) +
➥ dy*(-ACx*Az + ACx*Pz + ACz*Ax - ACz*Px) +
➥ dz*(ACx*Ay - ACx*Py - ACy*Ax + ACy*Px) / div

v = dx*(-ABy*Az + ABy*Pz + ABz*Ay - ABz*Py) +
➥ dy*(ABx*Az - ABx*Pz - ABz*Ax + ABz*Px) +
➥ dz*(-ABx*Ay + ABx*Py + ABy*Ax - ABy*Px) /div
```

This already looks like something we can use in an application. We should respect the project's coding conventions, of course, so some further grooming may be necessary. But conceptually, that's it.

Now whenever we want to determine whether a ray hits a triangle, we run this code and compute t, u, and v. Then we check the conditions:

```
if t >= 0. and u >= 0. and v >= 0. and (u+v) <= 1.:
    return True
return False
```

Have I forgotten anything? Of course! Not all systems have a solution at a point. Before checking the conditions, even before computing the parameters, we should check whether our system has a solution. Certainly, if a ray is parallel to the triangle's plane, there's no solution. If a triangle ABC is degenerate, or if a vector **d** is zero-length, there's no solution.

Checking all these things one by one may be difficult, both computationally and cognitively. Luckily, we don't have to do the work. All of these things lead to our equations being linearly dependent, which automatically reflects on the divisor!

If the `div` is 0, there's no solution, whatever the geometric reason is. So add this code before the parameters' computation, and it will do the trick:

```
if div == 0.:
    return False
```

NOTE Generally, it's not a good idea to compare floating-point numbers with some constants exactly. Even if the divisor was supposed to be 0, it could be computed with a small error, and this error will help the divisor squeeze past the check. Then it would go into the parameters' computation and corrupt their results. Avoiding that scenario properly would take us into a whole new field of computational mathematics that has little to do with geometry—namely, error analysis—and adding an arbitrary epsilon value instead of doing a solid error analysis would be lazy and wrong. So let's be lazy but mathematically correct, and leave the check as it is.

SEE ALSO If you want to learn more about dealing with floating point errors, the classic work to start with is *Accuracy and Stability of Numerical Algorithms*, 2nd ed., by Nicholas J. Higham (Society for Industrial and Applied Mathematics, 2002).

3.8.4 Section 3.8 summary

Solving a small-dimensional problem with a system of linear equations involves stating the problem, expanding the coordinates, writing it in Python, and feeding it to SymPy. SymPy results may need some grooming, as the code SymPy produces isn't necessarily human readable. Luckily, SymPy provides tools for improving readability.

3.9 Exercises

Exercise 3.1 Which of these systems has a point as a solution?

System 1

$$6x + 4y = 0$$

$$3x + 2y = 3$$

System 2

$$x + 2y + z = 1$$

$$2x + 3y + z = 2$$

System 3

$$x + 2y = 0$$

$$3x + 4y = 1$$

$$6x + 8y = 2$$

Exercise 3.2 Which systems from the previous exercise have a solution?

Exercise 3.3 This is a matrix multiplication algorithm:

```
def multiply_square_matrices(A, B, N):
    C = [[0. for j in range(N)] for i in range(N)]
    for i in range(N):
        for j in range(N):
            for k in range(N):
                C[i][j] += A[i][k] * B[k][j]
    return C
```

What is its complexity estimation in big-O?

Exercise 3.4 Suppose that we don't know how to compute a zero. We want to find it with an algorithm. To do so, we use this recurrent formula iteratively:

$$x_{i+1} = x_i^2 - x_i$$

Now if we start from $x_1 = 0.5$, the formula will converge slowly but steadily into 0:

1. $x_1 = 0.5$
2. $x_2 = -0.25$
3. $x_3 = 0.3125$
4. $x_4 = -0.21484375$

```
5 x5 = 0.2610015869140625
6 x6 = -0.19287975854240358
7 ...
```

But for some starting points, it diverges instead, as for $x = -1.5$:

```
1 -1.5
2 3.75
3 10.3125
4 96.03515625
5 9126.716079711914
6 83287819.6835923
7 ...
```

Please do the convergence analysis for this algorithm. Define the interval of x for which it converges and the values for which it neither converges nor diverges, but stalls. You don't have to do this exercise with pen and paper; you're more than welcome to use the language of your choice to conduct computational experiments.

Exercise 3.5 Here's something to challenge your hyperspatial intuition. Suppose that a ray in 4D is given as a point vector to its origin P and a vector of its direction **d**. Also suppose that a 3D singular cube is in this space. Because it's 4D space and only a 3D singular cube, it has to be flat in one of the dimensions, so let's say it's the fourth. Then cube C in parametric form is $C = (1, 0, 0, 0)\ i + (0, 1, 0, 0)\ j + (0, 0, 1, 0)\ k$, where i, j and k are in range [0, 1]. Note that there's no (0, 0, 0, 1). The last coordinate of this cube is always 0.

Write a function—or better yet, make SymPy write one for you—that for every ray defined by P and **d** answers whether it intersects the singular cube C.

3.10 Solutions to exercises

Exercise 3.1 System 3. The first one is well-specified, but it's two parallel lines. There's no point of solution or any solution at all. The third one is overspecified, but it has two identical lines, so all the lines intersect at a single point.

Exercise 3.2 System 2 and System 3. The second one is underspecified; two 3D planes intersect at a line. The whole line is a solution.

Exercise 3.3 $O(N^3)$. For each element of the resulting N × N matrix, it has to do N multiplications and additions.

Exercise 3.4 The algorithm is surprisingly tricky for such a simple formula. It converges in three separate intervals: (–1, 0), (0, 1), and (1, 2). In –1, 0, 1, and 2, it neither diverges nor converges. And for all x below –1 and above 2, it diverges.

Exercise 3.5 Write the ray in parametric form, and form an equation:

$$\mathbf{P} + t\mathbf{d} = (1, 0, 0, 0)\ i + (0, 1, 0, 0)\ j + (0, 0, 1, 0)\ k$$

The equation expands as follows:

$$td_{x1} - i = P_{x1}$$

$$td_{x2} - j = P_{x2}$$

$$td_{x3} - k = P_{x3}$$

$$td_{x4} = P_{x4}$$

This system of linear equations is easy to solve with SymPy. A few tests for the functions are

- P = (0.5, 0.5, 0.5, 1), **d** = (0, 0, 0, –1) – true
- P = (0.5, 0.5, 0.5, 1), **d** = (0, 0, 0, 1) – false
- P = (0.5, 0.5, 0.5, 1), **d** = (1, 0, 0, 0) – false
- P = (0.5, 0.5, 0.5, 1), **d** = (1, 0, 0, –2) – true

Summary

- A system of linear equations is geometrically a bunch of hyperplanes, which are lines in 2, and planes in 3D.
- The solution is the point where all the hyperplanes intersect.
- Overspecified systems rarely have a real point as a solution, but they may have a point that's close enough to all the hyperplanes to count as one pragmatically.
- Underspecified systems never have a single point of solution. They may have an infinite amount of solutions or no solution at all.
- Even seemingly well-defined systems may not have a solution if some of their equations are linearly dependent except for the free coefficient. Geometrically speaking, some of the hyperplanes are parallel.
- These systems may also have a hyperplane as a solution if some of the equations are completely linearly dependent. In other words, some of the hyperplanes are the same.
- Midsize problems are best solved with elimination-based algorithms.
- Use sparse-matrix representation when applicable for better performance.
- Large problems could be solved with iterative algorithms.
- When choosing an iterative algorithm, always check its convergence criteria. The matrix may need to be preconditioned.
- Small problems could be solved symbolically with SymPy and then turned directly into code, avoiding data preparation and function calling overhead.

4 Projective geometric transformations

This chapter covers

- Generalizing popular geometric transformations
- Using homogeneous coordinates to turn transformations into matrix multiplications
- Making a composition out of generalized projective transformations
- Finding the inverse transformation
- Making a transformation matrix from several displaced points

Geometric transformations, projective space, and homogeneous coordinates are associated concepts that not only enable, but also explain one another. The first one is the most applicable to real-world problems, so it usually gets the most attention. To exploit it in the most effective way, however, you should also know a little about the other two.

In this chapter, you'll learn how to do geometric transformations such as translation, rotation, scaling, and shear. You'll learn how to generalize them into a matrix multiplication. But you don't need a book to do all that. Any good framework has transformation routines; you can learn a few functions and get the job done.

More important, in this chapter you'll learn how to avoid code duplication by understanding and exploiting projective space. You'll see how to save processor cycles on the inverse transformation by knowing how homogeneous coordinates work. You'll understand how to create your own transformations from whatever bits of data you have. You'll learn not only how your framework of choice works, mathematically, all frameworks are the same, but also how to use it most effectively, as well as how to write your own routines when the framework doesn't provide enough for your particular problem.

4.1 Some special cases of geometric transformations

A *geometric transformation* is a function with both its domain and codomain in the same geometric space. In other words, it's something that accepts a point as input and results in a point as output.

Unless the domain and codomain are deliberately restricted, the transformation concerns every point in space, so it transforms the whole space. Even if we transform some specific object, and we're doing that by applying the transformation to each of its points, the transformation itself still describes how all the space changes.

All the transformations we will look into in this chapter are unrestricted. That is, they transform the whole $\mathbb{R}^n$.

The transforming function can be written as a vector equation, matrix operation, or anything else. The exact representation isn't essential. The simplest way to describe a transformation is to use explicit real-number functions for every coordinate separately:

$$f_i : \mathbb{R} \to \mathbb{R}$$

The transformation that is easier to program is called translation, so let's start with it.

4.1.1 Translation

Translation is simply moving points of space around. In its explicit per-coordinate form, a 2D translation looks like this:

$$x_t = x_i + dx$$

$$y_t = y_i + dy$$

Here x_i, y_i are the coordinates of the input point; x_t, y_t are the coordinates of the translated point; and dx and dy are the coefficients of the transformation. They're numbers that indicate how far the point is transported along the respective axes. The coefficient dx corresponds to moving a point along the *x*-axis, and the dy corresponds to moving the point along the *y*-axis. Using numbers instead of dx and dy results in specific translations (figure 4.1).

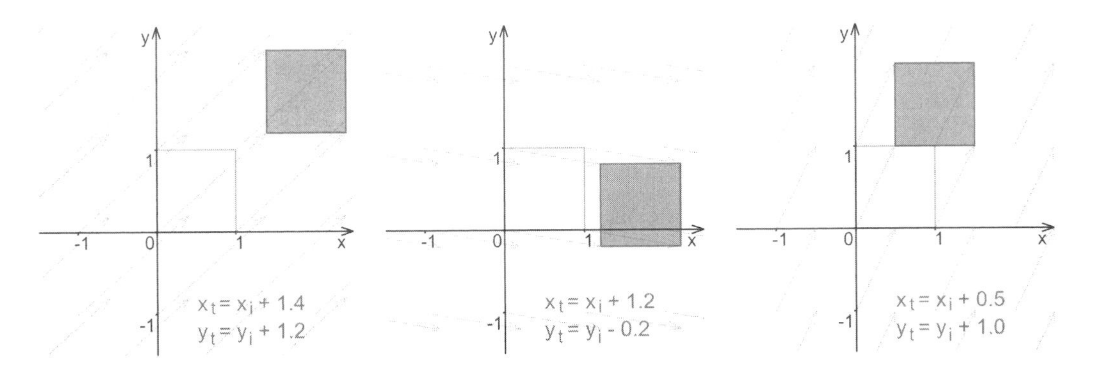

Figure 4.1 Three examples of standard square translations. A standard 1 × 1 square (transparent) moves and becomes the opaque square according to the coefficients.

Before we move on to other kinds of transformations, let's do one more thing. It's clear that transformations are functions of coordinates with some numeric coefficients in them. But what if coefficients themselves were functions? What if we change coefficients as time passes, following some predefined formula? Let's do that to a square:

$$x_t = x_i + \sin(t)$$

$$y_t = y_i + \cos(t)$$

This operation is still a translation, but now the whole translation is the function of t. This translation is called a *parametric transformation,* and it looks like figure 4.2.

Simply by incrementing our parameter t, we made a square go in circles. The concept of parametric transformation is the key to animation.

4.1.2 Scaling

Not all transformations are easy to write in vector notation. Let's take a look at another transformation that can be written in vector form only when it's in its special case. The new transformation is called *scaling*, and it goes like this:

$$x_s = ax_i$$

$$y_s = by_i$$

As before, x_i, y_i are the coordinates of the input point; x_s, y_s are the coordinates of the scaled point; and a and b are coefficients of the transformation. Now the transformation itself is different: a corresponds to stretching the space along the x-axis, and b corresponds to stretching the space along the y-axis (figure 4.3).

If scaling coefficients coincide, the scaling by one axis is proportionate to the scaling by the other. Because it works the same in any direction, this kind of scaling is

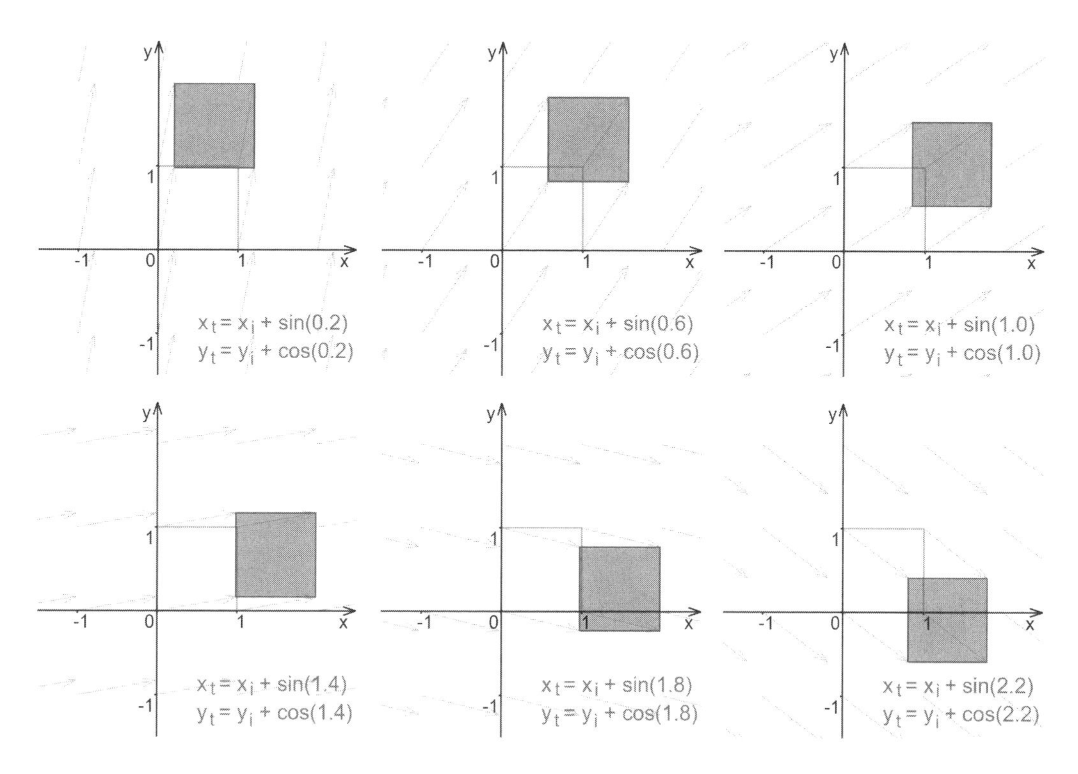

Figure 4.2 An example of parametric translation. A square goes around its original center.

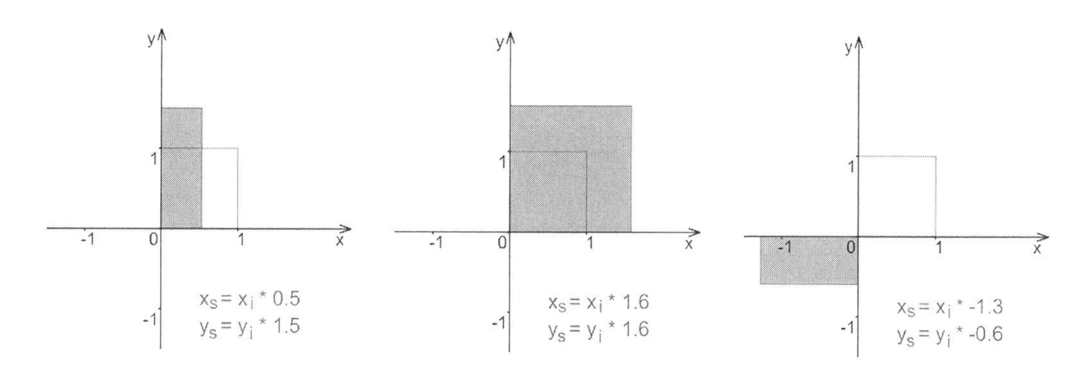

Figure 4.3 Three examples of standard square scaling

called *uniform,* which means "same shape" in Latin, or *isotropic,* which means "same turn" in Greek.

Now that we know about two different transformations, let's take a look at another important concept: a transformation composition. Scaling as a transformation always scales the whole space, but always with respect to the zero point (0, 0). If you want to scale an object with respect to its center, you should do a transformation composition. First, you translate the object so that its center coincides with the zero point; next, you do the scaling; and then you translate it back (figure 4.4).

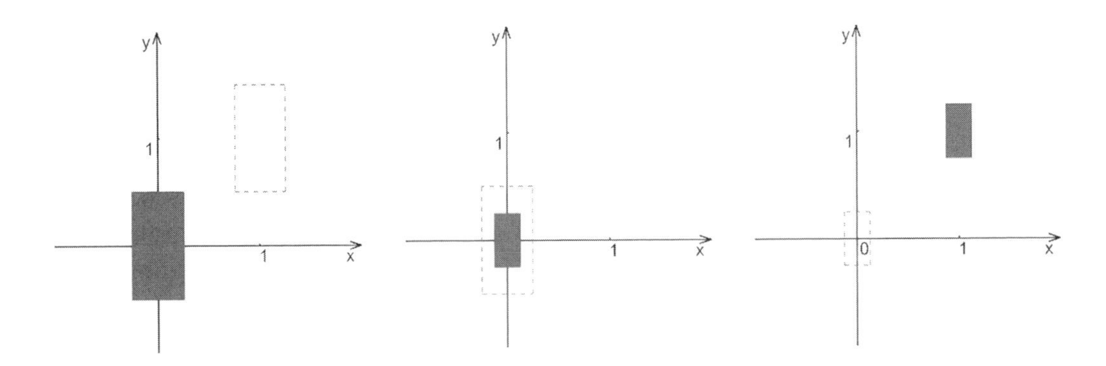

Figure 4.4 Scaling from an object's center is a composition of transformations.

In compositions such as this one, the order of operations matters. "Translate, rotate, translate back" is quite obviously not the same as "Translate, translate back, and rotate." In practice, you should understand the context of a transformation before trying to replace one with any other. If you alter some transformation when debugging, and the result becomes unpredictable, you should check the transformations before and after the one you're tinkering with.

Translation and scaling are two of the three most common transformations in computer graphics and design. They're intuitive, predictable, and simple to program.

4.1.3 Rotation

Rotation is the third of the three most common transformations in computer graphics and design. The following operation rotates the space around its zero point:

$$x_r = \cos(t)\ x_i + \sin(t)\ y_i$$

$$y_r = -\sin(t)\ x_i + \cos(t)\ y_i$$

As usual, x_i, y_i are the coordinates of the input point, and the coordinates of the rotated point are now x_r and y_r. The parameter t is the angle we want to rotate the space, written in radians (figure 4.5).

Fun fact: I've been writing about projective transformations for about ten years, and in my last tutorial, I wrote this very formula. But I wrote *sine* instead of *cosine* and *cosine* instead of *sine*. This mistake is a common one. I'll show you how to avoid making it.

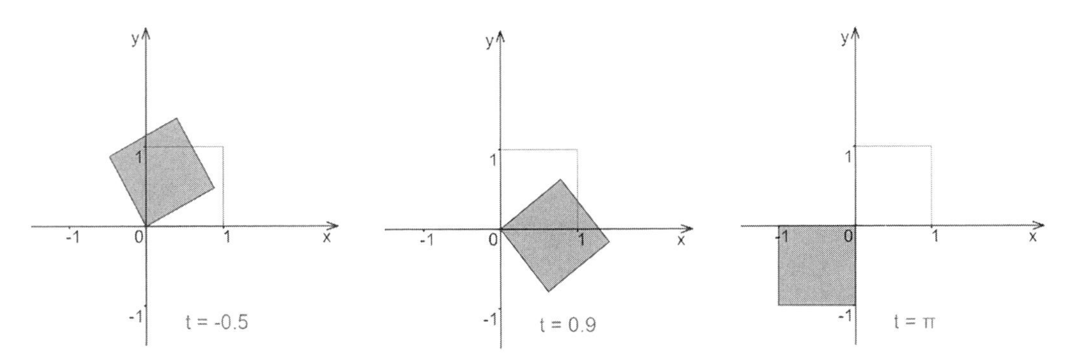

Figure 4.5 Three examples of standard square rotations. All rotations go around the zero point (0, 0).

But first, I should probably give you a small reminder of what sine and cosine functions are. Figure 4.6 shows a sine.

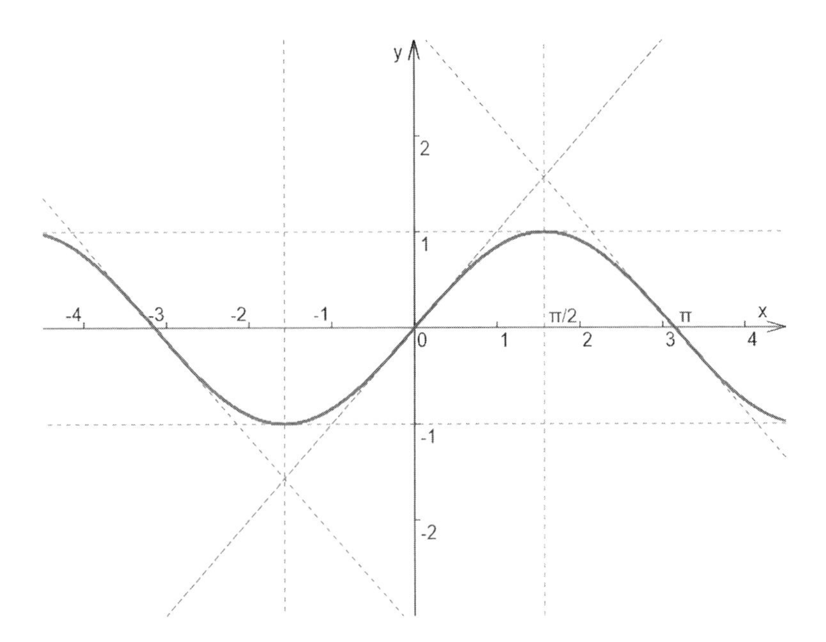

Figure 4.6 Sine function

A sine is a continuous function, with the domain being $\mathbb{R}$ and the codomain being the interval $[-1, 1]$. It's periodic, which means that it copies itself every 2π as it goes along the *x*-axis.

The sine is *odd*, meaning that its left half, where all the *x*s are negative, is exactly the opposite of the right half, where all the *x*s are positive:

$$\sin(-x) = -\sin(x)$$

As a continuous odd function, the sine goes through the zero point. It has to. If $x = 0$ is anything else but 0—let's say c—the function would have to jump immediately to $-c$ on the other side, which would break its continuity (figure 4.7). The only number that survives negation without change is zero: $0 = -0$.

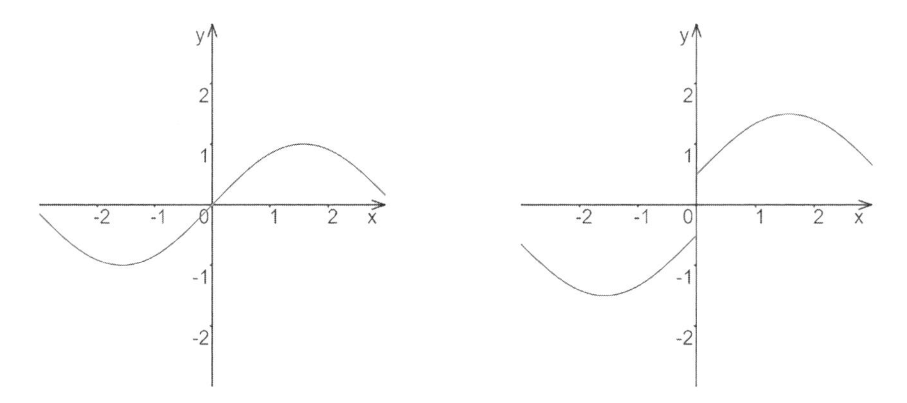

Figure 4.7 An odd function has to go through the zero point or lose its continuity.

Here's a little bonus for performance enthusiasts. Not only sine of zero is zero, but also, for small values of x, the sine of x is approximately x itself: $\sin(x) \approx x$. This means that for the small xs, the $\sin(x)$ function can be replaced by the much cheaper $f(x) = x$. For $x = 0.244$, $\sin(x) \approx 0.242$, so the relative difference is about 1%, which also means that for $x < 0.244$, the relative difference is even less than that.

Figure 4.8 shows the cosine.

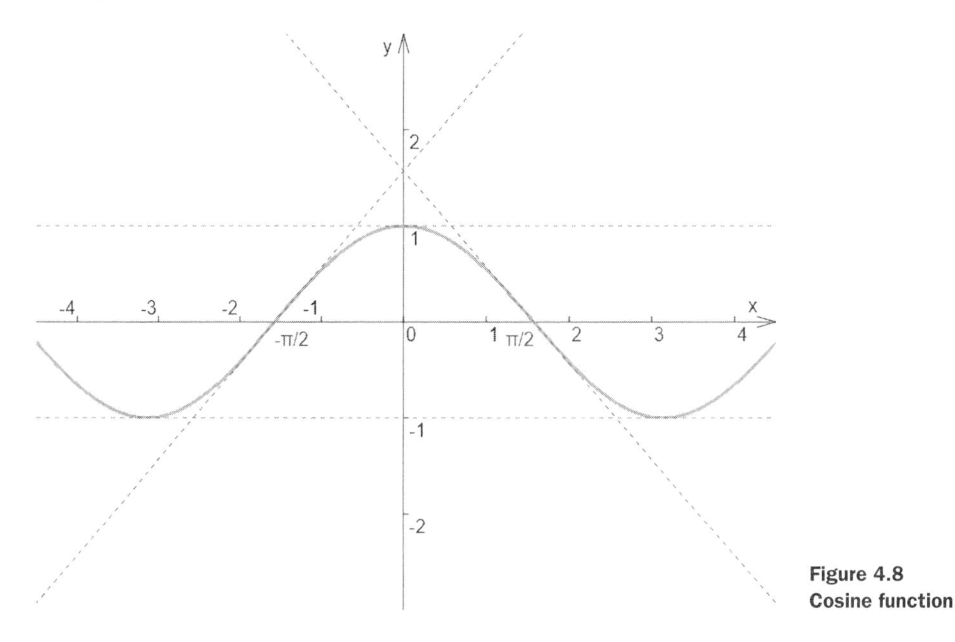

Figure 4.8 Cosine function

The cosine is almost the same as the sine except that it runs $\frac{\pi}{2}$ ahead of it:

$$\cos(x) = \sin\left(x + \frac{\pi}{2}\right)$$

The cosine is continuous and periodic, as the sine is. The major difference is that the cosine is even. Its x-negative half is the mirrored copy of its x-positive half:

$$\cos(-x) = +\cos(x)$$

In close proximity to 0, it could be approximated by the constant 1. For $x = 0.14$, $\cos(x) \approx 0.99$, so for $x < 0.14$, the relative error is less than 1%.

Now for the most important thing: how to never mistake a sine for a cosine. If you rotate the graphs 90 degrees counterclockwise, the sine looks like the letter *S*, and the cosine looks like the letter *C* (figure 4.9).

Figure 4.9 The way to remember sine and cosine functions

Yes, this is a silly way to remember your functions, but it works, doesn't it? Now let's get back to the transformations. The rotation formula rotates all the space around the zero point. If you want to rotate an object around its center, you should (as with scaling from the last section) do a composition. Translate the object, do the rotation, and translate the object back (figure 4.10).

Composition is the key to complex transformations. Merely by combining scale, rotation, and translation, we can already do quite a lot. But one thing we can't do yet is transform a pair of parallel segments into nonparallel segments. We can't do shearing.

Our basic transformations constrain the way we operate on the whole space. To go beyond these constraints, we need to generalize and expand our tool set.

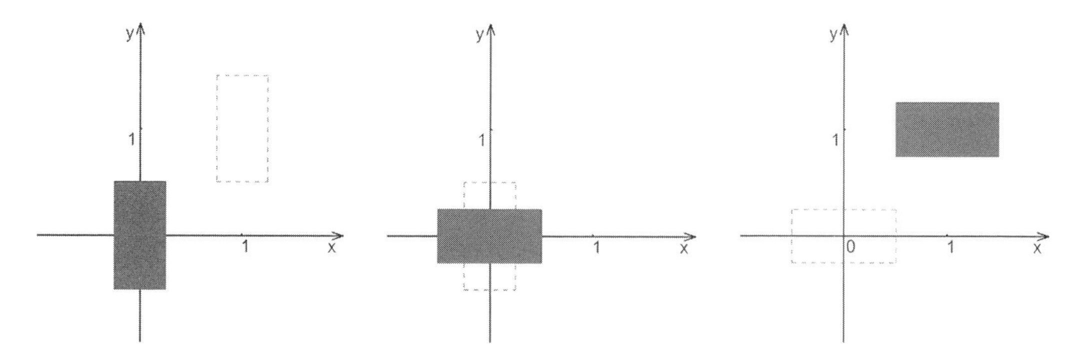

Figure 4.10 Rotation around the center is a composition of transformations.

4.1.4 Section 4.1 summary

All the geometric transformations we've seen so far operate on the whole space. The translation shifts not only the geometric object, but also the whole space, along the axes. The scaling also scales the object together with the whole space. If it scales the space proportionally—enlarges everything 2x by the *x*-axis and 2x by the y-axis, for example—we call this scaling *isotropical*. If the scaling works by axes differently—scales everything 2x by the *x*-axis, but 3x by the y-axis, for example—we call it *anisotropical*. The rotation rotates the object, and the whole space, around the zero point.

To use all three transformations effectively, we need a way to compose them. And to add new transformations such as shearing to the toolbox, we need some generalization.

4.2 Generalizations

There are multiple ways to generalize the transformations we've learned. Some expand our possibilities geometrically, some simplify computations, and some allow for easy composition. The trick is to get all the good properties in one package.

4.2.1 Linear transformations in Euclidean space

A *linear transformation* is a transformation in which every new coordinate is a linear combination of old coordinates. A *linear combination* is a sum of coordinates each multiplied by its dedicated constant, like this:

$$x_t = a_{11}\ x_i + a_{12}\ y_i$$

$$y_t = a_{21}\ x_i + a_{22}\ y_i$$

Obviously, it generalizes the rotation:

$$x_t = \cos(t)\ x_i + \sin(t)\ y_i$$

$$y_t = -\sin(t)\ x_i + \cos(t)\ y_i$$

And maybe slightly less obviously, it generalizes the scaling as well:

$$x_t = a_{11}\ x_i + 0\ y_i$$

$$y_t = 0\ x_i + a_{22}\ y_i$$

What's more important, a linear transformation is the same as a square matrix multiplication:

$$\begin{pmatrix} a_{11} & a_{12} \\ a_{21} & a_{22} \end{pmatrix} \times \begin{pmatrix} x_i \\ y_i \end{pmatrix} = \begin{pmatrix} x_t \\ y_t \end{pmatrix}$$

Remember the row-on-column rule?

$$\begin{pmatrix} a_{11} & a_{12} \\ a_{21} & a_{22} \end{pmatrix} \times \begin{pmatrix} x_i \\ y_i \end{pmatrix} = \begin{pmatrix} a_{11}x_i + a_{12}y_i \\ a_{21}x_i + a_{22}y_i \end{pmatrix}$$

A linear transformation being a square matrix multiplication is a great property, because now we can steal the matrix algebra (a set of rules and operations) and apply it to geometric transformations.

NOTE In matrices, the first index of an element indicates the row, and the second one indicates the column. The indices start from 1. In C-like arrays, indices start from 0, which comes from a half-century-old optimization. Accessing an array in C essentially involves adding up the array's pointer with the offset computed as the type size multiplied by the index. If array indices started with 1 in C, this would have added one more decrement to the offset computation and made accessing an array a tiny bit slower. Many matrix libraries, especially those that come from C and C++, also use 0-based indexing.

Now we can compose the transformations simply by multiplying matrices. We can invert the transformations by computing an inverse matrix. We know what the identity matrix looks like, which does nothing to the space but is still nice to have around:

$$\begin{pmatrix} 1 & 0 \\ 0 & 1 \end{pmatrix} \times \begin{pmatrix} x_i \\ y_i \end{pmatrix} = \begin{pmatrix} x_t \\ y_t \end{pmatrix}$$

Unfortunately, linear transformation doesn't include translation, so the most important compositions, such as translate-scale-translate and translate-rotate-translate, aren't accounted for. You can scale your space, and you can even turn squares into parallelograms, but without translation, you can't transform a zero point into anything else, so all your transformations remain kind of zero-bound. Linear transformations look like figure 4.11.

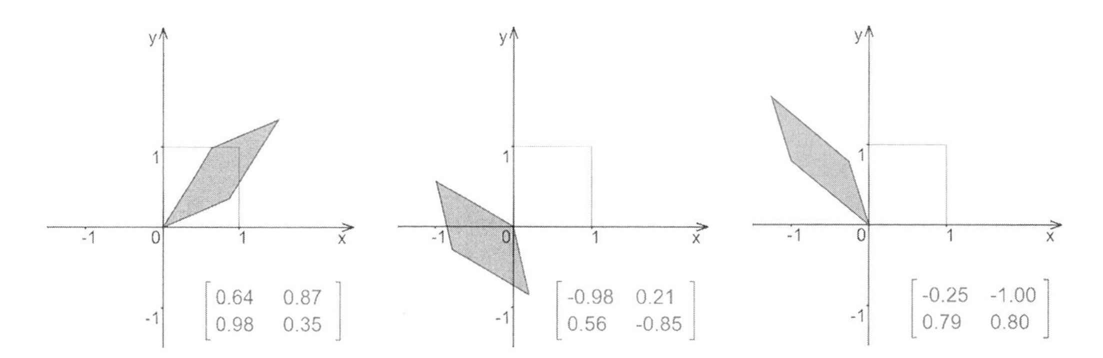

Figure 4.11 Three examples of linear transformations in Euclidean space. You can't transform (0, 0) into anything else.

The linear transformation retains parallelism, and it always transforms the zero point into itself. It's also symmetrical regarding both the *x* and *y* axes, but to see this effect, you need a symmetrical object to transform (figure 4.12).

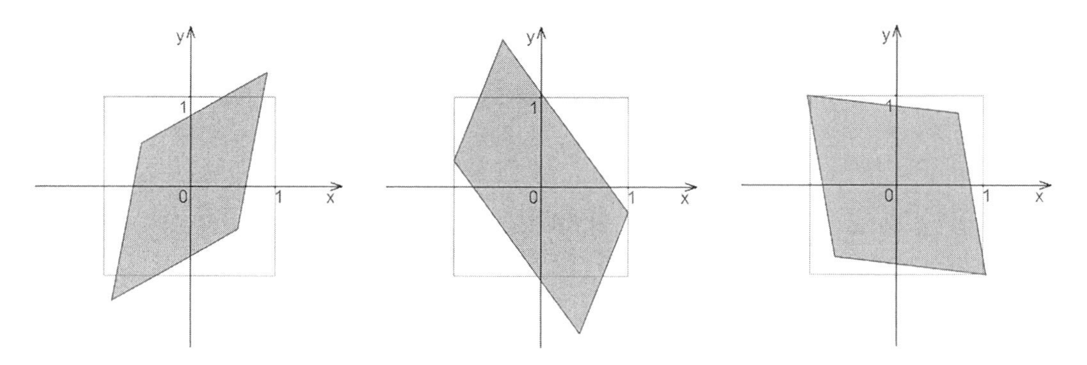

Figure 4.12 Linear transformation in Euclidean space is x- and y-symmetrical.

Because the linear transformation is bound to the zero point and is also symmetrical, it generalizes only the rotation and the nonuniform scaling—not the translation or anything else. The whole transformation could be guided by moving only two points in space, and they can be any points, not special ones. Let's see how to create a transformation by using a pair of transformed points as the input.

Suppose that we have a pair of points, (x_1, y_1), and (x_2, y_2). After the transformation, they're supposed to become (x_{t1}, y_{t1}) and (x_{t2}, y_{t2}), respectively. We can find a linear transformation that does this by solving a system of linear equations.

For this problem, we need SymPy. Let's start by putting all we know about the transformation in the form of equations. The first point being transformed one coordinate at a time looks like this:

```
[a11 * x1 + a12 * y1 - xt1,
a21 * x1 + a22 * y1 - yt1]
```

The second point is the same:

```
[a11 * x2 + a12 * y2 - xt2,
a21 * x2 + a22 * y2 - yt2]
```

The problem of finding matrix elements for the linear transformation translates into the small SymPy program in the following listing (ch_04/linear_transformation.py in the book's source code).

Listing 4.1 Computing the matrix for the linear transformation symbolically

```
from sympy import *

a11, a12, a21, a22 = symbols('a11 a12 a21 a22')
x1, y1, x2, y2 = symbols('x1 y1 x2 y2')
xt1, yt1, xt2, yt2 = symbols('xt1 yt1 xt2 yt2')

solution = solve([
    a11 * x1 + a12 * y1 - xt1,
    a21 * x1 + a22 * y1 - yt1,
    a11 * x2 + a12 * y2 - xt2,
    a21 * x2 + a22 * y2 - yt2
], (a11, a12, a21, a22))

print(solution)
```

Four coefficients for the linear transformation

Two points before the transformation

Two points after the transformation

This line is the "$x_t = a_{11}\ x_i + a_{12}\ y_i$" formula applied to point (x1, y1).

Here's the solution:

```
a11: (xt1*y2 - xt2*y1)/(x1*y2 - x2*y1),
a12: (x1*xt2 - x2*xt1)/(x1*y2 - x2*y1),
a21: (-y1*yt2 + y2*yt1)/(x1*y2 - x2*y1),
a22: (x1*yt2 - x2*yt1)/(x1*y2 - x2*y1)
```

The exact solution isn't important but please note how easily we could establish it symbolically for every pair of transformed points by solving a system of linear equations. A matrix of 2D linear transformation consists of four entries, so we have to have exactly four equations for our system—two points and two coordinates each. If there were three points, our system would be overspecified and normally wouldn't have a solution. If there was only one point, the system would be underspecified; there'd be not a single solution, but an unpragmatic continuum.

In this regard, this transformation could be called a *two-point transformation*. You can set the whole transformation by an example of where a pair of points should go in the result. Let's see how to go from a two-point transformation to a three-point transformation.

Row vectors exist, too

Some frameworks and engines prefer another way of doing matrix multiplications: they treat vectors as rows and keep matrices transposed. The row-on-column rule still stands, of course, so to achieve the same effect, you should multiply your vector by matrix from left to right:

$$\begin{pmatrix} x_i & y_i \end{pmatrix} \times \begin{pmatrix} a_{11} & a_{12} \\ a_{21} & a_{22} \end{pmatrix} = \begin{pmatrix} a_{11}x_i + a_{21}y_i & a_{12}x_i + a_{22}y_i \end{pmatrix}$$

This operation is essentially the same as the one we do in the book. One is no worse than the other. But when you're making two libraries work together, this ambiguity may become a problem. If one library treats vectors as rows and the other as columns, their transformation matrices have to be transposed when passing from one library to another.

Please stick to a single convention wherever you can. Consistency in working with matrices makes your code easier to read and reason about.

4.2.2 Bundling rotation, scaling, and translation in a single affine transformation

In the preceding section, we saw how to find coefficients for a two-point linear transformation. To make a three-point transformation in 2D, we need more coefficients in the formula. Let's add a pair of constants to our linear transformation:

$$x_t = a_{11}\ x_i + a_{12}\ y_i + a_{13}$$

$$y_t = a_{21}\ x_i + a_{22}\ y_i + a_{23}$$

This operation is called *affine transformation*, and it looks like figure 4.13.

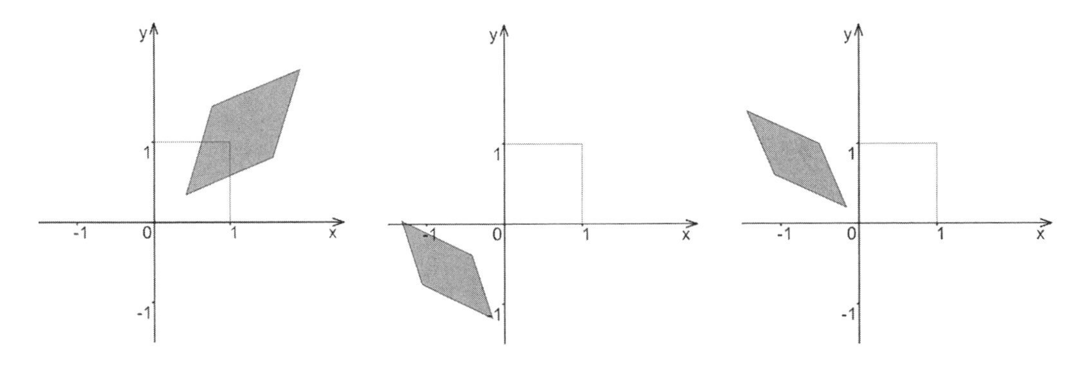

Figure 4.13 Three examples of affine transformations, which look a lot like linear ones but now have translations

The affine transformation still preserves parallelism. Any two lines that were parallel before the transformation remain parallel after, but the transformation is no longer bound to the zero point. An affine transformation is like a linear transformation we can carry around the plane. It's also symmetrical but this time relative to the (a_{13}, a_{23}) point.

This transformation still generalizes rotation and scaling, but now it also generalizes the translation:

$$x_t = 1\ x_i + 0\ y_i + a_{13}$$

$$y_t = 0\ x_i + 1\ y_i + a_{23}$$

Let's see how we can find the coefficients for the whole transformation by moving three points in space. We'll use the same program but add more coefficients and equations, as shown in the following listing (ch_04/affine_transformation.py in the book's source code).

Listing 4.2 Computing the coefficients for the affine transformation symbolically

```
from sympy import *

a11, a12, a13, a21, a22, a23 = symbols('a11 a12 a13 a21 a22 a23')
x1, y1, x2, y2, x3, y3 = symbols('x1 y1 x2 y2 x3 y3')
xt1, yt1, xt2, yt2, xt3, yt3 = symbols('xt1 yt1 xt2 yt2 xt3 yt3')

solution = solve([
    a11 * x1 + a12 * y1 + a13 - xt1,
    a21 * x1 + a22 * y1 + a23 - yt1,
    a11 * x2 + a12 * y2 + a13 - xt2,
    a21 * x2 + a22 * y2 + a23 - yt2,
    a11 * x3 + a12 * y3 + a13 - xt3,
    a21 * x3 + a22 * y3 + a23 - yt3
], (a11, a12, a13, a21, a22, a23))

print(solution)
```

Affine transformation in 2D looks like this: $x_t = a_{11}x + a_{12}y + a_{13}$; $y_t = a_{21}x + a_{22}y + a_{23}$. We need six coefficients to define it.

It's a three-point transformation, so we need to define three starting points.

Because we're going to infer the coefficients from the actual transformation of three points, we need to define the three ending points as well.

This line is "$x_t = a_{11}\ x_i + a_{12}\ y_i + a_{13}$" applied to the point (x1, y1).

The resulting solution provided by SymPy is a bit heavy; all the equations are supplied with a common divisor. If we write the solution separately, the equations become much more compact:

```
divisor = (x1*y2 - x1*y3 - x2*y1 + x2*y3 + x3*y1 - x3*y2)
a11 = (xt1*y2 - xt1*y3 - xt2*y1 + xt2*y3 + xt3*y1 - xt3*y2)/divisor,
a12 = (x1*xt2 - x1*xt3 - x2*xt1 + x2*xt3 + x3*xt1 - x3*xt2)/divisor,
a13 = (-x1*xt2*y3 + x1*xt3*y2 + x2*xt1*y3 -
➥ x2*xt3*y1 - x3*xt1*y2 + x3*xt2*y1)/divisor,
a21 = (-y1*yt2 + y1*yt3 + y2*yt1 - y2*yt3 - y3*yt1 + y3*yt2)/divisor,
```

```
a22 = (x1*yt2 - x1*yt3 - x2*yt1 + x2*yt3 + x3*yt1 - x3*yt2)/divisor,
a23 = (x1*y2*yt3 - x1*y3*yt2 - x2*y1*yt3 +
➥ x2*y3*yt1 + x3*y1*yt2 - x3*y2*yt1)/divisor
```

Once again, the geometry boils down to simple arithmetic. Two coordinates per three points is six. Six equations. Six equations require six coefficients, and six coefficients come from three coefficients per equation in the resulting formula. In the next section, we'll lose this simplicity, but we'll get something more important.

4.2.3 Generalizing affine transformations to projective transformations

To generalize the affine transformation, to turn it into a four-point transformation, we should add more coefficients to the equations. The trouble is that we have no place to put new coefficients if we want to keep linearity. If we don't care about linearity, we could add an x^2 member:

$$x_t = a_{11}\, x_i^2 + a_{12}\, x_i + a_{13}\, y_i + a_{14}$$

$$y_t = a_{21}\, x_i^2 + a_{22}\, x_i + a_{23}\, y_i + a_{24}$$

We can add even more members on a whim. Do we want a five-point transformation? Add y^2! Do we want a six-point transformation? Go for the cubes, or do x^2y^2. All these transformations are valid and interesting. They're called *polynomial transformations*, and we'll get back to them in chapter 8. They're not linear, though. In this context, *linearity* means that the transformation changes a straight line into a straight line, which is important in engineering graphics and applied geometry in general.

If we want to render a 3D picture on a 2D screen, for example, we want to do a projection. We take a piece of the 3D space, project it into screen space, and then draw all the objects one by one from back to front. We need a transformation that does the projection for us—the one that allows the perspective but doesn't do an unwanted fish-eye effect on everything that's otherwise straight and lined up.

Such a transformation exists, but it's not purely polynomial: it's rational. As with rational numbers, the word *rational* comes from *ratio,* as in ½ or ¾. The formula is

$$x_p = \frac{ax_i + by_i + c}{gx_i + hy_i + i}$$

$$y_p = \frac{dx_i + ey_i + f}{gx_i + hy_i + i}$$

This transformation doesn't preserve parallelism anymore, but on the other hand, it can turn a square into an arbitrary convex quadrilateral (figure 4.14). It's a four-point *projective transformation.*

NOTE I turned indexed *a*s into letters to shorten the equations and make them more readable. In programming, especially for polynomial transformations (as

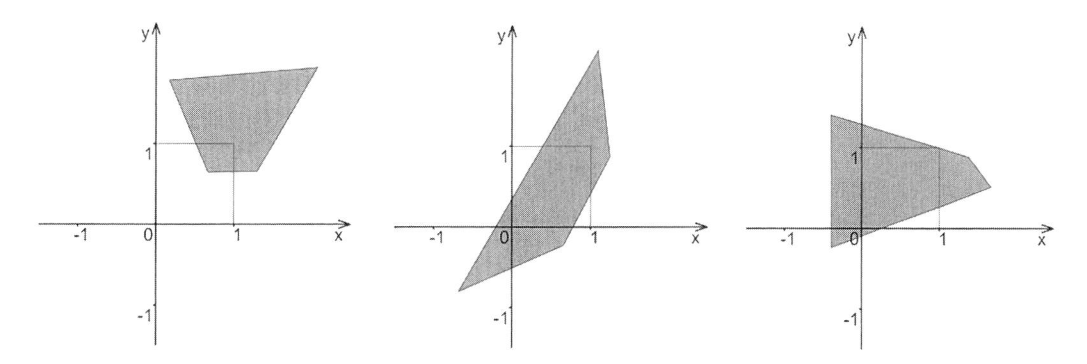

Figure 4.14 Three examples of projective transformations. They're like the affine ones, but for the projective ones, parallelism isn't mandatory anymore.

in chapter 8), sticking with indexes or even keeping your coefficients in arrays is preferable. But it's much easier to read formulas with graphically distinct coefficients in them.

If we count the coefficients in the formula, however, we'll see that the numbers don't add up. We have nine coefficients, and four points give us only eight equations. So how is that supposed to work?

Let's see the equations in SymPy. In this context, we're reverse-engineering a four-point transformation, so x_i, y_i, x_{pi}, and y_{pi} are our data, and coefficients *a*, *b*, *c*, *d*, *e*, *f*, *g*, and *i* are our variables we want to establish.

Because now *x*s and *y*s are coefficients for our equations, and *a*s, *b*s, and *c*s are variables, we can multiply both parts of every equation by its divider and get ourselves a nice set of linear equations for SymPy to process. The equation

$$x_p = \frac{ax_1 + by_1 + c}{gx_1 + hy_1 + i}$$

turns into

```
xp1 * (x1 * g + y1 * h + i) - x1 * a - y1 * b - c
```

and then into

```
xp1 * x1 * g + xp1 * y1 * h + xp1 * i - x1 * a - y1 * b - c
```

so that the whole system looks like this:

```
xp1 * x1 * g + xp1 * y1 * h + xp1 * i - x1 * a - y1 * b - c,
yp1 * x1 * g + yp1 * y1 * h + yp1 * i - x1 * d - y1 * e - f,
xp2 * x2 * g + xp2 * y2 * h + xp2 * i - x2 * a - y2 * b - c,
yp2 * x2 * g + yp2 * y2 * h + yp2 * i - x2 * d - y2 * e - f,
```

```
xp3 * x3 * g + xp3 * y3 * h + xp3 * i - x3 * a - y3 * b - c,
yp3 * x3 * g + yp3 * y3 * h + yp3 * i - x3 * d - y3 * e - f,
xp4 * x4 * g + xp4 * y4 * h + xp4 * i - x4 * a - y4 * b - c,
yp4 * x4 * g + yp4 * y4 * h + yp4 * i - x4 * d - y4 * e - f
```

Because our system is underspecified, we have only eight equations for nine variables; there's no single solution. But there is a whole class, a continuum of solutions of which each and every one is good. Every set of coefficients from this continuum describe the same four-point transformation.

Pragmatically, this means that we can pick any variable we like and assign it a value on a whim. We can pick 1 for *i*, for example, and add this decision to the equations list so that we have a nice unambiguous system:

```
[
xp1 * x1 * g + xp1 * y1 * h + xp1 * i - x1 * a - y1 * b - c,
yp1 * x1 * g + yp1 * y1 * h + yp1 * i - x1 * d - y1 * e - f,
xp2 * x2 * g + xp2 * y2 * h + xp2 * i - x2 * a - y2 * b - c,
yp2 * x2 * g + yp2 * y2 * h + yp2 * i - x2 * d - y2 * e - f,
xp3 * x3 * g + xp3 * y3 * h + xp3 * i - x3 * a - y3 * b - c,
yp3 * x3 * g + yp3 * y3 * h + yp3 * i - x3 * d - y3 * e - f,
xp4 * x4 * g + xp4 * y4 * h + xp4 * i - x4 * a - y4 * b - c,
yp4 * x4 * g + yp4 * y4 * h + yp4 * i - x4 * d - y4 * e - f,
i - 1
], (a, b, c, d, e, f, g, h, i))
```

Don't try to solve this system yet. Due to the way the solver's complexity rises with every new equation, it takes an improbable amount of time for the solver to finish a system of even nine equations. We'll find a way to deal with these equations when we get to the practical examples.

Pragmatism aside, how come a single transformation has a whole class of coefficients? Is this normal? Why is this happening? There's some exciting math behind this nonuniqueness, but to understand it, you'll have to rewire your whole intuition about space. Brace yourself; you'll need all the imagination you have.

4.2.4 An alternative to projective transformations

But before we get to the hard part, we have one more thing to settle. The projective transformation is a four-point linear transformation, but wouldn't this one also be linear?

$$x_t = a_{11}\, x_i\, y_i + a_{12}\, x_i + a_{13}\, y_i + a_{14}$$

$$y_t = a_{21}\, x_i\, y_i + a_{22}\, x_i + a_{23}\, y_i + a_{24}$$

Not really. This transformation is called *bilinear transformation,* but it isn't linear. On the plus side, it's even more potent than a projective transformation. For one example, it can transform a square into a curvy concave thing while preserving every straight line (figure 4.15).

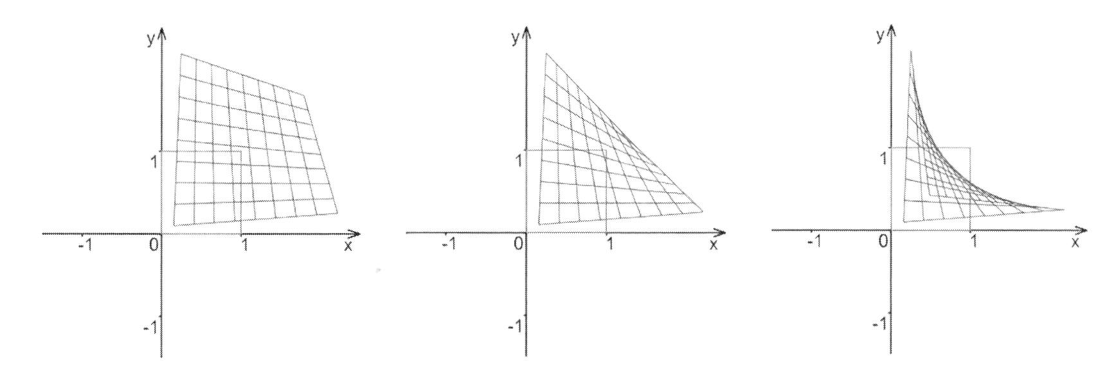

Figure 4.15 A bilinear transformation makes a square into a curvy concave object.

Bilinear transformation generalizes rotation, translation, and scale—all the affine transformations. But it doesn't generalize itself to 3D as nicely as a projective transformation does. A projective transformation is an (n+2)-point transformation in n-dimensional space, whereas bilinear is (2^n-1)-point transformation, meaning that in 3D, it suddenly becomes an eight-point transformation.

Most important, unlike linear transformation, the bilinear one doesn't correspond to matrix multiplication. We'll get back to bilinear, trilinear, biquadratic, and even cubic-linear transformations in chapter 8, but for now, let's get back to the projections to see why this transformation-to-matrix correspondence is so important.

4.2.5 Section 4.2 summary

There are different ways to generalize transformations. Some ways are more pragmatic than others. Linear transformation in Euclidean space generalizes only rotation and scaling. The affine transformation adds translation, and the projective transformation adds shear.

4.3 Projective space and homogeneous coordinates

This book started by discussing how people around the world expanded their numbers when they had a new need. They made up negative numbers to write debt, rational to write fractions, real numbers to deal with the continuum, and complex to get square roots of negative numbers. Now we'll go through a similar but cooler experience: we'll expand the space.

> **WARNING** This topic isn't easy, and this section is probably the hardest one in the whole book. You don't need special knowledge to understand it, but expanding the space will challenge your geometrical intuition. You'll have to develop a new intuition—a hard mental exercise that has nothing to do with book knowledge per se.

Please treat this space expansion as a mathematical investment. We'll pay for it in complexity, but it will return all the investment with a lucrative interest. Exploiting this expansion will make our code simpler and less error-prone in the end.

4.3.1 Expanding the whole space with homogeneous coordinates

Normally, when we add a coordinate to a coordinate tuple, we're adding a dimension. A two-dimensional (x, y) becomes three-dimensional (x, y, z), a three-dimensional becomes four-dimensional, and so on. Now we'll add a new number to the coordinate tuple, but it won't be a new coordinate and won't denote a new dimension; it'll be something new, a denominator. So in 2D, we now have three numbers per point: (x, y, w).

The expanded space we're getting close to is called *projective,* so let's give the letter p to its coordinates: (x_p, y_p, w_p). Our good old Euclidean space will have the letter e: (x_e, y_e). Because the new projective space is the extension of the Euclidean one, all the points from the Euclidean space reside in our new projective space, too, and there's a way to convert a point from p to e:

$$x_e = \frac{x_p}{w_p}$$

$$y_e = \frac{y_p}{w_p}$$

This conversion isn't one-to-one. Multiple coordinate tuples from projective space correspond to a single point in Euclidean space. This is a feature, not a bug.

We can also convert a point from Euclidean space to projective space. But instead of getting a single tuple, we'll get a continuum of tuples:

$$x_p = w_p \, x_e$$

$$y_p = w_p \, y_e$$

$$w_p = w_p$$

We saw a similar effect with underspecified linear systems. Sometimes, our solution isn't a single tuple of numbers, but a whole continuum. Just like that, a Euclidean point $(1, 0)$ is the same as projective point $(1, 0, 1)$ or $(2, 0, 2)$ or $(505, 0, 505)$ or anything $(w_p, 0, w_p)$. All these tuples describe the same point: $(1, 0, 1) = (w_p, 0, w_p)$.

For us programmers, it's not ideal, because managing continuums is harder than keeping points as tuples of coordinates. Usually, we keep w_p as 1. Then the translation becomes simple:

$x_p = x_e$ $y_p = y_e$ $w_p = 1$	$x_e = x_p$ $y_e = y_p$

But could w_p be 2 instead? Sure! But it's not pragmatic because we'll have to do extra division:

$x_p = 2x_e$ $y_p = 2y_e$ $w_p = 2$	$x_e = \frac{x_p}{2}$ $y_e = \frac{y_p}{2}$

Could w_p be any other number? Definitely! 3, –5, 12.34, 2π. It could be any real number. Well, almost. There is one exception. A point from Euclidean space can't have zero as w_p. A point with $w_p = 0$ can't even be converted to Euclidean, because $\frac{x_p}{w_p}$ wouldn't work.

But the points where $w_p = 0$ still exist—not in Euclidean space, but in projective space. They extend our Euclidean space because they complement it with an additional shell of points. These non-Euclidean points also make a lot of computations easier, but to understand them, we'll have to massage our imagination a little.

Imagine a point in Euclidean space: $(x_e, y_e) = (1, 0)$. Let's write it in projective space with $w_p = 1$: $(x_p, y_p, w_p) = (1, 0, 1)$. Now let's divide the w_p by 10 a few times and see what happens to the point:

$$(1, 0, 1)_p \equiv (1, 0)_e$$

$$(1, 0, 0.1)_p \equiv (10, 0)_e$$

$$(1, 0, 0.01)_p \equiv (100, 0)_e$$

$$(1, 0, 0.001)_p \equiv (1000, 0)_e$$

$$(1, 0, 0.0001)_p \equiv (10000, 0)_e$$

You see the pattern, right? The smaller the w_p is, the farther the point is from the zero point. It goes away, and away, and away.

Euclidean space is infinite. You can place a point as far away as you like, and you can always place another one further. The space doesn't end anywhere.

Projective space extends the Euclidean one, meaning that it contains all its points, the whole infinity of them, but it extends in with a "belt" of points that are truly infinitely far from the origin. There's nothing behind this belt, and all the Euclidean space lies within it.

This belt is $(x_p, y_p, 0)$. It doesn't convert to Euclidean space, because there are no such points in that space. Every point from this belt is farther from the zero point than any point in Euclidean space. Every attempt to measure the distance between an extension point and a Euclidean point is futile, because the distance is always infinite. No matter how far the Euclidean point travels from an origin, it can't get any closer to the extension.

I wish I could help you with an illustration here, but no one can illustrate a projective space on a piece of paper, which is mathematically impossible. That's why projective geometry is the most challenging subject for students and young professionals. You have to develop a new intuition to understand this space, and sadly, you can't rely on much help. You have to use your own imagination.

Imagine an infinite watermelon. It doesn't end anywhere; it goes on and on and on through the whole universe. The whole universe is a watermelon. Now imagine its skin.

This example is like a Zen Buddhist koan, but it has a pragmatic, applicative sense. If you want, take a break and meditate on it. Rebuilding your intuition doesn't happen only by reading.

This way of notation, supplying a denominator to a coordinate tuple, is called *homogeneous* or *projective coordinates.*

Why homogeneous?

Homogeneous means "same kind" in Greek. It usually indicates that something is made of similar parts with the same properties. Here is a homogeneous polynomial:

$$p(x) = ax^2 + bxy + cy^2$$

All the members are at the same degree. x^2 is of the second degree, and xy is also of the second degree because there is a first-degree x multiplied by a first-degree y. y^2 is of the second degree, too, of course.

By contrast, this isn't a homogeneous polynomial:

$$g(x) = ax^2 + bx + c$$

This polynomial, when it's put as an equation, describes a parabolic curve:

$$ax^2 + bx + c = 0$$

In homogeneous coordinates, all the polynomic equations are homogeneous:

$$ax^2 + bx + c = 0$$

$$\frac{ax_p^{\,2}}{w_p^{\,2}} + \frac{bx_p}{w_p} + c = 0$$

$$ax_p^{\,2} + bx_pw_p + cw_p^{\,2} = 0$$

Remember that we can safely multiply all the point coordinates by the same nonzero number. (1, 0, 1) and (2, 0, 2) are the same point in homogeneous coordinates.

Pragmatically, this simplifies the surface classification in 3D because, for example, a conic surface in homogeneous coordinates now has the same formula as an ellipsoid. Only the coefficients differ.

Consequently, this makes the code smaller and less error-prone. You don't have to keep a lot of different functions for different surfaces; you can have one neat polynomial expression that covers all of them. Nice!

This expansion may not seem to be practical yet, so consider an example. In Euclidean space, there are two kinds of projections: a central projection, where projection rays come from a point in space (figure 4.16), and a parallel projection, where they all go in parallel (figure 4.17).

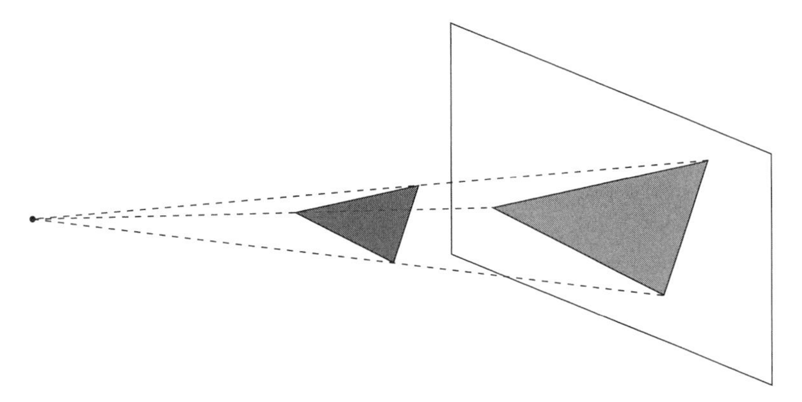

Figure 4.16 An example of central projection. Projection lines intersect in the center of the projection.

In projective space, these two types of projections are the same type! The center projection is the same, but for the parallel one, the center lies on the infinite belt. You surely know that parallel lines never ever intersect. A Euclidean postulate says so. Well, what's

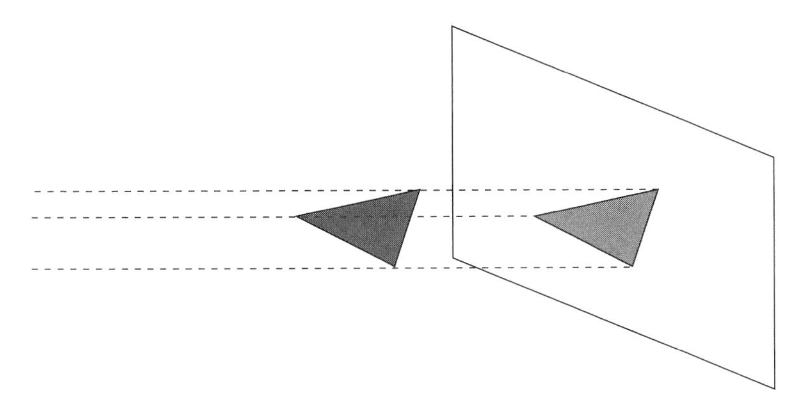

Figure 4.17 An example of parallel projection. In projective space, this example is the same as figure 4.16.

fair in carpooling is also fair in geometry: your space = your rules. The projective space is no longer Euclidean, so now the parallel lines intersect on the extension (figure 4.17).

These projection types are also used to model light in 3D. There is a point light and a directional light. Again, in projective space (and all GPUs operate in projective space anyway), these two types of light are the same.

This example isn't the sole benefit of projective coordinates. But to appreciate all the benefits, we have to get back to projective transformations, so let's wrap up with the coordinates.

To reiterate, we can extend our space with a belt (or, in 3D, a skin) of infinitely far points that ruin our intuition about how space works. There are no points farther than this extension, and the parallel lines intersect somewhere on it.

In return, the way we denote surfaces and projections becomes generalized, making code smaller and less error-prone. This also applies to transformations, which is the topic of the next section.

4.3.2 Making all the transformations a single matrix multiplication: Why?

As we've seen, the formula for a projective transformation is

$$x_p = \frac{ax_i + by_i + c}{gx_i + hy_i + i}$$

$$y_p = \frac{dx_i + ey_i + f}{gx_i + hy_i + i}$$

In homogeneous coordinates, all the polynomial components become homogeneous, too, and the denominator could be 1:

$$x_p = \frac{ax_i + by_i + cw_i}{gx_i + hy_i + iw_i}$$
$$y_p = \frac{dx_i + ey_i + fw_i}{gx_i + hy_i + iw_i}$$
$$w_p = 1$$

Unless the denominator is 0—and in real projective transformations, it isn't—we can multiply all the coordinates by the denominator. Remember that (1, 0, 1) and (2, 0, 2) are the same point:

$$x_p = ax_i + by_i + cw_i$$

$$y_p = dx_i + ey_i + fw_i$$

$$w_p = gx_i + hy_i + iw_i$$

If you remember the row-on-column rule, you probably see the resemblance to the matrix multiplication already. Indeed, in homogeneous coordinates, we can rewrite our projective transformation as matrix multiplication:

$$\begin{pmatrix} x_p \\ y_p \\ w_p \end{pmatrix} = \begin{pmatrix} a & b & c \\ d & e & f \\ g & h & i \end{pmatrix} \times \begin{pmatrix} x_i \\ y_i \\ w_i \end{pmatrix} = \begin{pmatrix} ax_i + by_i + cw_i \\ dx_i + ey_i + fw_i \\ gx_i + hy_i + iw_i \end{pmatrix}$$

This transformation generalizes translation, rotation, scale, shear, and perspective. All the basic transformations could be made into a single matrix multiplication. This is the translation matrix:

$$\begin{pmatrix} 1 & 0 & d_x \\ 0 & 1 & d_y \\ 0 & 0 & 1 \end{pmatrix}$$

This is the rotation:

$$\begin{pmatrix} \cos(r) & \sin(r) & 0 \\ -\sin(r) & \cos(r) & 0 \\ 0 & 0 & 1 \end{pmatrix}$$

And this is the scaling:

$$\begin{pmatrix} s_x & 0 & 0 \\ 0 & s_y & 0 \\ 0 & 0 & 1 \end{pmatrix}$$

The generic affine transformation that generalizes all three looks like this:

$$\begin{pmatrix} a & b & c \\ d & e & f \\ 0 & 0 & 1 \end{pmatrix}$$

In terms of code, this generalization is yet another simplification. You don't have to keep separate functions for all the partial cases anymore. You can implement them all—translations, rotations, and scalings—with a single matrix multiplication.

As an important but slightly overused reminder, (1, 0, 1) and (2, 0, 2) are the same point. You can multiply your transformation matrix by a scalar without changing the transformation mathematically, because applying the multiplied matrix to a point simply results in a multiplied point in homogeneous coordinates. The earlier translation could be rewritten this way:

$$\begin{pmatrix} 1 & 0 & d_x \\ 0 & 1 & d_y \\ 0 & 0 & 1 \end{pmatrix} \text{ becomes } \begin{pmatrix} 2 & 0 & 2d_x \\ 0 & 2 & 2d_y \\ 0 & 0 & 2 \end{pmatrix}$$

The matrix is different, and the resulting coordinates (in numbers in the coordinate tuple) will be different, but because each point in a projective space is a linear continuum, the transformation remains the same. Each transformation has not a single matrix, but a continuum of matrices that represent it. We'll use this property to gain some performance when computing inverse transformation later.

This matrix generalization works in 3D as well. In homogeneous coordinates, all your 3D points and vectors are four-number tuples, and the transformation matrix is a 4×4 array. As a nice bonus, this four-in-a-row arrangement works nicely with modern superscalar architectures.

But treating transformations as matrices isn't only about simplifying the computation; it's also about something more. As we saw with linear transformations, being able to turn a transformation into a matrix operation lets us steal the matrix algebra (the set of rules and operations). So what is matrix algebra, and what useful properties does it have?

COMPOSITION

First, every square matrix multiplication results in a square matrix. That's fairly trivial. When we multiply a 3×3 matrix by a 3×3 matrix using the row-on-column rule, the result will be a 3×3 matrix no worse than any of the inputs.

This means that we can compose our transformations, and the result will be a transformation as well. We discussed this topic with linear transformations, but composability was limited because the linear transformation didn't include the translation. We have the translation now, so let's see how it works! Figure 4.18 shows a rectangle.

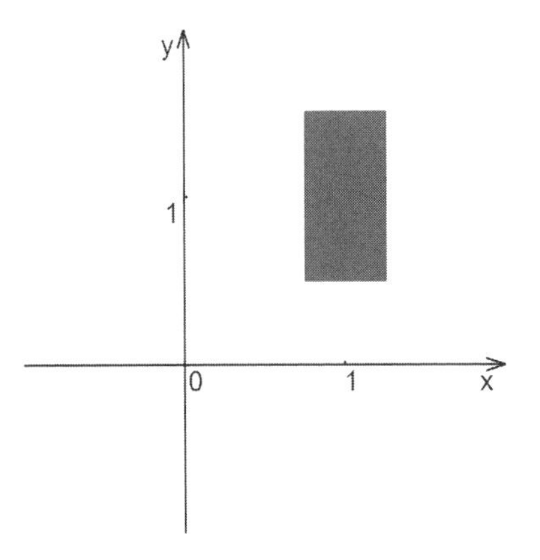

Figure 4.18 A rectangle in Euclidean 2D space. Its corners are (0.75, 0.5), (1.25, 0.5), (1.25, 1.5), (0.75, 1.5). With projective transformations, transforming its corners is the same as transforming the whole rectangle.

We want to turn it 90 degrees clockwise around its center. Normally, this operation consists of three different transformations (figure 4.19):

1. We translate it so that its center becomes the zero point.
2. We rotate it 90 degrees clockwise.
3. We translate it back.

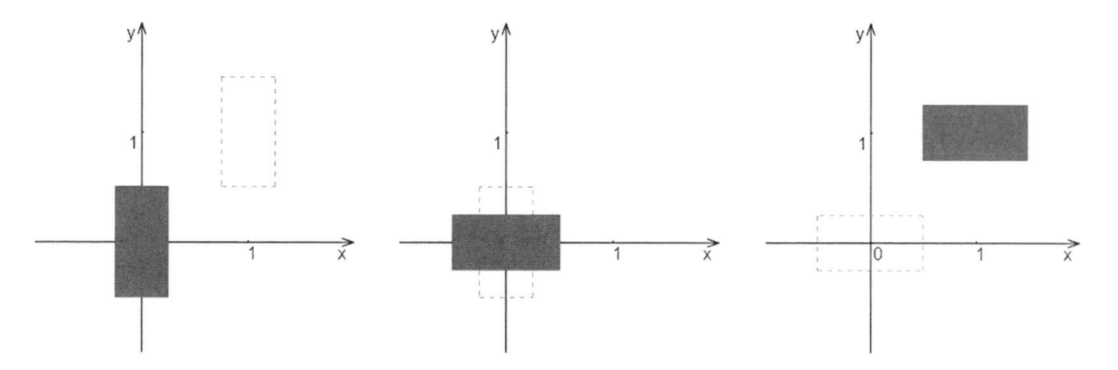

Figure 4.19 This composition could be made into a single transformation using one multiplication instead of three.

The respective matrices are

$$\begin{pmatrix} 1 & 0 & -1 \\ 0 & 1 & -1 \\ 0 & 0 & 1 \end{pmatrix}, \begin{pmatrix} 0 & 1 & 0 \\ -1 & 0 & 0 \\ 0 & 0 & 1 \end{pmatrix}, \begin{pmatrix} 1 & 0 & 1 \\ 0 & 1 & 1 \\ 0 & 0 & 1 \end{pmatrix}$$

Let's make a composition in SymPy (see ch04/matrix_composition.py in the book's source code) and see whether it does what we want it to. These are the matrices:

```
a1 = Matrix([[1, 0, -1], [0, 1, -1], [0, 0, 1]])
a2 = Matrix([[0, 1, 0], [-1, 0, 0], [0, 0, 1]])
a3 = Matrix([[1, 0, 1], [0, 1, 1], [0, 0, 1]])
```

This is the composition,

```
c = a3*a2*a1
```

which, as we previously discussed, is also a matrix:

```
Matrix([[0, 1, 0], [-1, 0, 2], [0, 0, 1]])
```

These are the rectangle's corners:

```
p1 = Matrix([0.75, 0.5, 1])
p2 = Matrix([1.25, 0.5, 1])
p3 = Matrix([1.25, 1.5, 1])
p4 = Matrix([0.75, 1.5, 1])
```

These are the transformations:

```
print(c * p1)
print(c * p2)
print(c * p3)
print(c * p4)
```

The transformation makes points of the vertical rectangle into the points of the horizontal one as though it were three separate transformations (translation, rotation, and another translation):

```
Matrix([[0.500000000000000], [1.25000000000000], [1]])
Matrix([[0.500000000000000], [0.750000000000000], [1]])
Matrix([[1.50000000000000], [0.750000000000000], [1]])
Matrix([[1.50000000000000], [1.25000000000000], [1]])
```

So instead of doing 12 matrix multiplications, 3 per point, we do only 6—2 to compute the composition and 4 to apply it to the points. That's economical.

In practice, this approach allows us to position much more complex geometric models consisting of millions of points in a scene by applying all the possible compound transformations, rotations, and scales, and then project them on the user's screen for rendering—all by multiplying each point by the corresponding matrix only once per frame.

PARALLELIZATION

The next useful property we can borrow from matrices is associativity. For any three matrices A, B, and C, $(AB)C = A(BC)$. This doesn't seem much, and to be fair, in computational geometry it isn't. But in programming in general, associativity is a much-prized property because it makes parallel computation possible.

Consider this: if you're not allowed to move parentheses, and then you have to compute a long composition, you have to start from the first pair of matrices, multiply their product by the third matrix, multiply that result by the fourth, and so on. For an eight-matrix composition, seven operations would take exactly seven steps:

$$A \times B \times C \times D \times E \times F \times G \times H =$$

$$(AB) \times C \times D \times E \times F \times G \times H =$$

$$(ABC) \times D \times E \times F \times G \times H =$$

$$(ABCD) \times E \times F \times G \times H =$$

$$(ABCDE) \times F \times G \times H =$$

$$(ABCDEF) \times G \times H =$$

$$(ABCDEFG) \times H =$$

$$(ABCDEFGH)$$

But if we can choose where to put the parentheses, if we can choose which pair to compute first, we can do several operations in parallel, and the same computation will take only three steps:

$$A \times B \times C \times D \times E \times F \times G \times H =$$

$$(AB) \times (CD) \times (EF) \times (GH) =$$

$$(ABCD) \times (EFGH) = (ABCDEFGH)$$

Associativity is great! Only one thing makes it less great for projective transformation: matrix multiplication itself is so cheap in computation terms that you normally wouldn't think about parallelizing its composition.

So with matrix algebra comes the ability not only to do several transformations in one go, but also to compute such compositions in parallel. Are any useful properties left? Sure!

An idle transformation

There is a matrix E so that for every matrix A, EA = AE = A. It's a nonmodifying element. It's like 0 for the addition of real numbers or 1 for their multiplication. For square matrices, it's an *identity matrix,* a diagonal matrix consisting of 0s and 1s:

$$\begin{pmatrix} 1 & 0 & 0 \\ 0 & 1 & 0 \\ 0 & 0 & 1 \end{pmatrix}$$

It works as a zero-length translation, as a 1-to-1 scale, and as a zero-degree rotation, too. This transformation does nothing, and it's helpful because now we can generalize both the objects we want to transform, and the objects we want to keep static as a single type of objects. The objects we want to transform will have a real transformation matrix, and the static objects will also have a transformation matrix but equal to the identity. Then whenever we decide to move them after all, we won't have to hack their type; all we have to do is fill in the matrix coefficients.

A transformation that puts things back as they were

The last useful property for us is the inverse element. For every invertible matrix A, there is another element A^{-1} so that $A^{-1} A = AA^{-1} = E$. In plain words, for every transformation that turns point x to point y, there's a transformation that turns y back to x.

This type of transformation is called an *inverse transformation.* Luckily for us, a matrix inversion is a well-known operation, and every framework or library that works with matrices has one.

NOTE What if a matrix isn't invertible? Surely, we can fill a matrix with all zeroes, and then no matter how we multiply it, we won't get anything but zeros as a result. Don't worry; all real-world transformation matrices are invertible. It wouldn't hurt to add an assertion for it, though, because a transformation matrix's losing invertibility might indicate a bug in the code.

Counterintuitively, using an inverse matrix for an inverse transformation may not always be effective. Remember that we can multiply our matrix by a nonzero scalar, and the matrix will correspond to the same transformation. In practice, we can use this property to compute an inverse transformation without computing the real inverse matrix. Here's how you're supposed to compute a matrix inverse:

$$A^{-1} = \frac{1}{\det(A)} C^T$$

The matrix C is the cofactor matrix of A. It's the same shape matrix in which every element is computed from the elements of the original matrix. How to compute one isn't important yet.

The function *det* is the matrix determinant. It's a relatively heavy operation computationally that turns a matrix into a number. The determinant is like a matrix's weight, a kind of measure. I haven't talked about it yet and won't discuss it further, because to get an inverse transformation, we don't need to compute the determinant. An inverse matrix multiplied by the determinant—or, rather, not yet divided—will be a matrix of the inverse transformation, too.

SEE ALSO If you want to learn more about the determinant, nice visualizations on the Intuitive Math website explain what it is and how it connects with transformations. See http://mng.bz/71Oe.

You don't have to learn this formula by heart; all the libraries that work with matrices have their inversion functions anyway. But let's look into a trick we can use to make the computation faster. We can use the cofactor transposition instead of the real matrix inversion. Let's get back to our rectangle from before to see how it works. First, let's find an inversion for our composition as a matrix inversion. SymPy has a method we can use:

```
c_inv = c.inv()
```

Now let's apply the inverted matrix to the rectangle's corners after the composed transformation:

```
pt1 = c * p1
pt2 = c * p2
pt3 = c * p3
pt4 = c * p4
...
print(c_inv * pt1)
print(c_inv * pt2)
print(c_inv * pt3)
print(c_inv * pt4)
```

This is the printout:

```
Matrix([[0.750000000000000], [0.500000000000000], [1]])
Matrix([[1.25000000000000], [0.500000000000000], [1]])
Matrix([[1.25000000000000], [1.50000000000000], [1]])
Matrix([[0.750000000000000], [1.50000000000000], [1]])
```

These coordinates are the original coordinates before the transformation, although with more zeroes at the end. So we transformed the rectangle back to its original position. Now let's do a cofactor transposition instead (see ch04/matrix_inversion.py in the book's source code):

```
def minor(M, i, j):
    M_copy = M[:, :]
    M_copy.col_del(j)
    M_copy.row_del(i)
    return M_copy

def cofactor(M):
    C = M[:, :]
    for i in range(M.shape[0]):
        for j in range(M.shape[1]):
            determinant = minor(M, i, j).det()
            sign = 1 if i+j%2 == 1 else -1
            C[i, j] = sign*determinant
    return C

def not_a_real_inverse(M):
    return cofactor(M).T

print(not_a_real_inverse(c) * pt1)
print(not_a_real_inverse(c) * pt2)
print(not_a_real_inverse(c) * pt3)
print(not_a_real_inverse(c) * pt4)
```

An i-j minor is a matrix without i-row and j column.

That's how you compute the cofactor matrix, if you're interested.

The determinant of the minor in SymPy

Matrix transposition in SymPy

The result is

```
Matrix([[-0.750000000000000], [-0.500000000000000], [-1]])
Matrix([[-1.25000000000000], [-0.500000000000000], [-1]])
Matrix([[-1.25000000000000], [-1.50000000000000], [-1]])
Matrix([[-0.750000000000000], [-1.50000000000000], [-1]])
```

This isn't the same matrix, and it doesn't produce the same coordinate tuple. But it's the same up to scalar multiplication, and in homogeneous coordinates, that's all we need. Now we have a transformation that turns point x to point y and the inverse transformation that turns y to kx, where k is a scalar. If we want to keep our tuples ending with +1, we can divide them all by their last element, which would be exactly k, instead of computing the whole determinant beforehand.

Now you see how knowing the math behind projective transformations enables you to write more efficient, faster code. You can not only save on composition, but also throw away whole chunks of computation and get away with it.

4.3.3 Section 4.3 summary

In homogeneous coordinates, a projective transformation that also generalizes translation, rotation, scale, and shear is the same as matrix multiplication. These multiplications are *composable*, meaning that you can encode a whole series of transformations in a single matrix. The composition itself is also parallelizable because of associativity, not that we'll ever have to resort to that. We also have a transformation that does nothing, and for every transformation that does something, there's a transformation that serves as its undoing.

4.4 Practical examples

It's time to do something practical. Putting objects in a scene seems like fun, but it's all about calling translations and rotations in the right order. Setting a camera for a 3D scene so that it follows an object is a little more interesting, but it involves a lot of irrelevant trigonometries, and you probably have a function for that task in your engine of choice anyway. Let's make our own transformation, then—a transformation that solves a practical problem and doesn't rely on any help from third-party libraries.

4.4.1 Scanning with a phone

In 2012, I made a simple camera app for Android that undoes the natural perspective of a picture (figure 4.20).

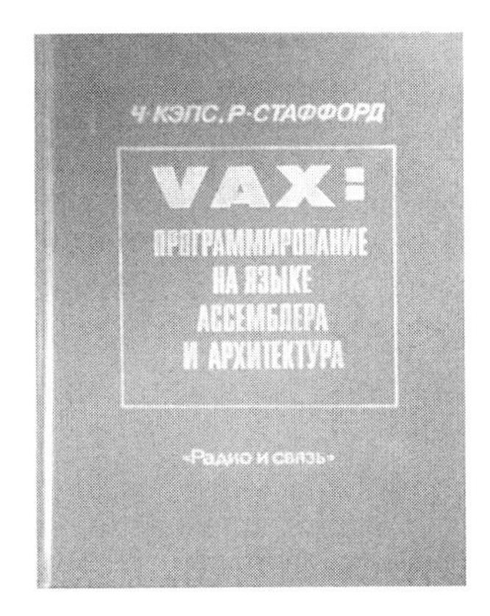

Figure 4.20 Before and after: undoing the natural perspective in a picture

Now we have all sorts of scanning apps on phones, but at that time, this app was a novelty. Let's dissect the application to see what's it made of.

We know that the projective transformation is a four-point transformation, so for input, we need a picture, which the camera API provides, and the four corner points of a rectangle that we'll ask the user to select.

Modern scanning apps do that point selection automatically, although not always with great success. This automatic selection is by itself an interesting problem, but it has little to do with projective transformations, so we'll set it aside.

We've already seen a system for an arbitrary four-point transformation, and it was a little problematic. It took forever to solve symbolically, and the solution also appeared to be unreasonably huge. We always have the option to solve the system numerically, but that option is the lazy way to go. Let me show you something else instead.

The scanning problem is as follows: for every pixel of a target rectangular picture (figure 4.20, right), find the original pixel color from a photographic one (figure 4.20, left), given that the four corner points of the rectangular picture have their corresponding points on the picture from the camera.

This example may seem to imply a convex quadrilateral–to–rectangular transformation, but in fact, it's the opposite. For every pixel of the flat picture, we want to find where it comes from. The transformation is rectangular–to–quadrilateral, not vice versa (figure 4.21).

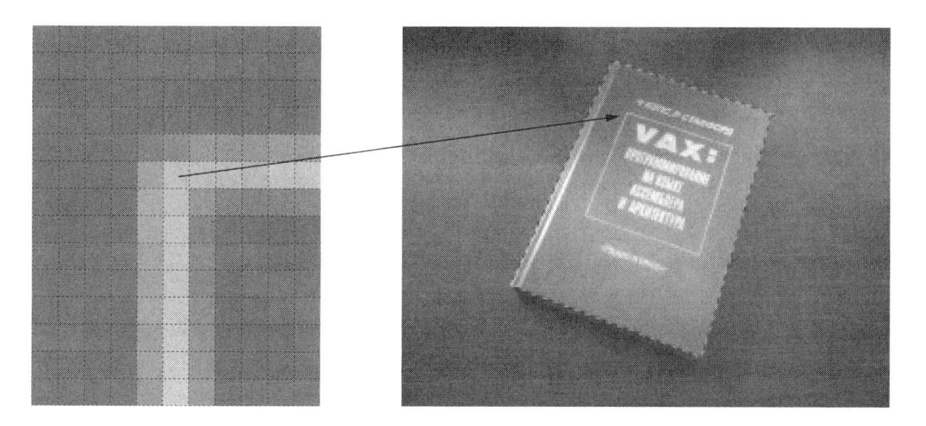

Figure 4.21 To undo a perspective, we use a four-point transformation from a rectangular.

Why is this important? Now we can use a standard square for an intermediate transformation. A *standard square* is a square with a side length of 1 that starts at the zero point (0, 0).

A rectangular–to–standard square transformation is scaling. Let's say that our rectangle has width w and height h. The transformation matrix that transforms its pixels to the standard square is

$$\begin{pmatrix} \frac{1}{w} & 0 & 0 \\ 0 & \frac{1}{h} & 0 \\ 0 & 0 & 1 \end{pmatrix}$$

The second transformation, from the standard square to the source picture, is a little trickier. Let's revise our system from section 4.3.2:

```
[
xt1 * x1 * g + xt1 * y1 * h + xt1 * i - x1 * a - y1 * b - c,
yt1 * x1 * g + yt1 * y1 * h + yt1 * i - x1 * d - y1 * e - f,
xt2 * x2 * g + xt2 * y2 * h + xt2 * i - x2 * a - y2 * b - c,
yt2 * x2 * g + yt2 * y2 * h + yt2 * i - x2 * d - y2 * e - f,
xt3 * x3 * g + xt3 * y3 * h + xt3 * i - x3 * a - y3 * b - c,
yt3 * x3 * g + yt3 * y3 * h + yt3 * i - x3 * d - y3 * e - f,
xt4 * x4 * g + xt4 * y4 * h + xt4 * i - x4 * a - y4 * b - c,
yt4 * x4 * g + yt4 * y4 * h + yt4 * i - x4 * d - y4 * e - f,
i - 1
], (a, b, c, d, e, f, g, h, i))
```

The system is large, sure, but as soon as we apply the coordinates of the standard square, it becomes much simpler. The exact order in which we pick the coordinates is irrelevant, so this will do:

```
(x1, y1) = (0, 0)
(x2, y2) = (1, 0)
(x3, y3) = (1, 1)
(x4, y4) = (0, 1)
```

By applying these specific numbers to the equations, we make them smaller:

```
[
xt1 * i - c,
yt1 * i - f,
xt2 * g + xt2 * i - a - c,
yt2 * g + yt2 * i - d - f,
xt3 * g + xt3 * h + xt3 * i - a - b - c,
yt3 * g + yt3 * h + yt3 * i - d - e - f,
xt4 * h + xt4 * i - b - c,
yt4 * h + yt4 * i - e - f,
i - 1
], (a, b, c, d, e, f, g, h, i))
```

This result is a leaner version of the same system; more important, it doesn't have to be a single system anymore! Now there are four pieces that we can solve one by one—much faster, and with more compact solutions, too. There's the obvious i:

```
[i - 1 ], (i)
```

Then, with the known i, the first two equations become independent:

```
[xt1 * i - c] , (c)
[yt1 * i - f,], (f)
```

Then, knowing the c and f, we can resolve the next two equations along with the two at the end in terms of g and h:

```
c = xt1
f = yt1
i = 1
[xt2 * g + xt2 * i - a - c,
yt2 * g + yt2 * i - d - f,
xt4 * h + xt4 * i - b - c,
yt4 * h + yt4 * i - e - f] , (a, b, d, e)
```

The result will be rather compact

```
a: g*xt2 - xt1 + xt2, d: g*yt2 - yt1 + yt2,
b: h*xt4 - xt1 + xt4, e: h*yt4 - yt1 + yt4
```

and reusable! Let's finish the system with these equations:

```
a = g*xt2 - xt1 + xt2
d = g*yt2 - yt1 + yt2
b = h*xt4 - xt1 + xt4
e = h*yt4 - yt1 + yt4
solution_3 = solve([
    xt3 * g + xt3 * h + xt3 * i - a - b - c,
    yt3 * g + yt3 * h + yt3 * i - d - e - f,
], (g, h))
```

The solution is in *x*s and *y*s only, with no other variables:

```
g: (xt1*yt3 - xt1*yt4 - xt2*yt3 + xt2*yt4 -
➥ xt3*yt1 + xt3*yt2 + xt4*yt1 - xt4*yt2)/
➥ (xt2*yt3 - xt2*yt4 - xt3*yt2 +
➥ xt3*yt4 + xt4*yt2 - xt4*yt3),
h: (xt1*yt2 - xt1*yt3 - xt2*yt1 + xt2*yt4 +
➥ xt3*yt1 - xt3*yt4 - xt4*yt2 + xt4*yt3)/
➥ (xt2*yt3 - xt2*yt4 - xt3*yt2 +
➥ xt3*yt4 + xt4*yt2 - xt4*yt3)
```

With these formulas, we can compute g and h numerically for every four points of the standard square transformation, and use that to compute a, b, d, and e. Computing c and f from their simple formulas and putting i as 1 will complete the solution. (For the full listing, see ch_04/projective_transformation_in_steps.py in the book's source code.)

Now we have a transformation that scales pixels to the points of the standard square and a transformation that transforms points from the standard square to the source picture. We can compose them into a single transformation and apply it to every pixel of the target rectangular picture to know where this pixel should take its color from the original picture.

We still have a few less-relevant problems to solve. First, how do we know the size of the rectangular picture? The answer is, we don't. There's a projective transformation for every rectangular picture to every possible quadrilateral. Most of these transformations are improbable in practice, though I'm not sure whether it's grammatically correct to apply the word *most* to an infinite continuum.

So let's solve this problem constructively. We don't want our picture to be overly pixelated, meaning that a lot of pixels on the target pixels would correspond to a single pixel on the source. But we don't want to lose any data either, as when a lot of pixels from the source picture are ignored by the transformation.

Here's the rule of thumb I came up with: the height of the target picture should be the average of the left and right segments' lengths from the quadrilateral defined by the user. Then the width is the average of the top and bottom segments' lengths.

This rule isn't math; it's engineering. It's not proved to be optimal, but it works most of the time. Please feel free to come up with your own solutions.

Another problem is how we compute colors between pixels so the target picture becomes nice and smooth, not pixelated. This problem is an interesting one, and we'll get to it in chapter 8.

4.4.2 Does a point belong to a triangle?

In chapter 3, we did a ray intersection with a spatial triangle. This time, the problem is even simpler: finding whether a point lies in a triangle on a plane.

There are practical applications for this problem, such as picking a triangle on a triangle mesh's projection or looking for a particular triangle when we're doing something on a mesh's texture. But what makes this problem fascinating is that it serves as an excellent interview question.

There are multiple ways to check whether a point lies in a triangle. They're all correct; all make sense in their own context. If you want to probe a person's experience in geometry, such as during a job interview, this simple problem gives them a lot of opportunities to discuss things all over the domain, including projective transformations.

Suppose that we want to probe a lot of points against a few triangles as fast as possible and that we have some free preprocessing time. One way to go would be to turn each triangle into a convenient transformation in its own basis and then apply this transformation to each point we want to check (figure 4.22).

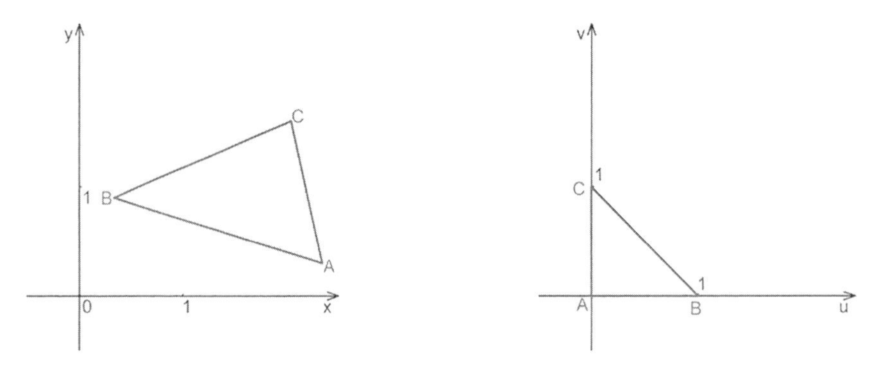

Figure 4.22 A triangle in space versus the same triangle in its own basis

How do we bring each triangle to its own basis, though? Well, we already know how to make a four-point transformation, and with that, we can transform triangles. So technically, we can give each triangle a fourth point by doing vector addition and revert the problem at hand to the known one (figure 4.23).

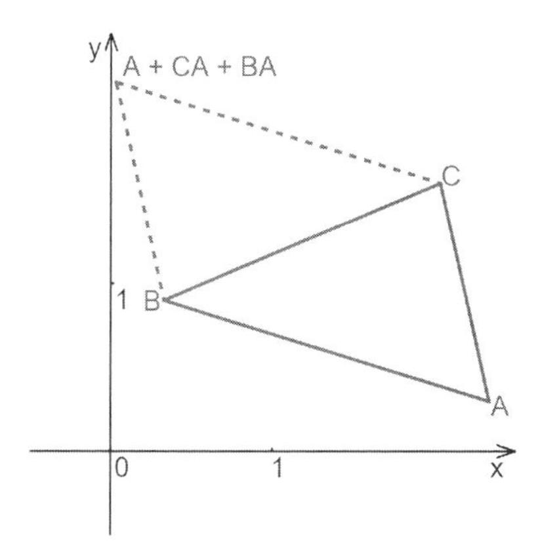

Figure 4.23 Adding a point to a triangle reverts the problem to the known one, but that's boring.

That approach is one 3 × 3 matrix multiplication per triangle, which is rather cheap, but it can be made cheaper still! We don't need a full-scale projective transformation to transform a triangle. We can use a special case that we already know about. For a three-point transformation in 2D, a mere affine transformation will work perfectly. An affine transformation is

$$x_t = ax_i + by_i + c$$

$$y_t = dx_i + ey_i + f$$

We want to find a transformation from an arbitrary triangle to its basis, so the three points should go to the corners of a standard square:

$$(x_1, y_1) \rightarrow (0, 0)$$

$$(x_2, y_2) \rightarrow (1, 0)$$

$$(x_3, y_3) \rightarrow (0, 1)$$

We can make this into a system of equations. Let's switch to SymPy notation because we're not going to solve it manually anyway:

```
a*x1 + b*y1 + c,
d*x1 + e*y1 + f,
a*x2 + b*y2 + c - 1,
d*x2 + e*y2 + f,
a*x3 + b*y3 + c,
d*x3 + e*y3 + f - 1,
```

That's six equations and six variables. Unlike the system we had with projective transformation, this solution has no ambiguity. But we didn't have to use homogeneous coordinates either. This transformation fits nicely in Euclidean space, so normally, the system also has a single Euclidean point as a solution—not a line, not a continuum, but a point. The solution is

```
a: (-y1 + y3)/(x1*y2 - x1*y3 - x2*y1 + x2*y3 + x3*y1 - x3*y2),
b: (x1 - x3)/(x1*y2 - x1*y3 - x2*y1 + x2*y3 + x3*y1 - x3*y2),
c: (-x1*y3 + x3*y1)/(x1*y2 - x1*y3 - x2*y1 + x2*y3 + x3*y1 - x3*y2),
d: (y1 - y2)/(x1*y2 - x1*y3 - x2*y1 + x2*y3 + x3*y1 - x3*y2),
e: (-x1 + x2)/(x1*y2 - x1*y3 - x2*y1 + x2*y3 + x3*y1 - x3*y2),
f: (x1*y2 - x2*y1)/(x1*y2 - x1*y3 - x2*y1 + x2*y3 + x3*y1 - x3*y2)
```

As before, the solution deserves a little grooming:

```
div = x1*(y2 - y3) + x2*(y3 - y1) + x3*(y1 - y2)

a = (-y1 + y3)/div,
b = (x1 - x3)/div,
c = (-x1*y3 + x3*y1)/div,
d = (y1 - y2)/div,
e = (-x1 + x2)/div,
f = (x1*y2 - x2*y1)/div
```

This is your code. With it, you turn each triangle into six coefficients on the preprocessing step. Note that this solution takes exactly the same memory as the coordinates of the triangles' vertices. Then you transform each point by doing only four multiplications and six additions:

```
xt = a*xi + b*yi + c
yt = d*xi + e*yi + f
```

Then you do the basis check as you did in chapter 3:

```
return True if (0 < xt) and (0 < yt) and (0 < xt+yt < 1) else False
```

That's it.

4.4.3 Section 4.4 summary

Being comfortable with projective transformations pays greatly in terms of performance. Your code runs faster, and you come up with solutions quicker. You have more options. Instead of opting for a numeric solution or a series of consequent transformations, you can solve problems symbolically and compose the transformation in one go, eventually turning a spree of quick wins into a fast, and robust application.

But although being fluent in projective transformations gives you options, it doesn't obligate you to use the most sophisticated one. Sometimes, picking a transformation that's enough for the job results in the fastest code, and it makes your job as a programmer easier, too.

4.5 Exercises

Exercise 4.1 Which of the following homogeneous coordinates tuples represent the same point in Euclidean space?

- A (18, 9, 1)
- B (18, 9, 3)
- C (10, 10, 0)
- D (6, 3, 1)
- E (6, 3, 0)

Exercise 4.2 Which of the following homogeneous coordinates tuples represent the same point?

- A (1, 1, 1)
- B (2, 2, 1)
- C (2, 2, 0)
- D (1, 1, 2)
- E (1, 1, 0)

Exercise 4.3 Which points lie exactly 1 unit apart in 3D?

- A A. (5, 4, 5, 1)
- B B. (5, 4, 5, 0)
- C C. (10, 10, 10, 2)
- D D. (10, 10, 10, 0)
- E E. (5, 4, 5, 2)

Exercise 4.4 Imagine that you're writing a 2D solitaire game. The cards are laid out on a rectangular canvas, so each card is rectangular, with its corners in the canvas's system of coordinates. The producer says that a flat view is boring and proposes to make a perspective view so that the card table looks more realistic (figure 4.24).

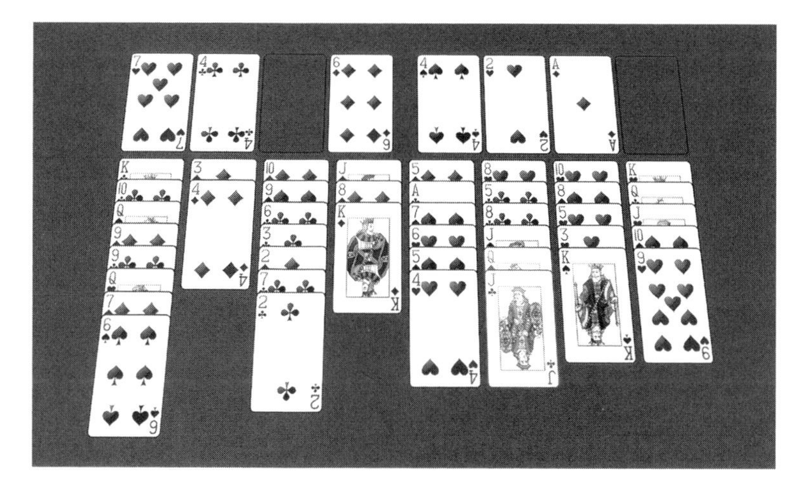

Figure 4.24 A FreeCell Solitaire game with added perspective. (Source: Mantas, CC BY-SA 3.0, via Wikimedia Commons)

How would you render such a table, and how would you process mouse clicks so that players can pick the cards they want from the projected table?

Exercise 4.5 Speaking of realistic views, let's say you want to position a camera in 3D. To do that, you need to specify the transformation that puts points from 3D space into the 3D standard cube. One of the cube walls would be the 2D screen on which we display the 3D scene, and the depth of the cube will be used to determine which objects stand before which.

You have four pieces of input data: a point the camera is looking at; a point of the camera's origin; a field-of-view angle; and one more vector, which we'll call *x*. How do you turn it into a single matrix transformation?

To solve this problem in full, you need to know a bit of vector algebra and a bit of trigonometry—quite a bit of both, to be honest. So don't solve it completely yet. For now, answer three questions:

1. What size will the transformation matrix be?
2. We know that in 2D, a projection is a four-point transformation. How many points are required for the transformation in 3D?
3. What's *x*?

4.6 Solutions to exercises

Exercise 4.1 B and D. A is the same as 2 scaled by three; C and E don't belong to Euclidean space.

Exercise 4.2 C and E. A, B, and D are the scaling of the same point, and they're all Euclidean, so they can't possibly represent the same as C or E.

Exercise 4.3 A and C. Brought to Euclidean space, C is (5, 5, 5), which is one unit apart from A, which is (5, 4, 5). Then E is (2.5, 2, 2.5), which is too far from any other Euclidean point we have. And B and D are both infinitely far from any point in Euclidean space, because they're on the projective extension.

Exercise 4.4 Surprisingly, picking and rendering are the same task. You need a color for each pixel of the projected table, and for this task, you need a projected-to-rectangular space transformation. The same transformation works for picking; you need to transform the mouse position from projected space to the original rectangular.

This transformation is a four-point transformation; we composed one in section 4.4.1. But it was easier there; we projected from a rectangular space to a standard square and then from the standard square to the arbitrary transformation. The zeroes from the standard square made our equations smaller and simpler. But computing the formulas for the reverse transformation—from arbitrary four-point projected space to the standard cube—might be too much for SymPy.

You can compute the transformation's coefficients numerically, using a direct method such as Gaussian elimination. The computation is only nine equations, and you have to do it only once per "camera" setup. But you can reuse the coefficients' formulas from the scanning-app example (see section 4.4.1) and do the inversion for the matrix they constitute, which will lead to the same result with less work.

Exercise 4.5 The answer to question 1 is a 4 × 4 matrix. It turns a 3D homogeneous coordinate into a 3D homogeneous coordinate, and they are tuples of 4 numbers: (x, y, z, w).

The answer to question 2 is a five-point transformation. It's not easy to see unless you try to evaluate the matrix size first. A 4 × 4 matrix gives you 16 variables. One variable is floating, because the result of the transformation is a linear continuum, not a single tuple, which leaves us 15. So we need 15 equations to fill the matrix with coefficients, and 15 equations may come only from 5 transformed 3D points.

Question 3 is an open question. There's no single correct answer, but there is a wrong one. You need an additional vector because two points and a field-of-view angle don't constrain the camera in space firmly enough. Imagine that camera is pierced by the line coming through the point of origin and the point of observation in kebab style. It becomes somewhat constrained, sure, but it can still rotate freely around this axis. So you need some kind of vector to constrain the camera in 3D space. Usually, this vector is from the point of origin pointing to the top of the camera. It's called an *up* vector, but it can be a vector to the right or to the bottom. It definitely shouldn't be pointing forward or back, because that axis contains the point of observation itself—the wrong answer. Otherwise, the constraint could be anything. You'll learn more about vectors and how to work with them in chapter 9.

Summary

- Translation, rotation, scaling, and shear can all be generalized as projective transformation.
- Homogeneous coordinates expand Euclidean space with a belt of infinitely far points.
- Euclidean coordinates and homogeneous ones are connected via $x_{ie} = \frac{x_{ip}}{w_p}$ proportion, but this doesn't work for the belt, where $w_p = 0$.
- The expansion simplifies the way we do math, allowing us to write a projective transformation as matrix multiplication. Then matrix multiplications and projective transformations are essentially the same. They're both composable; we can stack as many projective transformations on top of one other as we like, and the composition will still be a single projective transformation. Both matrix multiplications and projective transformations have their inverse operations.
- Due to the nature of homogeneous coordinates, an inverse transformation is wider than the inverse matrix and could be computed faster.
- You can create your own projective transformation—not necessarily composed of translations, rotations, and such—by solving a system of linear equations.

5 The geometry of calculus

This chapter covers

- Using the derivative as a measure of how fast a function grows
- Computing the derivative of a function not only as a number, but also as a function itself
- Using derivatives to compute tangent vectors of parametric curves
- Using second derivatives to assess the curvature of a curve
- Composing smooth piecewise functions and parametric curves

People usually have a pretty good intuition about the topics we're going to talk about. Does a function grow or decline? Is a surface smooth, or does it have some sharp edges? Is one curve curvier than another? Usually, we can answer all these questions intuitively. But we can't program intuition. We need formal things; we need formulas; we need equations. We need the mathematical basis on which to build our code. So this chapter is dedicated to calculus, its role in geometry, and the geometry behind it.

I know that calculus has a bad reputation for being dry and boring. But you'll learn it not by studying formulas, but by expanding the geometric intuition you already have. We'll leave the formulas to SymPy.

The concepts introduced in this chapter aren't necessarily geometrical, but they'll help you understand splines and nonlinear geometric transformations (chapters 7 and 8). They'll also enable some interesting practical examples in chapters 11 and 12, such as building a smooth contour out of square pixels and coining a continuous curve out of parabolas.

By the end of the chapter, you'll have a solid understanding of how to measure smoothness and curvature, and consequently, you'll know how to program your curves to be continuous, smooth, and curvy exactly as requested.

5.1 What is a derivative?

A *derivative* is a measure of how fast a function grows or declines. We can tell whether a function grows or declines by looking at its graph, of course. Only in mathematics do we usually use the verbs *increases* and *decreases*. But to turn an observation into a specific number, we need to do some math. For a linear function in real numbers, growth or decline correlates to the coefficient by the x in its explicit form:

$$y = ax + b$$

If a is positive, the function increases or grows. If a is negative, the function decreases or declines. If a is zero, the function isn't linear at all; it's a constant (figure 5.1). The larger the absolute value of a is, the more the function grows or declines.

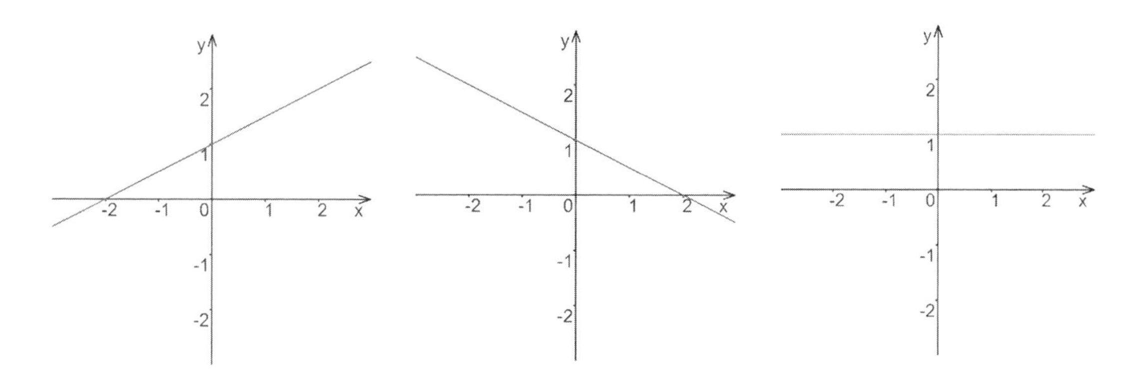

Figure 5.1 Increasing and decreasing linear functions, and a constant

For nonlinear functions, however, measuring growth isn't that easy. Nonlinear functions may increase in some places and decrease in others; they can even be periodical, like the sine, or discontinuous, like the tangent. They don't have a nice dedicated coefficient to look at if we want to see how they go. So what do we use instead?

5.1.1 Derivative at a point

Let's say we want to see how fast a car goes, and we don't have a speedometer. No problem. We'll drive the car on a piece of flat road, measuring the time it takes to go from A to B, and the distance from A to B divided by the time of the travel will be our average speed. Well, this approach isn't ideal; because the speed of a car isn't a constant value, we have to speed up and slow down to make the trip. But at least this would be some kind of measure.

We can take a similar approach to measuring functions. An average "speed" of a function on an interval [a, b] is something like this:

$$s = \frac{f(b) - f(a)}{b - a}$$

It's a number. It's some kind of characteristic. We can illustrate the formula graphically as a right-angled triangle where $(b - a)$ is the bottom leg, $(f(b) - f(a))$ is the right leg, and the hypotenuse works as a linear approximation of the function on the interval $[a, b]$. The hypotenuse is a line segment, so it also represents a function on its own: the linear function, the equation of the line it lies on. The "speed" of the function measured is the same as the incline of the hypotenuse (figure 5.2).

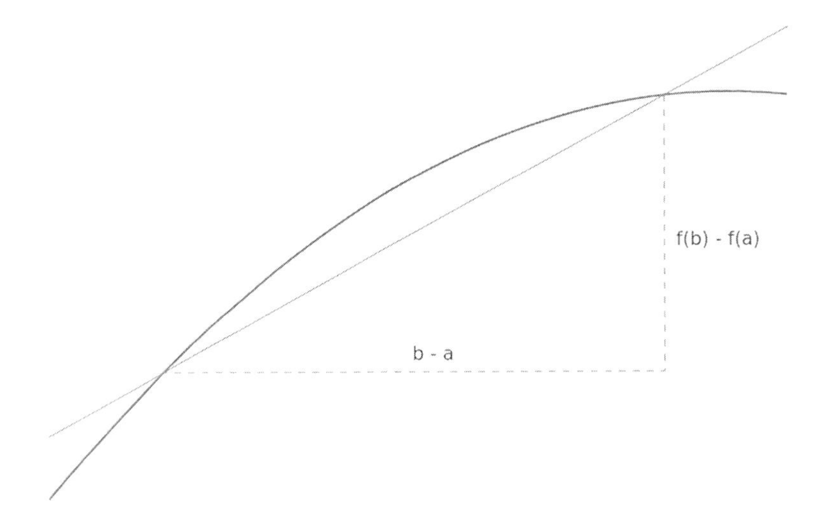

Figure 5.2 An average "speed" of a function on [a, b] is the same as the incline of the hypotenuse.

It's some kind of a measure, sure, but we usually don't want an average speed, especially in applied geometry. What we're interested in is how a function grows at a specific point. With this knowledge, we can make piecewise functions smooth because we can stop one function at some point and continue it with another. If both the values of functions and their growth measures coincide at the point of a cut, the transition appears to be smooth.

So we do need a "speedometer" after all—something that tells a "speed" of a function at a specific point. This mathematical speedometer was invented in the 17th century by Newton and Leibniz independently. To fully understand how it works, you need to get acquainted with infinitesimal analysis and limits in particular. But this book is about geometry, so we'll take a shortcut.

The main idea of infinitesimal quantities is that you can squeeze your interval into a point while keeping it an interval. This idea is mind-bending, but it's not entirely geometrical. To make it so, let's squeeze the interval further and further for our "average speed" measure and see what happens (figure 5.3).

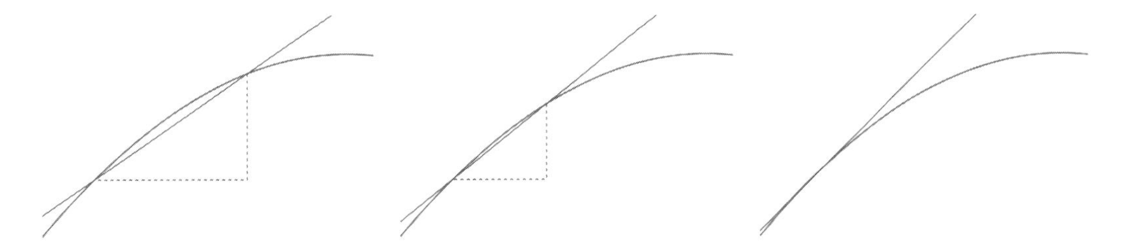

Figure 5.3 As the interval decreases, the hypotenuse line becomes more like the tangent line.

Geometrically speaking, the measure of the function's growth at a point is the incline (or decline) of the function's tangent line at this point. Now, when we decrease the $(b - a)$ interval, the hypotenuse becomes more like the tangent line. We already know how to measure a line's incline as a number: it's the coefficient by its x.

This measurement is what we call a *derivative* at a point. The process of measuring a function's derivative is called *differentiation.* The derivative at a point is a number, and if we can take this number at some point x_i, (and sometimes we can't), the function is called *differentiable* at x_i. We use geometry to build up our intuition about calculus. In practice, we usually go the other way, using the mathematical apparatus of calculus to solve our geometrical problems. By "we," I mean SymPy. But before we get there, we need to enhance our intuition a little more.

5.1.2 Derivative as a function

We know how to measure a function's growth at a point. Well, *function* is a broad term. So far, we've talked only about functions of real numbers:

$$f : \mathbb{R} \rightarrow \mathbb{R}$$

But that's okay; we'll stay in this area a little bit longer. For most functions discussed later in the book, we can safely do that measurement for any point of their domains.

So because a functional relation exists between x and the derivative of a function, the differentiation of a function, as a rule of putting a number to a point, becomes a function itself (figure 5.4).

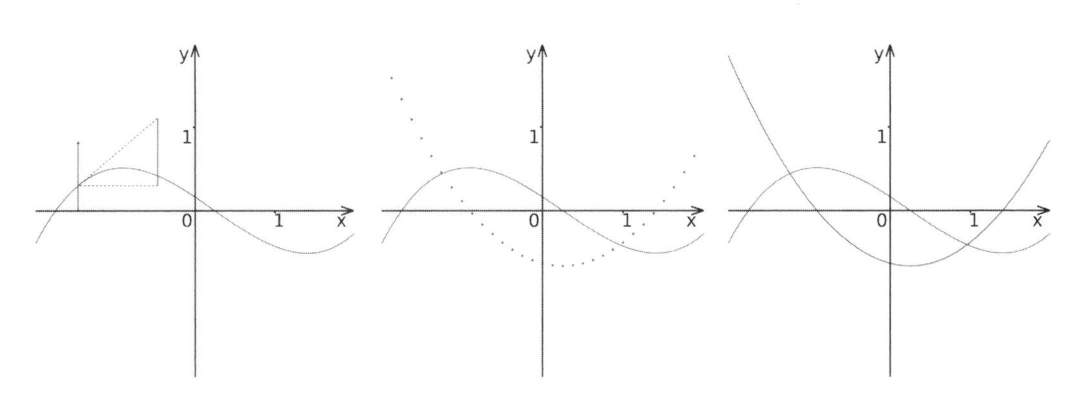

Figure 5.4 A derivative at a point, at a set of points, and as a continuous function

We need to know some of the "classic" functions by sight, including their derivatives. This knowledge will help us understand polynomials and series later in the book. We'll start with an old pal: a linear function (figure 5.5).

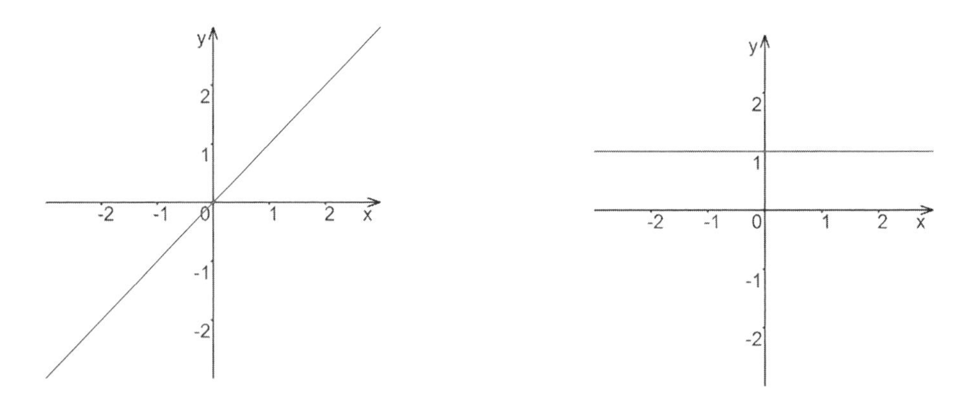

Figure 5.5 A linear function $y = x$ and its derivative $y = 1$

A derivative of a linear function is a constant. Remember that we turn the derivative into a number by approximating the function with a tangent line and taking one of the line's coefficients. Because the coefficient isn't influenced by the argument, it's constant.

Next, figure 5.6 shows a *quadratic function*, also known as a *quadratic parabola*.

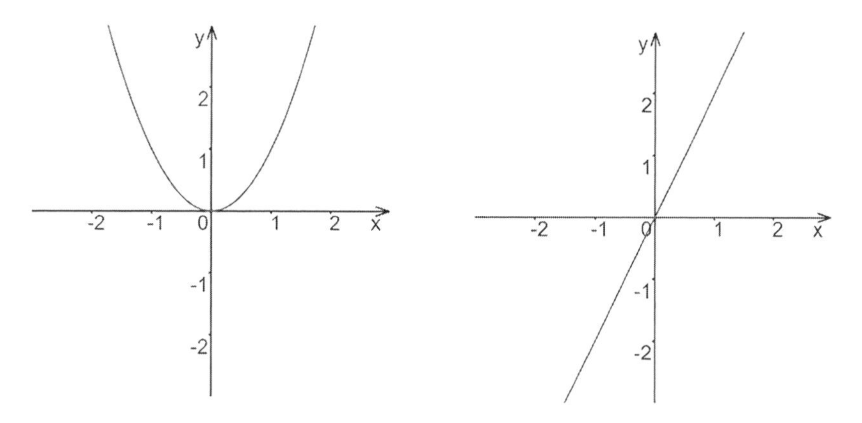

Figure 5.6 A quadratic function $y = x^2$ and its derivative $y = 2x$

A derivative of a quadratic function is a linear function. You probably can't tell from the figure yet, but you can use a formula to calculate this derivative:

$$(ax^2 + bx + c)' = 2ax + b$$

NOTE The prime here denotes that we take the derivative of $(ax^2 + bx + c)$ with respect to x. This notation is called *Lagrange's notation*, and its most valuable feature is its brevity. When the argument of a function isn't apparent, however, this notation may be insufficient. We might use an alternative *Leibnitz notation* in which the argument is stated explicitly. The same formula in Leibnitz notation looks like this:

$$\frac{d}{dx}(ax^2 + bx + c) = 2ax + b$$

Don't waste your time remembering the differentiation formulas yet; we'll get to something more valuable in a moment.

The derivative of a cubic function is a quadratic function (figure 5.7). Here is a formula for this differentiation:

$$f'(ax^3 + bx^2 + cx + d) = 3ax^2 + 2bx + c$$

Are you starting to see the pattern? The derivative of a cubic function is a quadratic one, and the derivative of a quadratic function is a linear one. The derivative of a linear function is a constant. In general, all this comes from differentiating the power of x:

$$f'(x^n) = nx^{n-1}$$

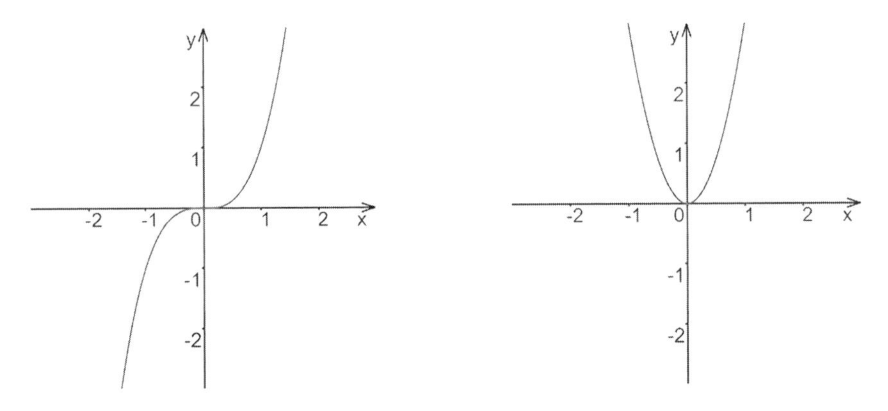

Figure 5.7 A cubic function $y = x^3$ and its derivative $y = 3x^2$

That's something worth remembering. Don't worry if you forget the exact appearance of the formula; SymPy will gladly give you a hint. Keep in mind that the differentiation of x^n lowers the n.

NOTE We normally use this formula to differentiate polynomials (the motivation for doing that is explained in chapter 6), so the n is usually an integer number. Interestingly, though, this formula works for noninteger powers as well.

Now let's go see our old friends sine and cosine. Figure 5.8 shows the derivative of the sine function.

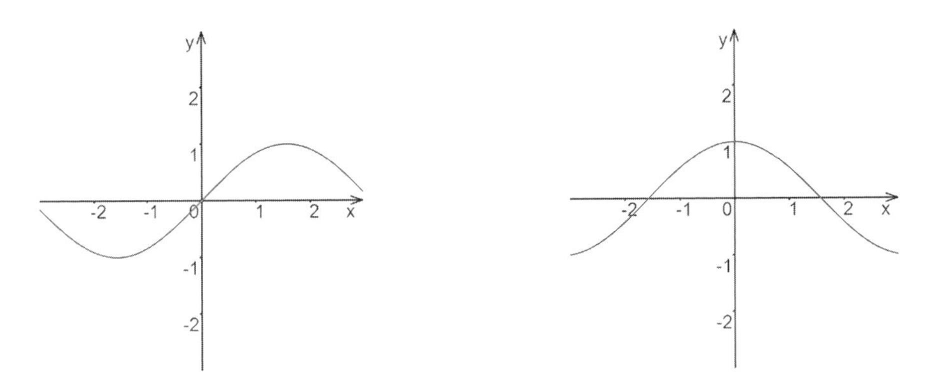

Figure 5.8 Sine function and its derivative cosine

Figure 5.9 shows the derivative of the cosine.

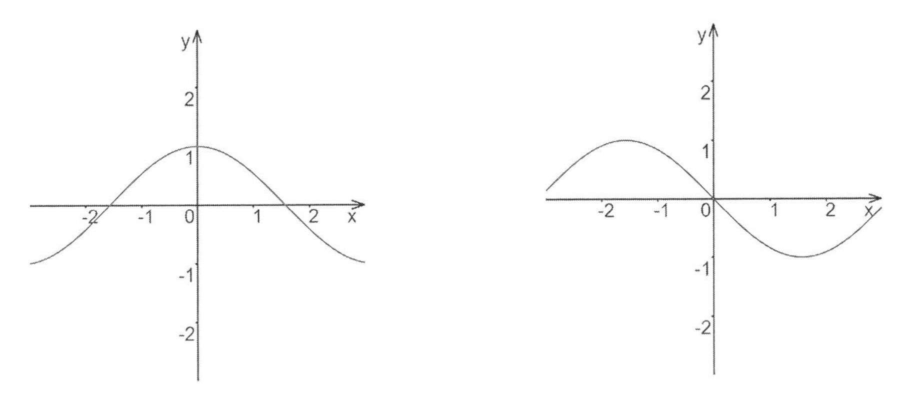

Figure 5.9 Cosine and the derivative of the cosine function, which is minus sine

Do you see the pattern? The derivative of sine is cosine. The derivative of cosine is almost a sine, but kind of inverted: $-\sin(x)$. This fact is fascinating, because it makes trigonometric functions infinitely differentiable. You turn sines into cosines and back. With a simple $f(x) = \sin(x)$, it takes four differentiations to get through $\cos(x)$, $-\sin(x)$, $-\cos(x)$, and back to the original state.

Another interesting function is the exponent:

$$f(x) = e^x$$

Figure 5.10 shows the exponent along with its derivative.

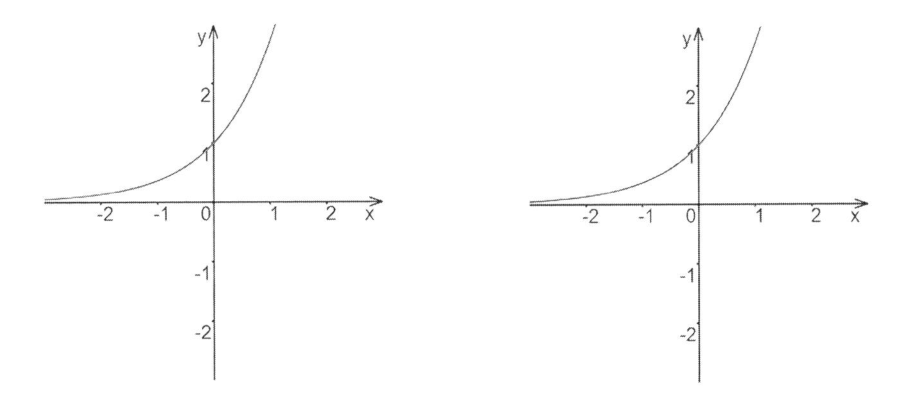

Figure 5.10 The exponent and its derivative

This figure isn't a glitch; I didn't copy the same plot twice. The exponent is the only function that always differentiates into itself—a fascinating property, to say the least.

In chapter 6, we'll use this property to explain the power series, which is a simple yet powerful mechanism to model complex functions with simple operations. Trigonometric functions in programming, for example, are almost exclusively modeled by their power series. These models are fast and precise, and at least for Intel x86/x64, they're even built in as CPU instructions. For now, though, let's accept exponent self-differentiation as a curiosity.

These functions are the most important ones you should know about. SymPy will help you with the rest. But before delegating your work to the machine you should also understand the rules of calculus. How do you differentiate the sum of functions, for example, and how do you go with their product and quotient?

You don't have to know the answers by heart; SymPy will help you with the differentiation anyway. Sometimes, however, even SymPy stalls on huge expressions. If you know the rules of differentiation, you can help SymPy with its work by making the expression simpler to deal with.

5.1.3 Rules of differentiation

If your function is a constant $f(x) = c$, it neither increases nor decreases. Its derivative, therefore, is always 0, which is also a constant. So you can derivate it again, and again, and it won't change. A derivative of $f(x) = c$ is $g(x) = 0$.

If a constant is a factor (in other words, a scaling coefficient for some other function), it remains that same constant after differentiation. The function differentiates, but the constant doesn't:

$$(af(x))' = af'(x)$$

A derivative of a nested function, or a composition of functions, is a derivative of the containing function times the derivative of the contained one:

$$f(g(x))' = f'(g(x)) \cdot g'(x)$$

A derivative of a sum is the sum of the derivatives:

$$(f(x)g(x))' = f'(x) + g'(x)$$

A derivative of a product is the sum of the following products:

$$(f(x)g(x))' = f'(x)g(x) + f(x)g'(x)$$

Here, the first function's derivative is multiplied by the second function, and the second function's derivative is multiplied by the first one. This property is important because it can be exploited for differentiating complex formulas. If you can factor a large expression into smaller ones, you can differentiate them piece by piece.

Similar to multiplication, a quotient of the function is the sum of products, but now it's also divided by the denominator squared:

$$\left(\frac{f(x)}{g(x)}\right)' = \frac{f'(x)g(x) - f(x)g'(x)}{g(x)^2}$$

This rule can also be used to simplify differentiable expressions. A special case of that rule is the derivative of a reciprocal:

$$\left(\frac{1}{f(x)}\right)' = \frac{f'(x)}{f(x)^2}$$

But in the context of polynomial interpolation, splines, and nonlinear transformations (we'll learn about them in chapters 6, 7, and 9), the most important rules for us are the constant factor rule and the sum rule. Luckily, these rules are the simplest ones, too. We also have to keep in mind the power function from before:

$$(x^n)' = nx^{n-1}$$

SymPy will help with the math, but these three rules will help you understand the material in chapter 6 better, so please try to keep them in your mental arsenal.

5.1.4 Using SymPy to do differentiation

In SymPy, the differentiation utility is called `diff`. Call it on everything you want:

```
>>> diff(x**3 + 2*x**2 + sin(x))
3*x**2 + 4*x + cos(x)
```

A function may have several symbols in it. Some symbols may be variables, and some may be coefficients. SymPy doesn't know which is which, so you may want to help by stating the variable explicitly as the second parameter of the `diff`:

```
>>> diff(a*x**2 + b*x + c, x)
2*a*x + b
```

We have to set x explicitly. From SymPy's perspective, a*x2 + b*x + c is a generic symbolic expression, not necessarily a function of x.**

Some functions are multivariable. A parametric surface, for example, is a function that connects pairs of coefficients and points of a 3D space. A single component of this connection is a function of two coefficients. You can differentiate this function by one or even both in one go by giving SymPy explicit instructions:

```
>>> diff(a*u**2 + b*v**2*u + c, u)
2*a*u + b*v**2

>>> diff(a*u**2 + b*v**2*u + c, v)
2*b*u*v

>>> diff(a*u**2 + b*v**2*u + c, u, v)
2*b*v
```

Both u and v are variables that we can use for differentiation.

As you can differentiate a function by u and then by v in one go, you can differentiate a formula twice by the same variable. The resulting formula is called a *second derivative*:

```
>>> diff(a*x**2 + b*x + c, x, x)
2*a
```

Note that the x is repeated.

Like addition, multiplication, or even projective transformation (chapter 4), differentiation has an inverse operation, called *integration*. Integration has no particular use in this book, so I won't talk much about it. But because it's so easy to play with these things in SymPy, why don't we try the integration as well?

```
>>> diff(a*x**2 + b*x + c, x)
2*a*x + b
>>> integrate(2*a*x + b, x)
a*x**2 + b*x
```

We differentiated a function and then integrated it back. But wait—something is missing.

Right: c, the constant! As a standalone function, a constant doesn't increase or decrease; it's stable by definition. That means that its derivative is always 0. Unfortunately, a zero doesn't leave traces in the resulting formula. We don't know which zero comes from which constant:

```
>>> diff(a*x**2 + b*x + c + d + e , x)
2*a*x + b
```

It's kind of 2*a*x + b + 0 + 0 + 0.

So the integration doesn't restore constants at all. At the same time, you have to keep in mind that there may have been a constant or two in the original formula, so this operation isn't the complete inverse of differentiation. It's complete up to the constant.

5.1.5 Section 5.1 summary

A derivative of a function at a point is a measure of how a function increases or decreases at the point. It's the same as the inclination of the function graph's tangent line.

A derivative of a function on its whole domain is also a function. Please use SymPy to compute these derivatives symbolically. All you have to do is to call a `diff` function and, if necessary, select the differentiation variables.

> **SEE ALSO** This book introduces calculus only in the context of geometry—mostly curves and, in later chapters, surfaces. But the second part of *Math for Programmers*, by Paul Orland (https://www.manning.com/books/math-for-programmers), is dedicated to calculus in a different context. It shows how calculus works for physical simulation. If you want a different perspective on differentiation and integration, you can find it in that book.

5.2 Smooth piecewise parametric curves

The world of technology is made of piecewise functions. Since the introduction of computer-aided design (CAD), all the parts and details, the mechanisms and engines, live in a form of parametric curves and surfaces. Cars, planes, and milling machines go through a stage of being a bunch of mathematical formulas. Before they become metal, they have to become geometry.

In this section, we'll see the basis of this kind of representation—not how to represent machines as formulas, which is rather tedious (and we have special software for that task anyway), but how the whole representation works conceptually. To learn that, we'll go from piecewise functions to smooth, continuous piecewise curves.

5.2.1 Piecewise functions

A *piecewise function* consists of several different functions, each defined on a specific interval of its domain. Here is a piecewise function (figure 5.11):

$$x < 0: f(x) = x^2 + 1$$

$$x \geq 0: f(x) = 1 - x^2$$

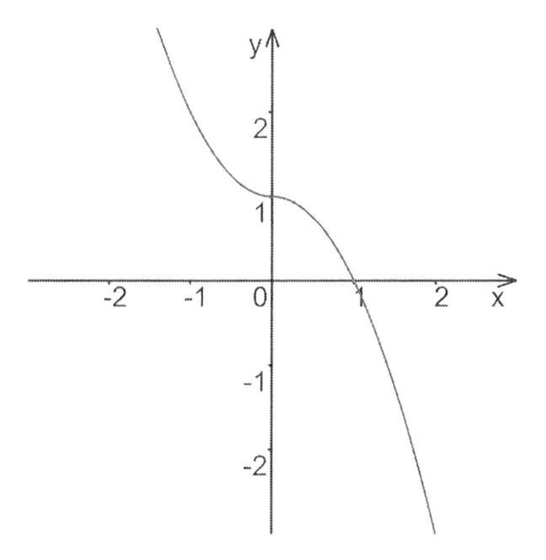

Figure 5.11 A piecewise function, smooth and continuous

This particular one is smooth and continuous, but this isn't a given. In fact, it's not easy to make a piecewise function continuous let alone smooth. *Continuity* means that the function on the left of a breaking point has the same value as the function on the right, and *smoothness* means that their derivatives are also equal.

For $x^2 + 1$ and $1 - x^2$, this works in $x = 0$, because their values are both 1 and their derivatives are both 0. But consider another example (figure 5.12):

$$x < 1: f(x) = x - 0.5$$

$$x \geq 1: f(x) = 1.5 - x$$

This example is still continuous, but it isn't smooth anymore. The smoothness of a function is the continuity of its derivative. Counterintuitively, smoothness is not only

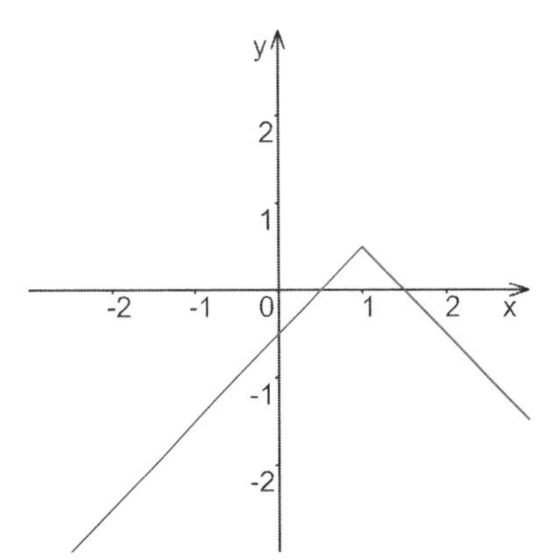

Figure 5.12 A piecewise function, continuous but not smooth

qualitative, but quantitative properly as well. The number that characterizes how far the smoothness goes is called the *order of continuity*.

- If a function is continuous, but its derivative is not, it has an order of smoothness, or an order of continuity 0. In math notation, we usually abbreviate this as C^0.
- If a function is continuous, and its derivative is, too, the notation is C^1.
- If a function is continuous, and its derivative and second derivative (the derivative of the derivative) are, too, the notation is C^2.
- If a function is continuous, and all its derivatives up to nth are continuous, too, the notation is C^n.

We already saw that unlike a graph of a C^0 function (figure 5.12), a graph of a C^1 smooth function (figure 5.11) has no sharp corners. In fact, it has no corners at all, as though the whole function has been filleted. This property can be exploited and extended to design smooth parametric curves in engineering.

5.2.2 Smooth parametric curves

As a reminder, a *parametric curve* is a function that maps a set of parameter's values to a set of points in space. In 2D, it's $f: t \rightarrow (x, y)$, or in per-coordinate form,

$$x = f_x(t)$$

$$y = f_y(t)$$

If both of the functions are C^n continuous at a point, the curve is also continuous. This means that if both coordinate functions' graphs are smooth, the resulting curve will be smooth as well (figure 5.13).

What's interesting is that a parametric curve may remain smooth even if the functions behind the curve's coordinates aren't. We already saw that a link exists between a function's graph's tangent and its derivative, which essentially represent the same

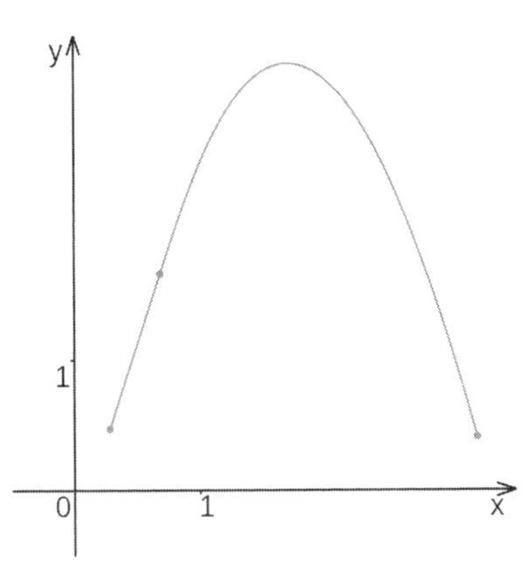

Figure 5.13 A continuous and smooth parametric curve made of two pieces: a linear one and a quadratic one

thing. For the parametric curve, the link remains, but now it's represented by something slightly different.

A tangent vector of a parametric curve is $(f_x(t)', f_y(t)')$. The curve looks smooth if a tangent vector remains continuous, so it looks as though the C^1 continuity for both f_x and f_y is mandatory. If we want to conjoint the curve: $(f_{1x}(t), f_{1y}(t))$ and $(f_{2x}(t), f_{2y}(t))$ smoothly, we may want their tangent vectors to be equal:

$$(f_x(t)', f_y(t)') = (f_{2x}(t)', f_{2y}(t)')$$

This equation will make the conjoint curve smooth indeed. But for a parametric curve to remain smooth, the only mandatory condition is that the tangent vectors of its conjoined pieces at the point of conjunction must share direction, not necessarily length! We want only f_x and f_y to be continuous up to a common factor:

$$(f_{1x}(t)', f_{1y}(t)') = c(f_{2x}(t)', f_{2y}(t)'), \text{ where } c > 0$$

If at some point, the tangent vector changes from (a, b) to $(c \times a, c \times b)$, the parametric function may change its shape, but it remains smooth (figure 5.14).

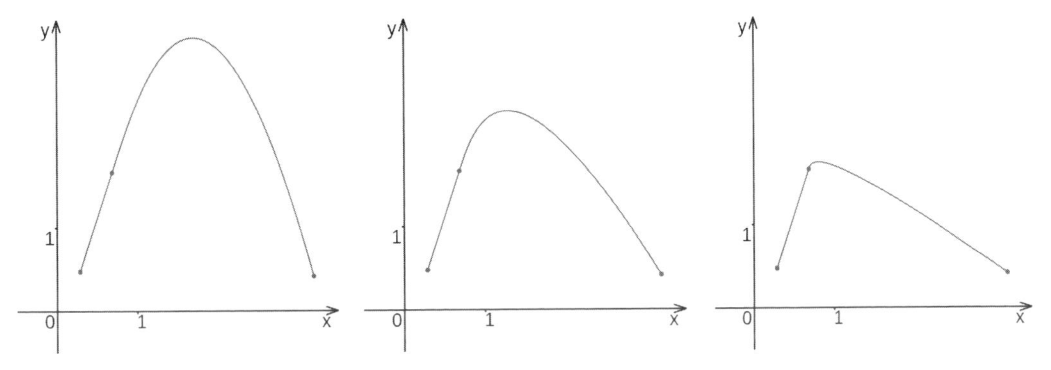

Figure 5.14 The same as in figure 5.13, linear and quadratic parametric curves joined together, but now both derivatives for the quadratic piece in a joint are being divided by the same number: 1, 2, 10.

This property may not seem interesting yet, but we'll use it to create a custom spline in chapter 7. The example comes from game development, and our custom spline will move a boat along the pathway selected by a user so that the boat will move smoothly and steadily. The property will reappear in chapter 12, where we'll dissect an algorithm that vectorizes a bitmap, or in other words, turns a picture into a set of curves.

5.2.3 Curvature

Curvature is a quantitive measure of how curvy a curve is. A straight line isn't curvy at all, so its curvature is 0 (figure 5.15).

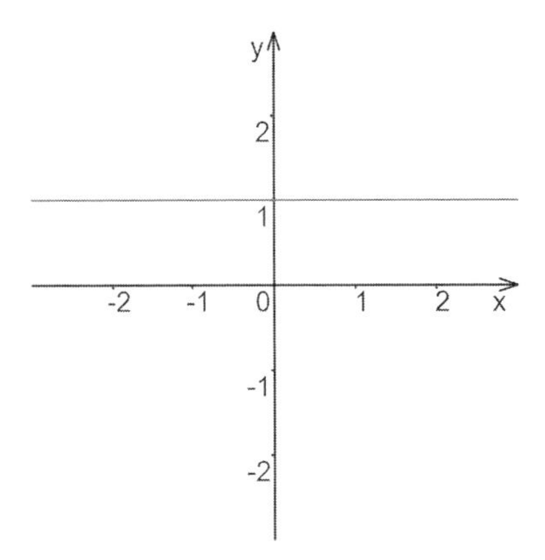

Figure 5.15 A straight line has a curvature of 0.

A circle has curvature, and the smaller its radius is, the larger the curvature is. The curvature is inverse to the circle radius (figure 5.16).

$$k = \frac{1}{R}$$

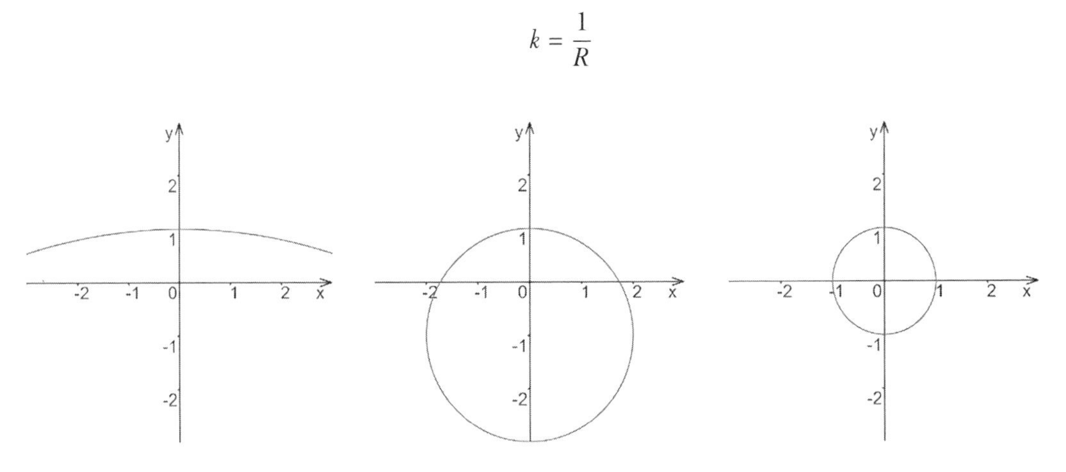

Figure 5.16 The smaller the radius, the larger the curvature.

If you imagine that a straight line isn't entirely straight, but is in fact an arc of a circle with a huge radius, the formula starts to make perfect sense. The smaller the circle is, the less resemblance between an arc and a line remains, and the larger the curvature is.

A straight line and a circle are the only two types of curves that have constant curvature everywhere. For the rest, the curvature changes from point to point; it can be positive, zero, or even negative. Yes, the idea of a negative radius sounds counterintuitive, but we want curvature to remain signed, so we wouldn't try to glue pieces with negative and positive curvatures together and expect them to flow smoothly from one to another—smoothly as in C^2 smooth.

We can visualize the curve's curvature at a point by fitting a circle (or a line, if the curve is flat) in the point's proximity (figure 5.17). The sign of the curvature corresponds to the side on which the fitting circle lies in relation to the function's graph. If it's above the graph, the curvature is positive; if it's below the graph, the curvature is negative.

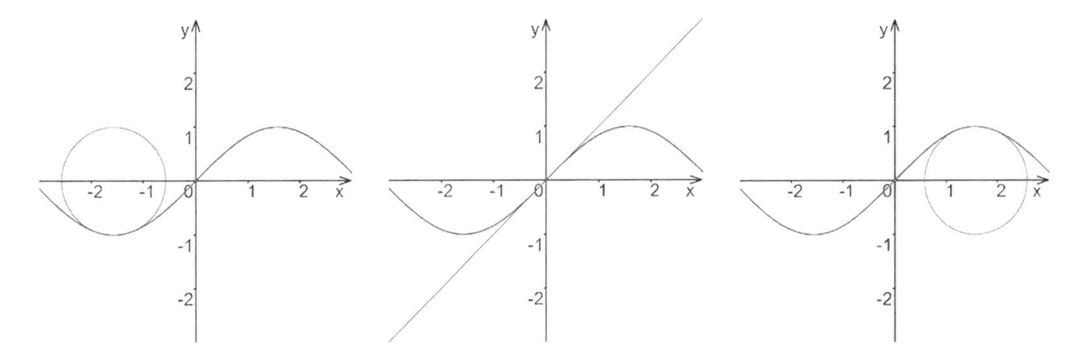

Figure 5.17 The sine graph has positive curvature in $-\pi/2$, zero in 0, and negative in $\pi/2$.

The visualizing circle has to touch the curve, so the circle and the curve should share not only a point, but also a tangent vector. That's not all. A circle may share a tangent vector with a function at any particular point and still change its radius. A point and a tangent vector aren't enough to determine the curvature, as we saw in figures 5.15 and 5.16. The point (0, 1) is shared by a straight line and three circles of different radii. The tangent vector in (0, 1) is always (1, 0), though.

To fit a circle of an appropriate radius, we have to account for how the tangent vector changes in the proximity of the point, which brings us back to the topic of this chapter. For parametric curves, the tangent vector is closely associated with the derivatives of the coordinate functions; it reflects the direction of the curve or the way a function changes.

The curvature is a measure of how the tangent vector changes in its own turn. As such, it's closely associated with the second derivatives of the coordinate functions. For a two-dimensional parametric curve, the formula for the curvature is

$$k = \frac{x'(t)y''(t) + x''(t)y'(t)}{\left(x'(t)^2 + y'(t)^2\right)^{\frac{3}{2}}}$$

Never mind the exact formula; you can search for it online. What you have to keep in mind is that for a parametric curve, the continuous curvature implies the continuous second derivatives.

Now, why is this important? Let's say you have to design a twisty road. You don't want drivers to slow down too much before each corner, so you want to make the cornering smooth. Your natural instinct is to make the corners round, to use circle arcs instead of sharp corners, and your natural instinct would be wrong.

Remember that straight lines and circles are the only two curves with constant curvatures. Although you can put a circular arc and a line segment together smoothly, you can't make their curvatures match. You'll always have a discontinuity in the curvature.

When you're driving, your car's position is the integral of your car's speed. I used the metaphor "Derivative is the speed of a function" before, and the metaphor works both ways. Your speed in turn is the integral of the acceleration. You push the gas pedal down, and your speed starts to rise; you release the gas pedal, and the speed starts to go down. The speed doesn't follow the pedal immediately; it integrates the pedal's input.

The same goes for the relationship between your angular speed—how fast your car turns in a corner—and the force you put on a steering wheel. A discontinuity in the curvature is the discontinuity in the force you apply. If you drive from a straight line to an arc, you'd have to turn your wheels immediately at the point of conjunction, in a single snap. Unless you're a cartoon character, this task is mechanically impossible.

That's why the smooth corners on the roads are not only smooth, but also curvature-continuous. The functions behind them are C^2. In chapters 6 and 7, we'll learn how to make functions such as these from polynomial building blocks.

Surface curvature

We can reuse the analogy with lines and circles and say that a plane has zero curvature and a sphere has some constant curvature. The trouble is that for an arbitrary point of an arbitrary surface, we can't necessarily fit a sphere in the point's proximity. Not all surfaces are locally spherelike.

What we do instead is fit 2D circles again. All the planes that go through the point of a surface and contain the surface's normal vector there intersect the surface differently. The curves where the surface intersects these planes are usually different, too. Every one of them has its own fitting circle; every one of them has its own curvature at the given point.

The smallest and the largest signed curvature values are called *principal curvatures*. Their half-sum is called the *mean curvature*, and the product is the *Gaussian curvature*. The corresponding planes are called *principal planes*.

Naturally, both the mean curvature of a sphere and its Gaussian curvature are constants, and for a plane, both are zero everywhere, which is consistent with 2D. The difference shows on a cylinder. The Gaussian curvature of any point of a cylinder is zero, but the mean curvature is some other nonzero number.

If you want to learn more about surface curvature, take a look at Bartosz Ciechanowski's interactive tutorial Curves and Surfaces at https://ciechanow.ski/curves-and-surfaces. The concept of curvature is explained there beautifully.

5.2.4 Section 5.2 summary

Piecewise parametric curves are continuous and smooth if the coordinate functions are continuous and smooth, too. In short notation, they're both C^0 and C^1.

The curvature is the concept associated with the second derivative. If you want to keep your curvature continuous, too, make sure that your coordinate formulas are C^2.

5.3 Practical example: Crafting a curve out of lines and circles

This example comes from a 2D game that I worked on in 2007. We used a simple but robust 2D engine that was perfect in every way except that it had limited capabilities for animation. We wanted a bug (not a programming bug, but an animated character) to crawl following some given trajectory at a constant speed. The trajectory, of course, should have been continuous so that the bug wouldn't jump through space and smooth so that its crawling wouldn't seem jerky.

Normally, I'd have used some kind of a polynomial spline, and we will learn about splines in the following chapters. But as it turns out, using polynomial splines to make a character go a proportional distance for every tick of time isn't trivial. It's possible—people have done it—but the task isn't strictly mathematical. There's no single formula for constant-speed movement along a polynomial parametric curve. You have to pick a different parametric pace for every step.

I started looking for a more elegant solution, and I think I found one. We can set up our trajectory as a series of lines and arcs.

Moving along the line at a constant speed is simple, given that the line is written in its parametric form, of course. We increase the parameter at a constant rate, and the trajectory also changes linearly.

Going in a circle isn't much more complicated. We know its radius; therefore, we know its circumference. A full circle corresponds to $2\pi rt$. If we increment the t in equal portions, the path will grow in equal steps, too, proportionately to the circle's circumference. This will make a bug go round and round at a constant speed.

How do we describe a whole complex trajectory with only arcs and lines? We split it into building blocks first.

5.3.1 The biarc building block

The *biarc* is a piece of a curve that consists of two arcs. It connects two points, holds the given tangent vectors in them, and makes sure that the transition from the first arc to the second is continuous and smooth (figure 5.18).

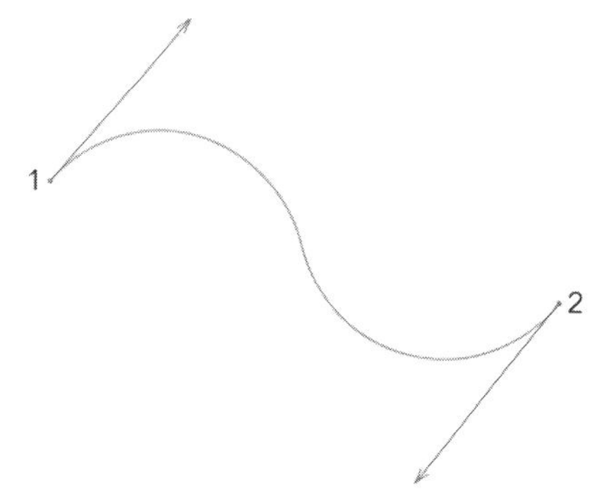

Figure 5.18 A biarc segment is two arcs smoothly joined together.

Let's make a program that computes the biarc that goes through a pair of points and follows the predetermined tangent vector. We'll use SymPy to get the formulas for this program.

The input for the SymPy code that generates the formulas is a pair of points, `(x1, y1)`, `(x2, y2)` in their symbolic form, and a pair of corresponding tangent vectors, `(dx1, dy1)`, `(dx2, dy2)`. The output is the formulas for the arcs' center points, `(ax1, ay1)` and `(ax2, ay2)`, and for the corresponding radii, `r1` and `r2`. Let's write it all down:

```
x1, y1, dx1, dy1 = symbols('x1 y1 dx1 dy1')
x2, y2, dx2, dy2 = symbols('x2 y2 dx2 dy2')

ax1, ay1, ax2, ay2 = symbols('ax1 ay1 ax2 ay2')
r1, r2 = symbols('r1 r2')
```

The beauty of arcs is that they come from circles, and the circles touch one another smoothly when the distance between their centers is equal to the sum of their radii. We can rewrite this as an equation and start forming a system:

```
(ax1-ax2)**2 + (ay1-ay2)**2 - (r1+r2)**2
```

Because we need to establish six variables—two circle centers, a pair of coordinates each, and two radii—we need five more equations to make the system well-defined. The first circle goes through the first point, so the distance between the circle's center and the point is equal to the circle's radius. This could become an equation, too:

```
(x1 - ax1)**2 + (y1 - ay1)**2 - r1**2
```

This equation, however, isn't the best to have in a system because it's a quadratic equation, and each quadratic equation may have up to two solutions. When you put quadratic equations in a system, the system's solutions start multiplying, too. So let's try to come up with better equations. Consider this: the vector between the point and the circle radius `(x1 - ax1, y1 - ay1)` is orthogonal to the tangent vector at the point `(dx1, dx2)`. Combined with the fact that the distance between `(x1, y1)` and `(ax1, ay1)` is `r1`, if you turn your normalized tangent vector 90 degrees, you can multiply it by the circle's radius and you'll have a connecting vector from the first point to the first circle's center (figure 5.19):

```
(x1 - ax1, y1 - ay1) = rotated(dx1, dx2) * r1
```

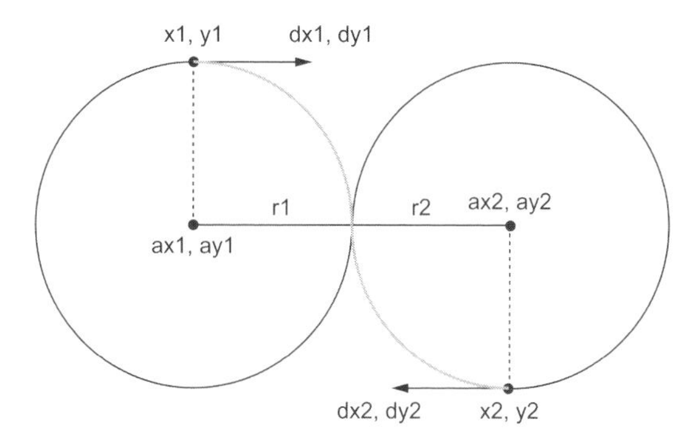

Figure 5.19 How to get from segment points to arc centers

Rotating 2D vectors 90 degrees is simple: swap their *x* and *y* coordinates, and invert one of the signs. So the first point along with its tangent vector gives us a pair of equations:

```
x1 - ax1 + r1*dy1,
y1 - ay1 - r1*dx1,
```

The second point is no different from the first, so we'll get a similar pair of equations from it:

```
x2 - ax2 + r2*dy2,
y2 - ay2 - r2*dx2,
```

Now we're lacking only one final equation. Usually, lacking one equation means that we have not a single solution, but a continuum of solutions, which is definitely true in our case. There isn't only one way we can put a biarc through a pair of points with given tangents. There could be different configurations, and the arcs of which the biarc consists of could have vastly different radii. But what if we enforce the radii of the arcs to be equal?

```
r2 - r1,
```

We can choose any other proportion. Or we can let users choose the proportion, making it an input parameter. Or we can add one of the radii to the input and compute the other one ourselves. The mathematics behind the biarc formulas will still work. But because there's no particular reason to add more parameters to the input, let's leave the system this way.

Where to start and stop an arc

With our equations, we're getting circles' centers and radii, but we don't get the angle intervals to turn our circles into actual arcs. Well, that's because we already have the starting and ending point for the piece from the input, and the intersection point is a midpoint between the circle's centers. Getting angles from points and vectors is a well-known trigonometrical problem that isn't exciting enough to feature in a practical example.

But if you want a hint, look for the `atan2` function in your programming language of preference. It's like the arctangent function, also known as an inverse tangent, but better because the arctangent works only for angles between -90 and 90 degrees. The `atan2` covers all the 360 degrees or all the 2π in radians.

The whole program is shown in the following listing (ch_05/biarc.py in the book's source code).

Listing 5.1 Solution to the conjoined arcs problem

```
x1, y1, dx1, dy1 = symbols('x1 y1 dx1 dy1')
x2, y2, dx2, dy2 = symbols('x2 y2 dx2 dy2')

ax1, ay1, ax2, ay2 = symbols('ax1 ay1 ax2 ay2')
r1, r2 = symbols('r1 r2')

solutions = solve(
    [
     r2 - r1,
     x1 + r1*dy1 - ax1,
     y1 - r1*dx1 - ay1,
     x2 + r2*dy2 - ax2,
     y2 - r2*dx2 - ay2,
     (ax1-ax2)**2 + (ay1-ay2)**2 - (r1+r2)**2
], (ax1, ay1, ax2, ay2, r1, r2))
```

- The input is a pair of points (x1, y1), (x2, y2) and a pair of corresponding tangent vectors (dx1, dy1), (dx2, dy2).
- The output is the arcs' center points (ax1, ay1) and (ax2, ay2) and corresponding radii r1 and r2.
- Radii are equal by design.
- Both radius vectors are orthogonal to tangent vectors.
- Circles touch at some point, which means that the distance between their centers is the same as the sum of their radii.

There's a caveat, though: because our last equation is quadratic, not linear, when we run our piece of code, we won't get a nice single solution, but two solutions. Each one is valid; each one could be easily groomed and turned into a wall of code by using SymPy functions, as we did in chapter 4. When you get formulas to put into your final program, the one that draws the arcs for the actual numeric input might look like this:

```
div = (dx1**2 - 2*dx1*dx2 + dx2**2 + dy1**2 - 2*dy1*dy2 + dy2**2 - 4)

squared = dx1**2*(-x1**2 + 2*x1*x2 - x2**2) +
    dx1*(2*dx2*x1**2 + 2*dx2*x2**2 + x1*
    (-4*dx2*x2 - 2*dy1*y1 + 2*dy1*y2 + 2*dy2*y1 - 2*dy2*y2) +
    x2*(-2*dy1*y2 + 2*dy2*y2) + y1*(2*dy1*x2 - 2*dy2*x2)) +
    dx2**2*(-x1**2 + 2*x1*x2 - x2**2) + dx2*(2*dy2*x2*y1 -
    2*dy2*x2*y2 + x1*(-2*dy2*y1 + 2*dy2*y2)) +
    dy1**2*(-y1**2 + 2*y1*y2 - y2**2) + dy1*(2*dx2*x2*y2 +
```

```
➥ 2*dy2*y1**2 + 2*dy2*y2**2 + x1*(2*dx2*y1 - 2*dx2*y2) +
➥ y1*(-2*dx2*x2 - 4*dy2*y2)) + dy2**2*(-y1**2 + 2*y1*y2 - y2**2) +
➥ 4*x1**2 - 8*x1*x2 + 4*x2**2 + 4*y1**2 - 8*y1*y2 + 4*y2**2

ax1_common = dx1**2*x1 - 2*dx1*dx2*x1 + dx1*dy1*y1 - dx1*dy1*y2 +
➥ dx2**2*x1 - dx2*dy1*y1 + dx2*dy1*y2 + dy1**2*x2 -
➥ dy1*dy2*x1 - dy1*dy2*x2 + dy2**2*x1 - 4*x1

ay1_common = dx1**2*y2 - dx1*dx2*y1 - dx1*dx2*y2 + dx1*dy1*x1 -
➥ dx1*dy1*x2 - dx1*dy2*x1 + dx1*dy2*x2 + dx2**2*y1 +
➥ dy1**2*y1 - 2*dy1*dy2*y1 + dy2**2*y1 - 4*y1

ax2_common = dx1**2*x2 - 2*dx1*dx2*x2 + dx1*dy2*y1 - dx1*dy2*y2 +
➥ dx2**2*x2 - dx2*dy2*y1 + dx2*dy2*y2 + dy1**2*x2 -
➥ dy1*dy2*x1 - dy1*dy2*x2 + dy2**2*x1 - 4*x2

ay2_common = dx1**2*y2 - dx1*dx2*y1 - dx1*dx2*y2 + dx2**2*y1 +
➥ dx2*dy1*x1 - dx2*dy1*x2 - dx2*dy2*x1 + dx2*dy2*x2 +
➥ dy1**2*y2 - 2*dy1*dy2*y2 + dy2**2*y2 - 4*y2

r1_common = dx1*y1 - dx1*y2 - dx2*y1 + dx2*y2 -
➥ dy1*x1 + dy1*x2 + dy2*x1 - dy2*x2

r2_common = -(-dx1*y1 + dx1*y2 + dx2*y1 - dx2*y2 +
➥ dy1*x1 - dy1*x2 - dy2*x1 + dy2*x2)
ax1 = (ax1_common - dy1*math.sqrt(squared))/div
ay1 = (ay1_common + dx1*math.sqrt(squared))/div
ax2 = (ax2_common - dy2*math.sqrt(squared))/div      The first
ay2 = (ay2_common + dx2*math.sqrt(squared))/div      solution
r1 = (r1_common - math.sqrt(squared))/div
r2 = (r2_common - math.sqrt(squared))/div

ax1 = (ax1_common + dy1*math.sqrt(squared))/div
ay1 = (ay1_common - dx1*math.sqrt(squared))/div
ax2 = (ax2_common + dy2*math.sqrt(squared))/div      The second
ay2 = (ay2_common - dx2*math.sqrt(squared))/div      solution
r1 = (r1_common + math.sqrt(squared))/div
r2 = (r2_common + math.sqrt(squared))/div
```

These two solutions differ only in signs before the squared part. Which one should we pick? The problem isn't really mathematical; both solutions will give us valid biarcs. So let's pick an arbitrary rule that seems pragmatic. How about computing both solutions numerically and picking the solution with the smallest sum of radii? That should work, and it kind of does.

We have our biarc, and it's nice and smooth. The only problem is that it's a little "bubbly" when the tangent vectors point in roughly the same direction (figure 5.20).

We might use this example as a building block for our smooth and continuous curve, but we need another kind of building block to avoid bubbles.

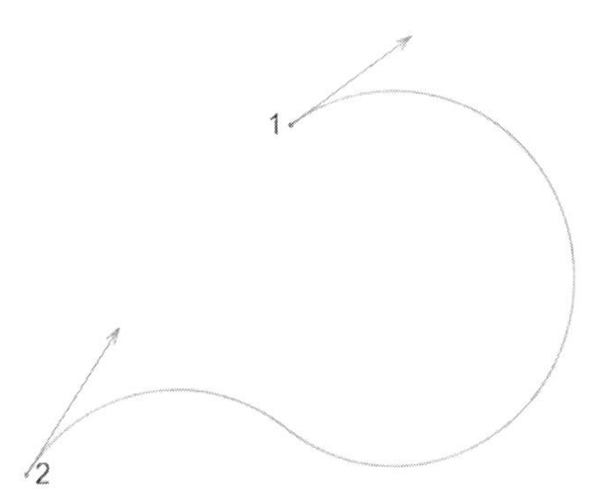

Figure 5.20 Biarc segments don't necessarily work well in all cases.

5.3.2 The line segment and arc building block

Another good building block is a curve that starts as an arc and ends as a straight line segment (figure 5.21). As with the biarc, its input would be a pair of points, `(x1, y1)`, `(x2, y2)`, and a pair of corresponding tangent vectors, `(dx1, dy1)`, `(dx2, dy2)`. But now we have to add an arc's radius to the input because we may want to reuse the radius we got from computing biarcs.

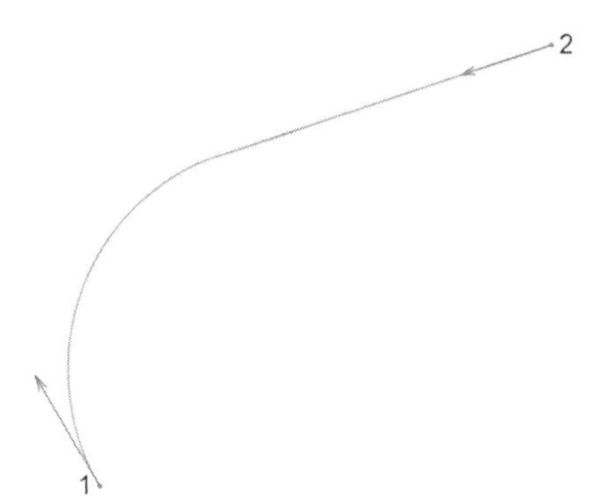

Figure 5.21 An arc can be joined smoothly with a line segment so that we have both straight and curvy pieces in one building block.

Once again, we won't write the actual code that computes the arc; we'll write a SymPy snippet that will generate this code for us. The input for the SymPy snippet, of course, is symbolic instead of numeric:

```
x1, y1, dx1, dy1 = symbols('x1 y1 dx1 dy1')
x2, y2, dx2, dy2 = symbols('x2 y2 dx2 dy2')
r1 = symbols('r1')
```

The output is even smaller than before, consisting of formulas for the center of the arc `(ax1, ay1)` and a formula for the parameter `t2`. A parameter is the length of the line segment from the second point and to the point where the arc and the line intersect:

```
ax1, ay1, t2 = symbols('ax1 ay1 t2')
```

We'll also need the point of intersection in our computation, although not necessarily in the output:

```
ix, iy = symbols('ix iy')
```

Figure 5.22 shows all the input and output data.

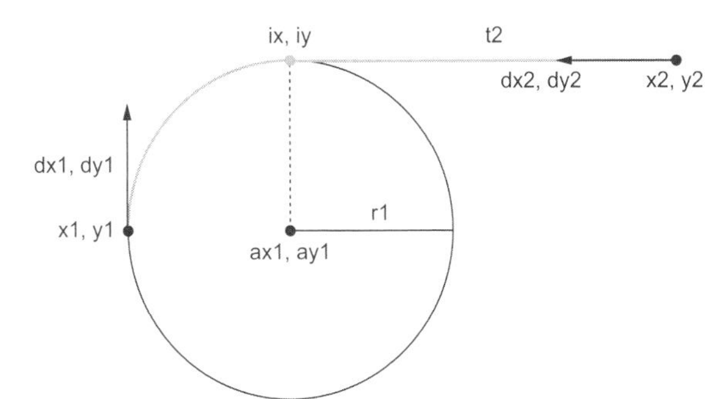

Figure 5.22 An arc and line segment piece

Now for the equations. We already know how to find an arc's center, its `r1` times the 90 degrees rotated tangent vector from the first point:

```
x1 - r1*dy1 - ax1,
y1 + r1*dx1 - ay1,
```

Similarly, we can set an intersection point, a tangent vector from the second point multiplied by `t2`:

```
x2 + dx2 * t2 - ix,
y2 + dy2 * t2 - iy,
```

Now for the difficult part. Because we want the intersection to be smooth, we want the arc to touch the segment gently, so the vector from the intersection point to the arc's center should be orthogonal to the line segment. The vector operation that establishes this orthogonality is called a *dot product*, but don't worry if you don't know what it is. We'll get to vector products and their applications in chapter 9. For now, let's try to solve our problem without vector algebra, using geometric reasoning alone.

> **NOTE** If you already know what a dot product is, you can safely skip the following three paragraphs.

A vector **a** is orthogonal to a vector **b**. We already saw that in 2D, we rotate a vector 90 degrees if we swap its coordinates and add a minus sign to one of them. Also, we scale vectors by multiplying them by a scalar. So if vector **a** is (x_a, y_a), vector **b** should be $(ky_a, -kx_a)$ or $(-ky_a, kx_a)$. Given that k could just as well be negative, both notations are essentially the same.

Now let's multiply **a**s coordinated by **b**s and sum them up, as we'd multiply row vector **a** by column vector **b**, thinking that they're matrices:

$$d = x_a x_b + y_a y_b, \text{ where } x_b = ky_a, \; y_b = -kx_a$$
$$d = x_a k y_a - k x_a y_a = 0$$

Isn't that fascinating! If vectors **a** and **b** are orthogonal, the sum of the products of their coordinates is 0:

$$x_a x_b + y_a y_b = 0$$

Great! Let's make the equation that says "The vector from the arc center to the point of intersection is orthogonal to the line segment":

```
(ix-ax1)*(ix-x2) + (iy-ay1)*(iy-y2)
```

Because the line segment follows its tangent vector, we can even simplify the equation:

```
(ix-ax1)*dx2 + (iy-ay1)*dy2
```

Let's recount our equations. We have two equations that establish the relationship between the arc center and a point of intersection, two equations that define the point of intersection, and the one that makes the intersection smooth. That's five equations per five variables. Now the system looks like listing 5.2 (ch_05/arc_and_segment.py in the book's source code).

Listing 5.2 Solution to an arc and line segment equation

```
x1, y1, dx1, dy1 = symbols('x1 y1 dx1 dy1')
x2, y2, dx2, dy2 = symbols('x2 y2 dx2 dy2')
r1 = symbols('r1')

ix, iy = symbols('ix iy')

ax1, ay1, t2 = symbols('ax1 ay1 t2')

solutions = solve(
    [
    x1 - r1*dy1 - ax1,
    y1 + r1*dx1 - ay1,
    x2 + dx2 * t2 - ix,
    y2 + dy2 * t2 - iy,
    (ix-ax1)*dx2 + (iy-ay1)*dy2
], (ax1, ay1, t2, ix, iy))
```

- **The input is the same as before, but now it's joined with the circle's radius.**
- **(ix, iy) is the point where the arc meets the line.**
- **The output is now only one arc's center (ax1, ay1), and we already know its radius from the input. The other piece of output is the parameter t2, which says where the line ends.**
- **The intersection point is on the tangent line of (x2, y2) + t2*(dx2, dy2).**
- **The intersection point is also on the arc.**
- **The arc's radius vector is orthogonal to the tangent vector (dx1, dy1).**

This time, the system is linear, so a single solution is available:

```
ax1 = -dy1*r1 + x1
ay1 = dx1*r1 + y1
```

```
div = (dx2**2 + dy2**2)
t2 = (dx1*dy2*r1 - dx2*dy1*r1 + dx2*x1 - dx2*x2 + dy2*y1 - dy2*y2)/div
ix = (dx1*dx2*dy2*r1-dx2**2*dy1*r1+dx2**2*x1+dx2*dy2*(y1-y2)+dy2**2*x2)/div
iy = (dx1*dy2**2*r1+dx2**2*y2-dx2*dy1*dy2*r1+dx2*dy2*(x1-x2)+dy2**2*y1)/div
```

The equations are simple, but they could be simpler! To make them so, we have to reevaluate our whole approach. You see, with SymPy, it's almost too easy to put facts in equations and get a solution. This approach is valid, of course, but it doesn't work best for every possible problem.

In our case, the solution doesn't have to come from solving a common system. Note that the first two equations aren't really equations but already a solution! We simply have to rewrite them this way:

```
ax1 = x1 - r1*dy1
ay1 = y1 + r1*dx1
```

Now all we have to do is solve a system of three equations with three variables. Even so, because our second pair of equations is effectively a way to compute *ix* and *iy* from *t2*, we aren't too interested in getting them as formulas from the system's solution. The equations we put in the system are already better to compute than the equations we got in the output:

```
ix = x2 + dx2 * t2
iy = y2 + dy2 * t2
```

The full solution, the formulas you can actually put into your arc and segment program, looks like this (ch_05/ better_arc_and_segment.py in the book's source code):

```
ax1 =  -dy1*r1 + x1
ay1 =  dx1*r1 + y1
t2 = (dx1*dy2*r1-dx2*dy1*r1+dx2*x1-dx2*x2+dy2*y1-dy2*y2)/(dx2**2+dy2**2)
ix = dx2*t2 + x2
iy = dy2*t2 + y2
```

Sorry for taking you through the extra step, I wanted to show that although SymPy is awesome at doing your math for you, you don't always have to follow the same procedure: gather all that you know about the problem and throw it into a machine. This approach works, but it doesn't always give you the best result possible.

But let's get back to geometry. The conjoined arc and segment piece work fine when the tangents are codirected-ish, which was a problem for the biarcs, but it doesn't work as expected when the tangents are counter-oriented (figure 5.23).

The equations are fine; we do have a circle and a line segment that touch properly. But no possible arc is both co-oriented with the second vector and counter-oriented with the first one. Nevertheless, we have a building block that doesn't work well in one case and another building block that doesn't work well in another. Because the cases

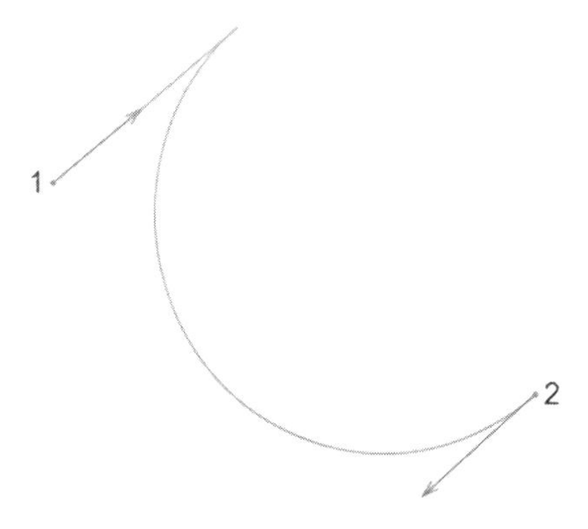

Figure 5.23 Technically, both ends are aligned with tangent vectors, but this alignment doesn't work for us.

are mutually exclusive, we can combine building blocks to build a curve with the properties we want and avoid unpleasant corner cases.

5.3.3 *The combination of both*

We have two building blocks. One doesn't work when the vectors are more or less co-oriented; the other doesn't work when they're close to being counter-oriented. As you may have noticed, *more or less* and *close* aren't mathematical terms. This is fine. For most practical applications, fuzzy nonmathematical criteria are unavoidable. That's the point where engineering steps in.

So let's define "more or less co-oriented," as the angle between the tangent vectors is less than 90 degrees, and use the arc and segment combination for these cases. For the rest, the biarc will do. This combination is potent, allowing us to build complex curves out of these pieces (figure 5.24).

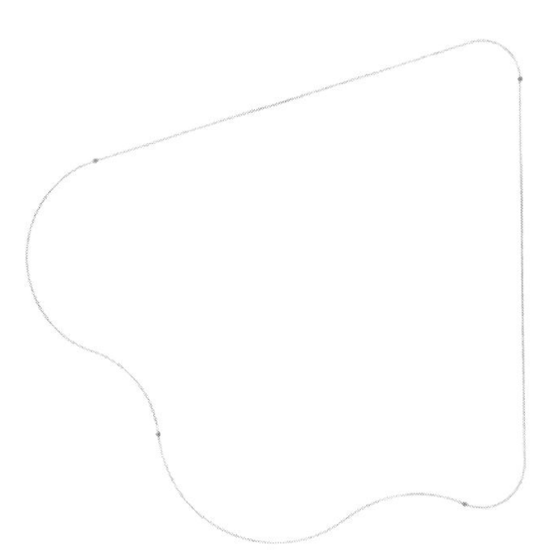

Figure 5.24 The combination of biarcs and arc and segment pieces in a complex curve

This combination of arcs and lines is rarely used in practice, however. We usually use polynomial and rational splines for curves, and we'll get to them soon enough. Our current approach, as primitive as it is, still has a property that polynomial splines struggle to achieve: the parameter growth is linear to the travel of the curve.

With this property, we can easily split the curve into equal pieces. As mentioned at the beginning of this section, this approach allows us to run a bug on a curve with a constant speed. For both parametric arcs and line segments, we do this by splitting a parameter into equal intervals; for polynomial splines, the task is far less trivial (figure 5.25).

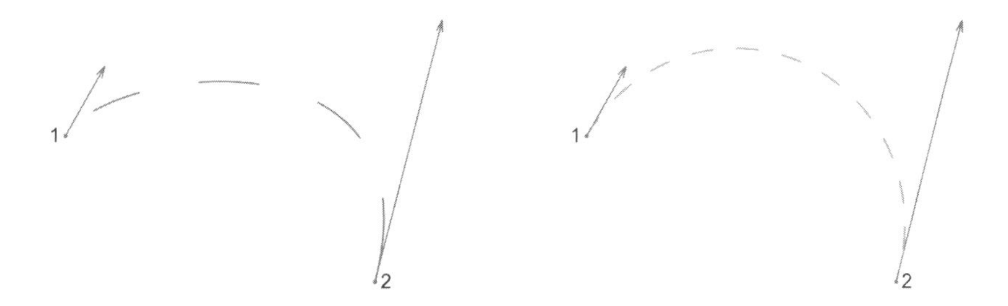

Figure 5.25 Cubic spline (on the left) with 8 pieces equal in parameter space, and the arc and line segment (on the right) with 16 pieces equal in both parameter space and 2D

I also like this approach because it's not entirely mathematical. You get to tinker with things, you get to pick the solutions you want, and you get to determine when and which piece to use. You can expand the approach further by introducing new pieces.

Although the pragmatic value of making curves from arcs and lines is limited, they're still fun to play with.

NOTE This chapter originates from an interactive Words and Buttons article. Immediately after it was published, Johnathon Selstad came up with an alternative approach. If we add a segment between two arcs, we can afford adjustable radii! What's even more awesome, he published his code in a form of a JavaScript sketch that anyone can tinker with. You can find it at http://mng.bz/51ea.

5.3.4 Section 5.3 summary

Knowing the notion of continuity and smoothness, you can model curves from the pieces you want and with the properties you need. The more pieces you know about, of course, the wider your choice is. We started with arcs and line segments; in the next two chapters, we'll learn about polynomials and rational splines.

5.4 Exercises

Exercise 5.1 What is the derivative of $ax^2 + bx + c$? The variable is x; a, b, and c are some arbitrary parameters.

1. $ax + b$
2. $2ax + b$
3. $2a + b$
4. 2

Exercise 5.2 What functions are the integrals of $2ax + b$? Again, the variable is x.

1. $ax^2 + bx$
2. $ax^2 + bx + c$
3. $ax^2 + bx + 1$
4. 4

Exercise 5.3 What is the 64th derivative of $\sin(x)$?

1. $\sin(x)$
2. $\cos(x)$
3. $-\sin(x)$
4. $-\cos(x)$

Exercise 5.4 When is a piecewise parametric curve smooth in a joint point?

1. When the coordinate functions are both C^1
2. When the proportion of coordinate functions is the same for both pieces in the joint point
3. When the coordinate functions are both C^2
4. When the curvature of the first piece coincides with the curvature of the second in the joint point

Exercise 5.5 Suppose that you want to model an escalator in a game. A character steps on a flat piece, which it moves horizontally and then elevates, and in the end, it again moves horizontally, all at a constant speed. How would you approach that task?

5.5 Solutions to exercises

Exercise 5.1 Answer 2. The derivative of an n-degree polynomial is an (n-1)-degree polynomial. Answers 3 and 4 are constants, so they're out. Answer 1 is closer, but it doesn't follow the rule of differentiation exactly.

Exercise 5.2 Answer 1 is correct when integrating with SymPy, but answers 2 and 3 are valid, too. That's the problem with integration; the symbolical operation we do in SymPy doesn't restore the constant, but the constant is still there. All three functions are valid integrals.

Exercise 5.3 Answer 1. At first, the sin(x) function is differentiated into cos(x), cos(x) becomes –sin(x), –sin(x) becomes –cos(x), and –cos(x) reverts back to sin(x), all in four operations. 64 is 16 times 4, so it reverts to sin(x) 16 times and settles there.

Exercise 5.4 The first condition is the classical condition for the parametric curve smoothness. If you want the single most important takeaway from this chapter, this answer is your takeaway. But the second condition is also enough to establish a smooth curve. The first is a special case of the second.

As for the third one, C^2 implies C^1, so it grants smoothness to the curve as well. Only the fourth one doesn't. C^2 establishes that the pieces have the same curvature in the joint point, but it doesn't work the other way around. Imagine Olympic rings. Every ring has the same smoothness at every point, yet the intersections of these rings are not smooth.

So the answers are 1, 2, and 3.

Exercise 5.5 This question is an open question. You can think of your own unique way to solve this problem. If you want a hint, here it is: we did something similar with arcs and line segments before, putting everything we know in a system and making SymPy solve it. The hint is *not* to do that.

What you want to do instead is find the coordinates of the arcs' centers by intersecting the offset lines of the top and bottom lines and the escalator slope (figure 5.26). Top and bottom offsets are top and bottom lines elevated or lowered along the y-axis, and the slope offset is the line we get by offsetting an origin point by the direction vector of the slope rotated 90 degrees. This approach creates simpler equations and results in shorter code.

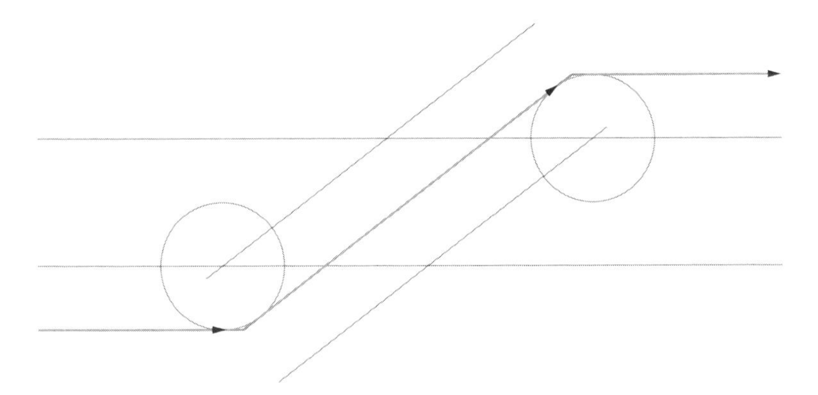

Figure 5.26 The escalator model is three lines and two circles.

Summary

- A derivative is a measure of how fast a function changes.
- Although a derivative at every particular point is a number, it's also a function with its own domain.
- You can differentiate a derivative function, resulting in a second derivative.
- Differentiation is recursive so you can always get an nth derivative out of (n-1)-th as long as the former is differentiable.
- Integration is almost inverse to differentiation. Differentiation eats up constants, and integration can't do anything about it.
- The smoothness of a function is a number saying how many of its derivatives are still continuous. A C^1 function has its first derivative continuous, a C^2—the second, and so on. The C^1 smoothness is closely associated with the smoothness of a parametric curve. The C^2 smoothness is associated with its curvature.
- You can build piecewise curves, smooth and continuous, by making sure that the values and derivatives of every two pieces put together coincide in the joint points.

6 Polynomial approximation and interpolation

This chapter covers

- Understanding polynomials and their properties
- Using polynomial interpolation and approximation to describe continuous phenomena
- Understanding power series and their balance between speed and accuracy
- Circumventing the limitations of polynomials for data representation

A *polynomial* is the simplest mathematical object malleable enough to present the concepts of approximation and interpolation well. You have an *approximation* when you have a complex mathematical object and want to represent it approximately with a simpler one, such as when you want to find a linear function that represents 1000 data points or when you want to emulate a trigonometric function using only multiplications and additions.

The approximation is important in data representation when we want to show a general trend behind some data, and we're fine with the approximating function missing the actual data points. But in a special case of approximation called *interpolation*, the approximating function goes through all the data points precisely.

Interpolation is often used in descriptive geometry for building curves and surfaces. In chapter 8, we'll learn to build surfaces using polynomial interpolation.

Moreover, understanding polynomials and polynomial interpolation is a necessary step toward understanding polynomial splines, which we'll discuss in chapter 7, and splines in general are staples of computer graphics and computer-aided design.

6.1 What are polynomials?

A *polynomial* is an expression that consists only of multiplications, additions, and non-negative exponentiations of variables and constant coefficients. In a *canonical representation,* the powers of the variable are sorted in reverse order, and every variable power n is multiplied by its coefficient. This is a canonical representation of a polynomial:

$$P(x) = a_n x^n + a_{n-1} x^{n-1} \dots + a_2 x^2 + a_1 x + a_0$$

We've already met a cubic function that's a good example of a polynomial:

$$P(x) = ax^3 + bx + cx + d$$

Some coefficients may equal 0, some equal 1, and some may be negative. This fact affects the notation but not the essence of an expression. A specific example of a cubic function might look like this:

$$C(x) = x^3 - 2x^2 + 4$$

In full canonical notation, it would be

$$1x^3 + (-2)x^2 + 0x + 4$$

The highest power of an argument in a polynomial is called the *degree of the polynomial.* A cubic is a third-degree polynomial.

There are other representations apart from the canonical representation shown here. This way of representing a polynomial saves us computational time:

$$P(x) = ((ax + b)x + c)x + d$$

This example is the same cubic function, cleverly rearranged to win us some performance. The original formula had six multiplications; this one has three. This way of computing a polynomial is called *Horner's scheme* or *Horner's method.*

SEE ALSO Horner's method is not only fast, but also a rather stable way to compute a polynomial. *Accuracy and Stability of Numerical Algorithms,* 2nd ed., by Nicholas J. Higham (Society for Industrial and Applied Mathematics, 2002), mentioned earlier in this book, contains the error analysis for it. Even more accurate algorithms exist, however. In 2009, Stef Graillat, Philippe Langlois, and Nicolas Louvet published a survey on this topic called "Algorithms for Accurate, Validated and Fast Polynomial Evaluation," which is available online at http://mng.bz/1MOj.

A cubic polynomial with the coefficient at x^3 being 1, could be also factored as this:

$$P(x) = (x - r_1)(x - r_2)(x - r_3)$$

The numbers r_1, r_2, and r_3 are called the *roots* of a polynomial. When $x = r_1$, $x = r_2$, or $x = r_3$, the polynomial itself becomes zero. Or, geometrically speaking, where the graph of a polynomial intersects the *x*-axis, there is a root. Let's elaborate on that.

6.1.1 Axis intersections and roots of polynomial equations

A linear equation, as we know from chapter 3, is also a kind of polynomial:

$$y = ax + b$$

$$P(x) = ax + b$$

A root of an equation $P(x)$ is the x for which $P(x) = 0$.

Imagine that the *x*-axis ($P(x) = 0$) is a linear equation itself. Then our initial linear equation and the axis together make a system:

$$y = ax + b$$

$$y = 0$$

As we saw in chapter 3, a system of linear equations may have infinite solutions, one solution, or no solutions at all. Infinite solutions occur when the equation lines coincide; no solutions occur when they are parallel. For our system, both cases can happen only if $a = 0$. But this means that our starting linear equation isn't a linear equation at all; it's a constant. Any real linear system of this kind—the one with nonzero a—always has one and only one solution (figure 6.1).

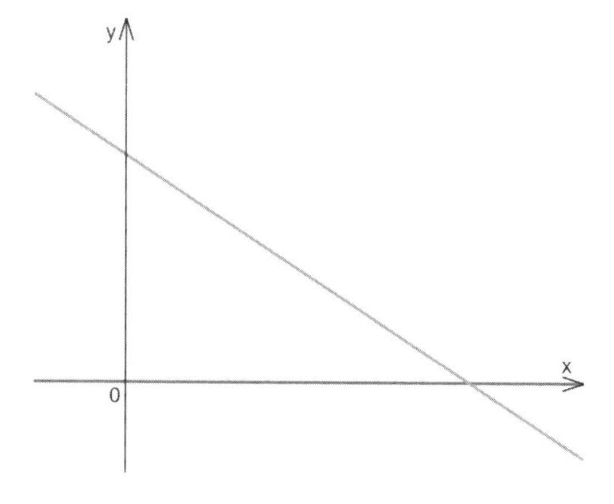

Figure 6.1 Linear equation with nonzero *a* intersects the *x*-axis one and only one time.

This also means that any linear equation $ax + b = 0$ also has one and only one root. Let's add a new member, a new degree for the x:

$$ax^2 + bx + c = 0$$

This type of equation is called a *quadratic equation*. As with a linear equation, we can make it into a function, build its plot, and use geometrical reasoning instead of algebra:

$$y = ax^2 + bx + c$$

The plot of this function makes a curve called a *parabola*. We can play with coefficients and see that a parabola may intersect the x-axis twice, touch it at a single point, or ignore it (figure 6.2).

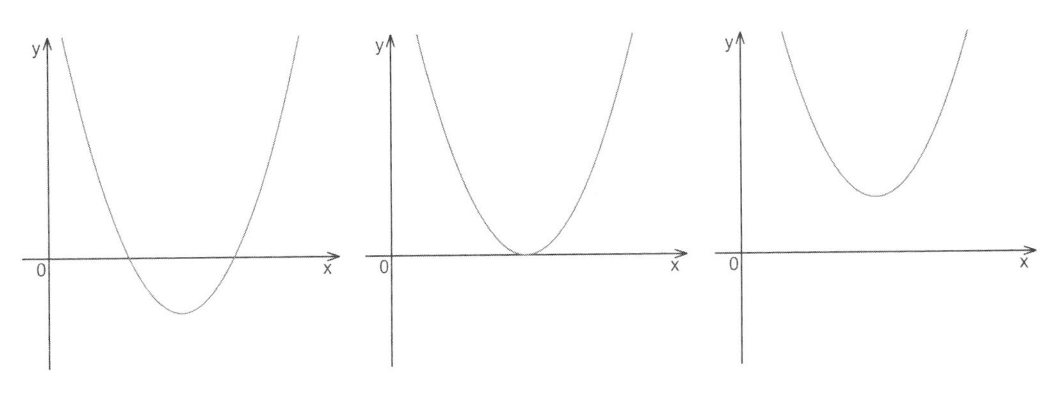

Figure 6.2 A quadratic polynomial may intersect the *x*-axis twice, touch it once, or ignore it.

So unlike a linear equation, a quadratic equation is allowed not to have roots, but it can also have one or two.

> **NOTE** In practice, the one-root solution is rare due to all kinds of computational and representational errors. A quadratic solver may not even bother with this particular case. If so, when an equation has only one root mathematically, a solver gives you a pair of roots that are identical. This is fine.

If we add another member, the equation becomes cubic:

$$ax^3 + bx^2 + cx + d = 0$$

As before, we can turn this equation into a function and build its plot. When we play with the coefficients, we'll see that it may have one root, two when the graph touches but doesn't cross the x-axis, or three (figure 6.3).

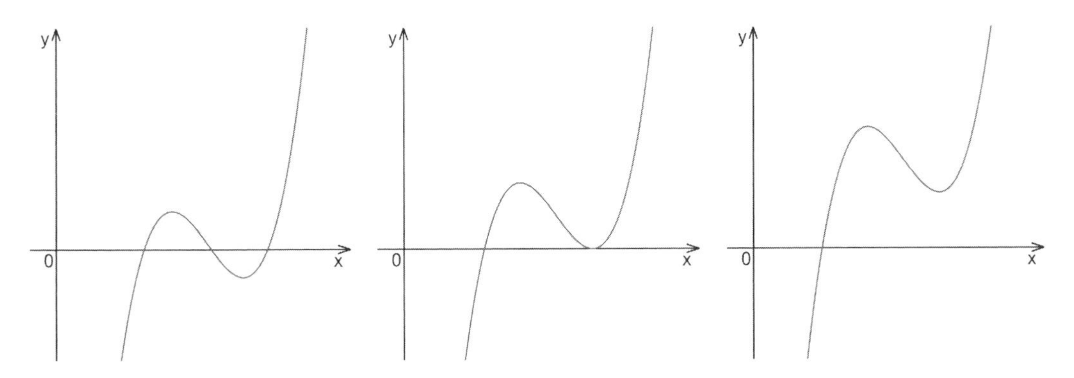

Figure 6.3 A cubic polynomial may have one, two, or three roots.

Interestingly, the graph can't escape the *x*-axis completely. Like the linear equation, it has to either go through the axis or stop being cubic. But as a quadratic equation, it's allowed to "choose" how many roots it has, and it can have up to three.

If we add a new member, the equation becomes quartic:

$$ax^4 + bx^3 + cx^2 + dx + e = 0$$

Somewhere around this point, we usually start using coefficients with indices instead of letters. It's not a rule, but you're unlikely to turn this equation into code directly; you'll probably use an array of coefficients to encode the polynomial anyway. Besides, you have to start numbering coefficients before you get to *x* anyway so why not start at *e*? The same polynomial equation with the numbered coefficients looks like this:

$$a_4x^4 + a_3x^3 + a_2x^2 + a_1x + a_0 = 0$$

Again, if you turn this expression into a function and see its graph, you see that the graph has zero, one, two, three, or four *x*-axis intersections (figure 6.4).

I hope you can see the pattern. The number of roots in a polynomial equation depends on the highest power of the *x* in the equation. Let's call it *n*. If *n* is even, the number of roots the equation has is less or equal to *n*. If *n* is odd, the number of roots is still less than or equal to *n*, but there's always at least one root. To understand where all these roots are coming from, let's see how polynomials are differentiated.

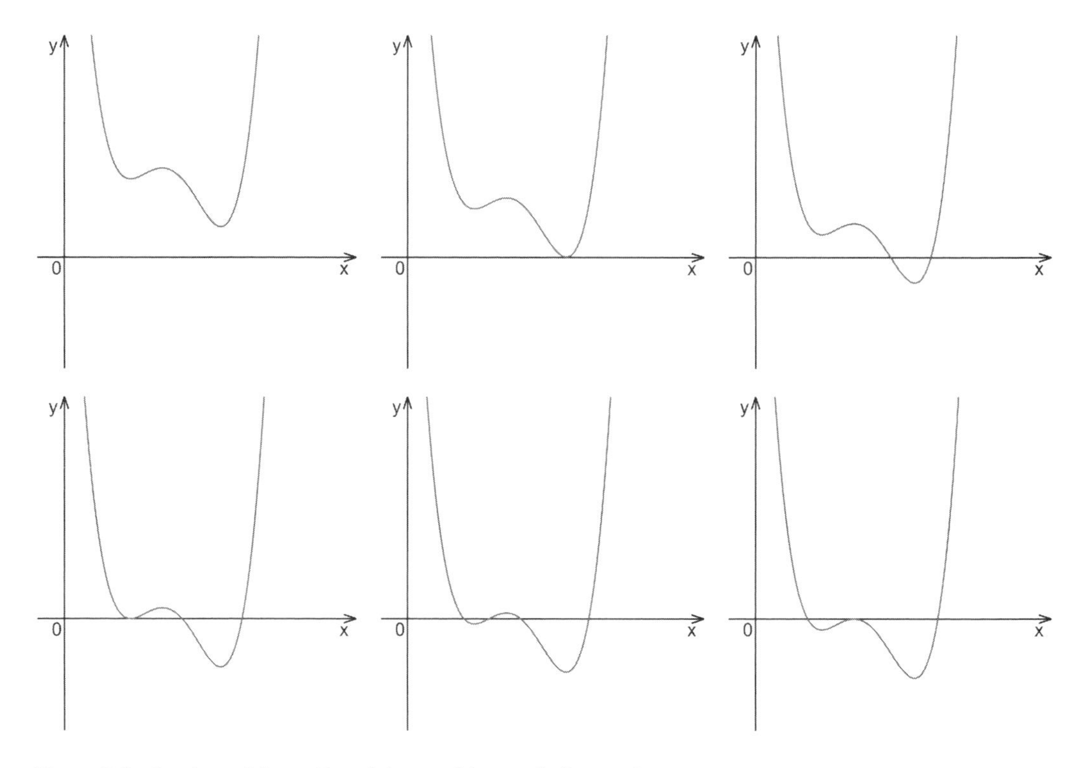

Figure 6.4 A polynomial equation of degree 4 has up to four roots.

6.1.2 Polynomial derivatives

The only three facts we need to remember when differentiating polynomials are

- How to differentiate the x^n function:

$$(x^n)' = nx^{n-1}$$

- How to differentiate a function multiplied by a constant:

$$(af(x))' = af'(x)$$

- How to differentiate a sum of two independent functions:

$$(f(x) + g(x))' = f'(x) + g'(x)$$

With this information, we can differentiate any polynomial we want by applying the differentiation of powers to each member one by one. Let's take a cubic polynomial for example:

$$P(x) = x^3 - x^2 - 2x + 1$$

$$P'(x) = 3x^2 - 2x - 2$$

As you can see, the derivative of our cubic polynomial is a quadratic polynomial. As a general rule, the derivative of a polynomial is a polynomial—or a constant if our polynomial is a mere linear function. Now let's put both functions on a single plot and look at the roots of the derivative (figure 6.5).

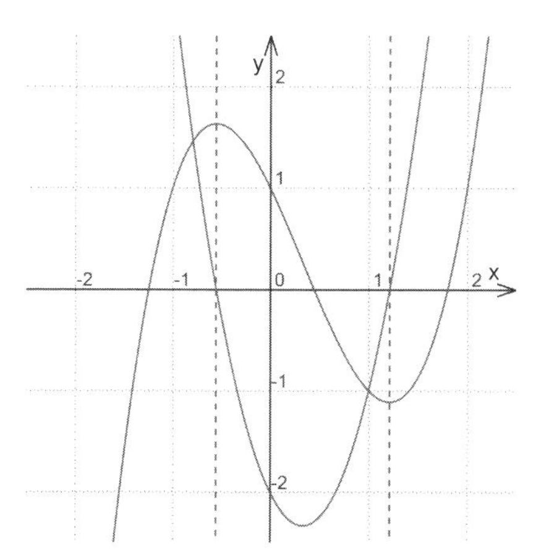

Figure 6.5 A cubic polynomial (two bumps, three roots) and its derivative, which is a quadratic polynomial (two roots)

Do you notice? Whenever the derivative has a root, the original function has a bump.

These bumps have proper names. The point on a graph where a function stops growing and prepares to decline is called the *local maximum point.* The point where it stops decreasing and prepares to increase is called a *local minimum point.* The values of a function in these points are called, respectively, *local maximum* and *local minimum* (no "point").

If a local maximum happens to be the highest value of the whole function, it can be called a *global maximum* or *absolute maximum.* The same goes for the *global* or *absolute minimum.* If it's the lowest value of the function ever, it's not only the *local minimum* but the *global minimum* as well. The word *extremum* means either *minimum* or *maximum,* and now I'm starting to regret not introducing it two paragraphs earlier.

Anyway, as you see from figure 6.5, a local minimum coincides with the derivative's root. Coincidence? Of course, not. Remember, the derivative indicates how fast a function increases or decreases. If the function goes up, the derivative is positive; if it goes down, the derivative is negative. If a function has a bump, an extremum, or specifically a local minimum, the derivative turns from negative to positive, and because it's continuous, it must go through the *x*-axis when changing the sign. A continuous function can't switch its sign without touching the *x*-axis; there must be a root somewhere.

The same applies to a local maximum. A growing function becomes declining, which means that its derivative changes from positive to negative, and it has to go through the axis once again.

So a local extremum of a function indicates a derivative's root. In practice, we often use this method in reverse to find extremums: differentiate the function, and find the roots of its derivative, and each root is expected to be a local maximum or minimum of a function. We do that, but it's technically wrong. A derivative has a root when a function stops growing or declining, but the function can change its mind and continue to grow or decline after a stop (figure 6.6).

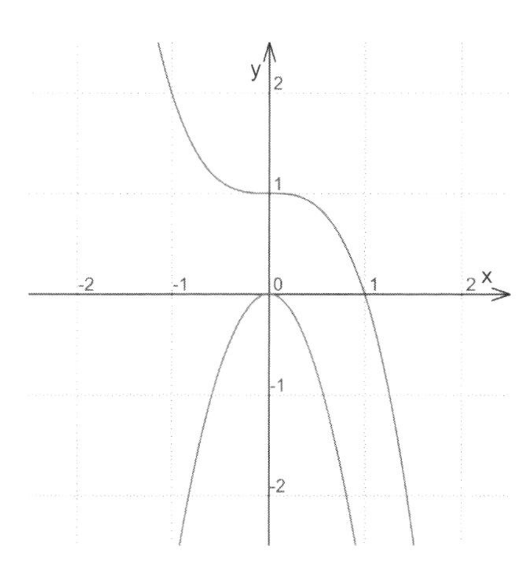

Figure 6.6 Still a cubic polynomial and its quadratic derivative. The derivative has a root, but it doesn't intersect the axis, so the cubic doesn't have an extremum.

Now let's get back to the "maximal number of roots" question. To have a root, a function has to go through the *x*-axis. To have two roots, a function has to make a U-turn and go back through the axis to its original half. If it makes a turn exactly on the axis, the two roots become the same root. To have three roots, a function has to cross the axis again, so it has to make two turns. The maximal number of roots a function can have, then, is the number of turns it makes, plus one.

In mathematical terms, each U-turn is an extremum, and an extremum can happen only when a derivative has a root.

A linear function has one root and no U-turns. Therefore, it has no extremums.

A quadratic polynomial has a linear function as its derivative. The linear function has one root, so the quadratic polynomial has one extremum, and one extremum allows it to have up to two roots.

A cubic polynomial has a quadratic function as a derivative. The quadratic function has up to two roots, so the cubic polynomial has up to two extremums and up to three roots.

The number of roots of a polynomial depends on the number of roots of its derivative. Because its derivative is also a polynomial, but one degree lower, each new degree of a polynomial gets at most one more root—one root per degree increment. The maximal number of roots a polynomial has, then, is the same as the polynomial degree.

> **NOTE** In complex numbers, it's even simpler: a polynomial of nth degree always has n roots.

6.1.3 Section 6.1 summary

A polynomial is a function that consists only of additions and multiplications of variables and constant coefficients. There's an economical way to store a polynomial: by storing only the coefficients of its canonical notation. There's also an economical way to compute a polynomial: by using Horner's scheme.

A derivative of a polynomial is also a polynomial, or a constant if the polynomial in question is a linear function. A polynomial of degree n has at most n roots. If n is odd, the polynomial must have at least one root; if n is even, it may have none.

6.2 Polynomial approximation

A polynomial consists solely of additions and multiplications. These operations are fairly easy to do by hand, but most important, they're also arithmetical operations that semiconductor chips are really good at doing, which makes polynomials cheap to compute.

In fact, all the mathematical functions our processors provide us—logarithm, exponent, sine, and cosine—are computed not as true functions, but as polynomials that resemble the originals well enough to be practical substitutes.

This substitution is called *approximation*. We have a function, we do some magic on it, and we get a copy of that function that's made of some other functions or operations. The copy may be imprecise, of course.

Polynomial approximation turns arbitrary functions into polynomials. On the hardware level, turning arbitrary functions into polynomials enables processors to compute non-polynomial functions they wouldn't be able to compute otherwise. And on the software level—our level—rewriting functions as polynomials of different degrees allows us to trade precision for performance. Let's see how it works, starting where we left off in the preceding section.

6.2.1 Maclaurin and Taylor series

A derivative of a polynomial is also a polynomial. That's nice to know, but what's even nicer to know is that the integral of a polynomial, a reverse of differential, is also a polynomial. We can exploit this feature to model whole functions solely with their derivative values. Let's see how.

First, let's take a cubic function and compute all of its derivatives:

$$
\begin{aligned}
P(x) &= x^3 - \frac{1}{2}x^2 - 2x + 1 \\
P'(x) &= 3x^2 - x - 2 \\
P''(x) &= 6x - 1 \\
P'''(x) &= 6
\end{aligned}
$$

One thing that attracts the eye is that for every derivative, the last coefficient is the same as the coefficient before x one line above. This makes sense, given that

$$a = (ax)'$$

But we can go further. When the last coefficient goes even one integration higher, it's divided by 2, which is maybe not so obvious but still consistent with what we know about differentiation:

$$b = \left(\frac{b}{2}x^2\right)''$$

With a cubic polynomial, we can do one step more and see that three lines above the coefficient get divided by 6:

$$c = \left(\frac{c}{6}x^3\right)'''$$

The 6 comes from multiplying 2 by 3. These divisors come from the second and the third derivation, respectively. As you can guess, we can easily add a few more derivations. For a quadric polynomial, its first coefficient is its fourth derivative divided by 24. For a fifth-degree polynomial, the divisor would be 120, and so on.

As a reminder, the product of all the natural numbers from 1 to n is called the *factorial of* n and is usually abbreviated as $n!$. The nth coefficient of a polynomial is the last coefficient of its nth degree divided by n factorial.

Okay, but what does this have to do with trading precision for performance? Hold on; we're getting there.

Let's compute our cubic along with all its derivatives in 0. This computation should be easy, because computing a polynomial in zero means essentially throwing away all the members but the constant:

$$P(0) = 1$$
$$P'(0) = 2$$
$$P''(0) = -1$$
$$P'''(0) = 6$$

Now, knowing how to compute the *n*th coefficient from the *n*th derivative's last coefficient, let's resurrect the $P(x)$. Its first coefficient from the left will be the third derivative divided by 3!, which is $\frac{6}{6} = 1$. Its second coefficient from the left is the second derivative divided by 2!, so $\frac{-1}{2}$. The third coefficient is –2 because it's the first derivative divided by 1. The fourth is the $P(0)$, so 1:

$$P(x) = x^3 - \frac{1}{2}x^2 - 2x + 1$$

Yay! We've disassembled a polynomial and assembled it back like a gun. How cool is that! Well, not too cool because it's not even practical. I mean, we ended where we started. But the good news is that this procedure also works with other source material. You can pick a function, compute a series of its derivatives at zero, and assemble a polynomial out of these values.

Let's try this trick on the exponent (figure 6.7). This function is the easiest one to differentiate because $(e^x)' = e^x$.

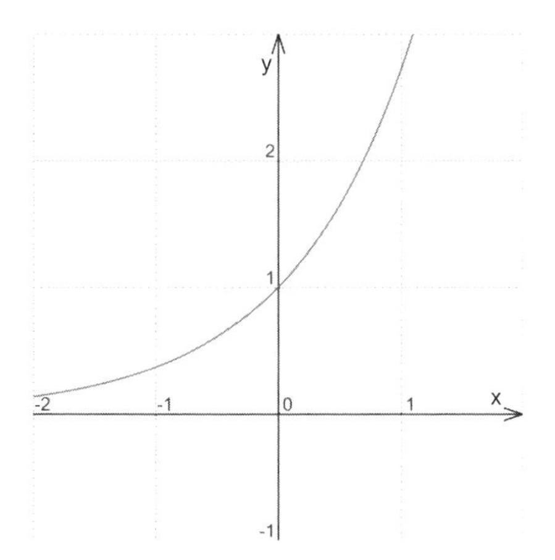

Figure 6.7 The exponent function

The exponent's value in 0 is 1. And because e^x differentiates into itself, its first derivative in 0 is also 1. So if we approximate it with a linear function (figure 6.8), both its coefficients will be 1s:

$$e_{\text{app}}(x) = x + 1$$

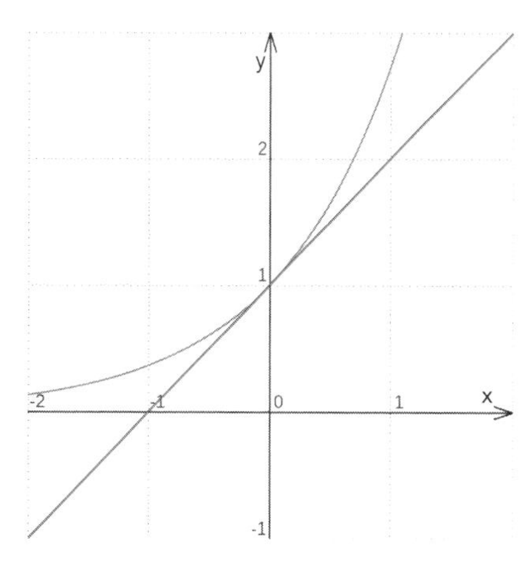

Figure 6.8 Simple but impractical approximation of the exponent with a linear function

Its second derivative in 0 is also 1. In its quadratic approximation (figure 6.9), this will turn into a coefficient near x^2 being $\frac{1}{2}$:

$$e_{\text{app}}(x) = \frac{1}{2}x^2 + x + 1$$

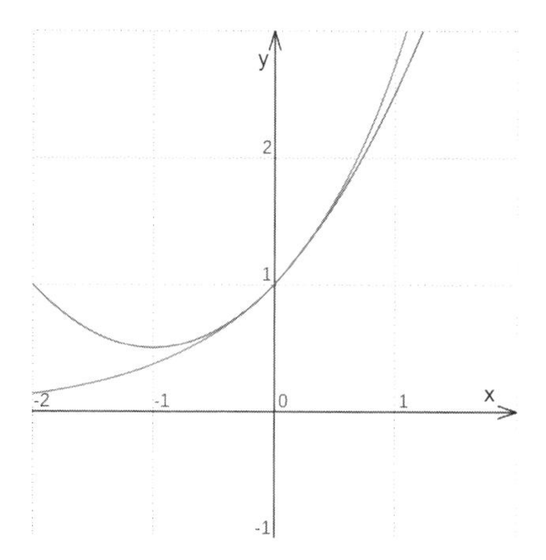

Figure 6.9 Slightly more practical quadratic approximation

Then the cubic approximation is the same as quadratic, but with one more term. The member is $\frac{1}{3!}$:

$$e_{\text{app}}(x) = \frac{1}{6}x^3 + \frac{1}{2}x^2 + x + 1$$

Following the same rule, we can easily add another coefficient, $4! = 2 \times 3 \times 4 = 24$,

$$e_{\text{app}}(x) = \frac{1}{24}x^4 + \frac{1}{6}x^3 + \frac{1}{2}x^2 + x + 1$$

and another, $5! = 5 \times 4! = 120$:

$$e_{\text{app}}(x) = \frac{1}{120}x^5 + \frac{1}{24}x^4 + \frac{1}{6}x^3 + \frac{1}{2}x^2 + x + 1$$

As it seems from figure 6.10, the approximation leans closer toward the original function but will never coincide with it. Let's once again turn to our intuitive understanding of infinity. The exponent infinitely rises toward positive infinity, but in negative infinity, the function "starts" at 0. Polynomials can't do that. The best a polynomial can do is to either go all the way from negative to positive infinity if its degree is odd or both start and end in positive infinity if its degree is even.

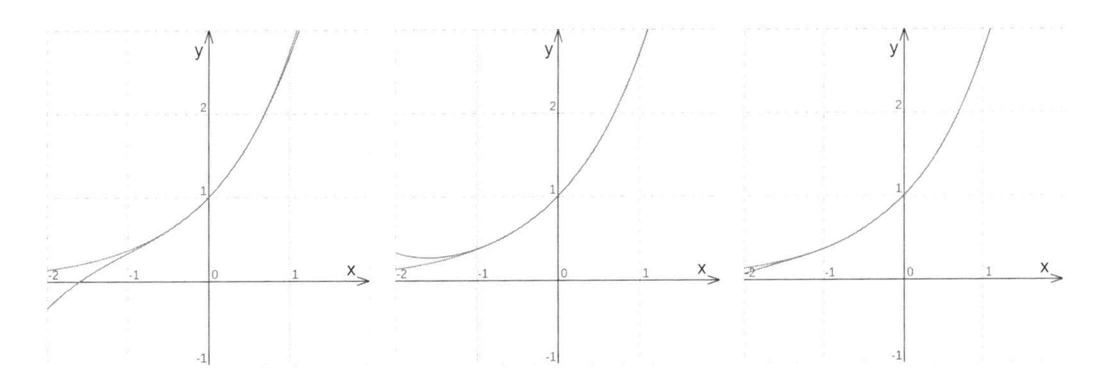

Figure 6.10 With more members coming, the approximation gets closer to the original function.

Still, in some locality, in some proximity of 0, this approximation works within some margin of error. And by adding members, we make the approximation error smaller. More members mean more computation, of course, so that's how we exchange speed for precision. We can have ultrafast but imprecise models, or we can make them as precise as our representation error allows, and they'll become relatively slow to compute.

So we can approximate a function by taking its derivatives in 0 and turning these derivatives into a polynomial. This approach is called the *Maclaurin series*. It's one example of a broader term, *power series*, and we do indeed need to broaden this approach.

What if a function has no derivatives in 0? What if it's not even defined in 0? Or what if a function we want to approximate is a natural logarithm, *ln*? It isn't defined in 0; it exists only for positive *x*s. Are we stuck?

Well, while $\ln(x)$ exists for $x > 0$, a $\ln(x + 1)$ exists for $x > -1$. So we can approximate the latter instead of the former and then, if we want, turn it into the conventional polynomial approximation with algebra:

$$\ln_{app}(x) = \ln_{x\_incremented}(x-1) = a_n(x-1)^n + \cdots + a_2(x-1)^2 + a_1(x-1) + a_0$$

$a_2(x-1)^2 + a_1(x-1) + a_0$ expands to this canonical polynomial:

$$a_2x^2 + (a_1 - 2a_2)x + (a_0 - a_1 + a_2)$$

So the coefficients of the resulting approximating polynomial in order of increasing degree are $(a_0 - a_1 + a_2)$, $(a_1 - 2a_2)$, and a_2.

You don't have to do the algebra by hand, of course. You have SymPy to do it for you:

```
>>> expand(a2*(x-1)**2 + a1*(x-1) + a0)
a0 + a1*x - a1 + a2*x**2 - 2*a2*x + a2
```

You can use expand to get rid of the brackets.

```
>>> collect(expand(a2*(x-1)**2 + a1*(x-1) + a0), x)
a0 - a1 + a2*x**2 + a2 + x*(a1 - 2*a2)
```

You can use collect to collect the coefficients before x.

When we take the derivatives in some other point than 0, the power series is called the *Taylor series*.

SEE ALSO The Better Explained website offers an awesome alternative explanation of the Taylor series. It appeals to your intuition rather than your memorization capacity, and it uses DNA analogy to do so. You can find it at http://mng.bz/Pxgn.

In practice, you wouldn't have to turn the Taylor series into the Mclaurin series by hand. That work has already been done. SymPy provides a function `series` that computes a finite number of polynomial members for you, starting at any given point. With this function, you can turn any function you want into its cheap polynomial approximation, and you can balance "cheap" and "accurate" with a single parameter! Let's take a look at a couple of examples:

We specify n = 6, but the highest degree of the polynomial is 5.

```
>>> series(E**x, x, x0 = 0, n = 6)
1 + x + x**2/2 + x**3/6 + x**4/24 + x**5/120 + O(x**6)
```

That's the exponent we have turned into a polynomial manually. The arguments for the `series` are

- The function itself
- The symbol for the function's argument
- The point to start derivating in x0
- The highest degree of a polynomial we want to get, plus one

If we can't take any derivatives in 0, we can easily change the x0, as with ln(x):

```
>>> series(ln(x), x, x0 = 1, n = 5)
-1 - (x - 1)**2/2 + (x - 1)**3/3 - (x - 1)**4/4 + x + O((x - 1)**5, (x, 1))
```

And if we want to increase a polynomial degree, we can use n. Note that n corresponds to the highest degree of a polynomial, not the amount of members. Case in point, $\sin(x)$ is an odd function, and its polynomial series contains only odd members so to get five members, we want the highest polynomial degree to be 9:

```
>>> series(sin(x), x, x0 = 0, n = 10)
x - x**3/6 + x**5/120 - x**7/5040 + x**9/362880 + O(x**10)
```

Approximating trigonometric functions with polynomials was a common optimization in game development a few decades ago. You don't need a precise sine model if you use it only to animate a ceiling fan in a dark room. Modern processors, however, are effective in computing trigonometry, so racing the native implementations for sine and cosine with polynomials is getting harder. You can still go this way, turning a simple but costly function into a cheap polynomial by using the Mclaurin or Taylor series, but you'll get performance benefits only if you allow considerable error for your model. Using power series as an optimization technique might still work if you want to approximate something much heavier than a sine, however.

Making a function into a theoretically infinite series of polynomial members is great if we can get a theoretically infinite amount of its derivatives. But what if we can't? What if all we have is a data set—a handful of points and some initial preconception of what a function might look like? Don't worry. This input is enough for our next approach: the least-squares polynomial approximation.

6.2.2 The method of least squares

So we have a set of N points, (x_i, y_i), $i = 1..N$. We want a function $f(x)$ to approximate them somehow. We have more than one way to build an approximating function, as shown in figure 6.11.

Visually, some functions fit the data set better than others. But how do we access this "betterness" programmatically? What is the metric—the measuring device for how good or bad an approximation is? Well, we can measure the quality of an approximation by measuring its collective error for every data point we have:

$$E = \sum_{i=1}^{N} |f(x_i) - y_i|$$

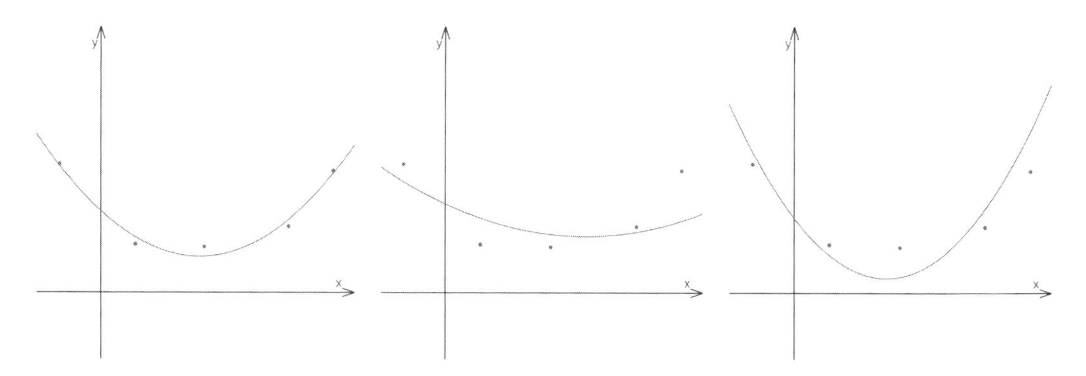

Figure 6.11 The same data set may be approximated with different functions to a different degree of success.

This is something. The larger the error, the worse the approximation is. The problem is if one point sticks out too far, its error gets diluted in the sum of all the other errors, and then the approximation works well for all the points but one (figure 6.12).

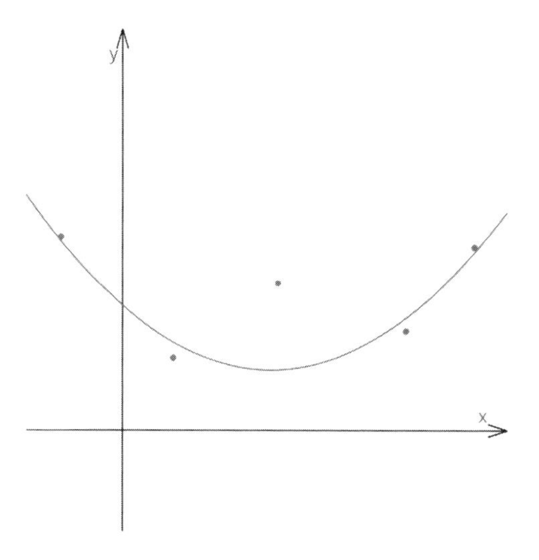

Figure 6.12 When mean error is a quality-of-approximation metric, some points tend to stick out.

To stop approximation from leaving outcasts, we usually put the error in the sum to squares. This metric is called *squared error*:

$$SE = \sum_{i=1}^{N} (f(x_i) - y_i)^2$$

This way, the sticking-out point will contribute to the error significantly more than any of the others, and the approximating algorithm will have to find a more evenly erroneous function (figure 6.13).

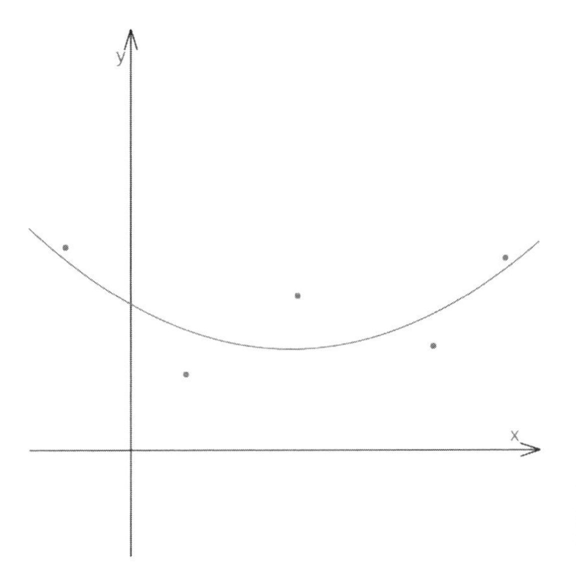

Figure 6.13 With squared mean error, the error is shared in a way that makes "sticking out" improbable.

Squared error or mean squared error?

Squared error is close to another popular metric often used in data analysis. It's called *mean squared error* or *mean squares deviation*, and it's the squared error divided by the number of points. This metric is better for human introspection because it scales for data sets of all sizes:

$$MSE = \frac{\sum_{i=1}^{N} (f(x_i) - y_i)^2}{N}$$

The approximating algorithm matches a set of points with a function of a certain kind in such a way that the error metric is guaranteed to be minimal. The error metric doesn't have to be human-readable, so usually, the approximating algorithm prefers squared error to mean squared error. Because the only difference between them is a division, minimizing one automatically means minimizing the other.

Theoretically, an approximating algorithm can take a function of any kind and find its coefficients by solving a minimization problem for a set of points numerically. In practice, this approach is rarely the way to go, because for several special kinds of functions, we already have much more efficient ways to get approximating coefficients. Polynomial is one of these functions.

The coefficients for an approximating quadratic polynomial $P(x) = a_2x^2 + a_1x + a_0$ come from this particular formula:

$$\begin{pmatrix} N & \sum_{i=1}^{N} x_i & \sum_{i=1}^{N} {x_i}^2 \\ \sum_{i=1}^{N} x_i & \sum_{i=1}^{N} {x_i}^2 & \sum_{i=1}^{N} {x_i}^3 \\ \sum_{i=1}^{N} {x_i}^2 & \sum_{i=1}^{N} {x_i}^3 & \sum_{i=1}^{N} {x_i}^4 \end{pmatrix} \begin{pmatrix} a_0 \\ a_1 \\ a_2 \end{pmatrix} = \begin{pmatrix} \sum_{i=1}^{N} y_i \\ \sum_{i=1}^{N} y_i x_i \\ \sum_{i=1}^{N} y_i {x_i}^2 \end{pmatrix}$$

Yes, this is a system of linear equations. We saw these systems in chapters 3 and 4, and you might expect that normally, a well-defined system has one and only one solution.

This particular approximation works for every set of points. It works for a single point resulting in a constant function $f(x) = y_1$. For a pair of points, it results in a straight line. For three points, it makes a perfectly fitting quadratic polynomial (figure 6.14).

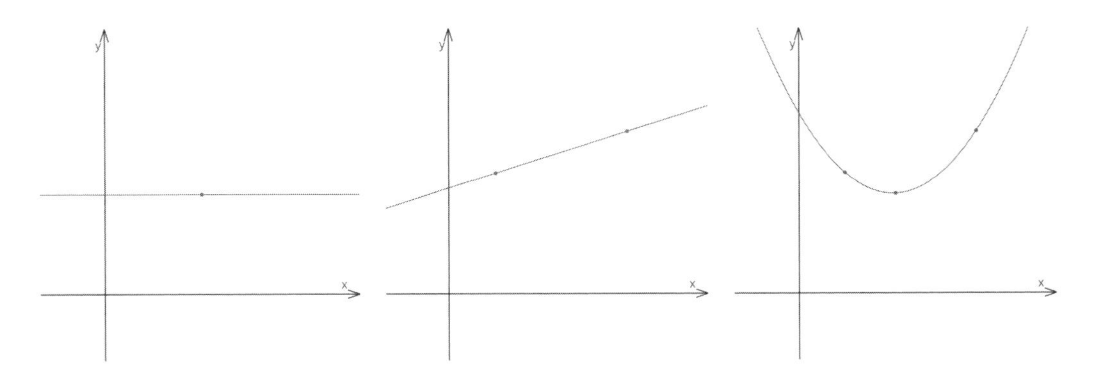

Figure 6.14 A polynomial of degree 2 approximates sets of 1, 2, and 3 points with no error.

For all the points more than three, it still results in a quadratic polynomial, but we compute only three coefficients so what else could it be? Due to the rather elaborate math, which we won't get into, the least-squares error metric is guaranteed to be minimal for this polynomial on any set of points, no matter how large the set is (figure 6.15).

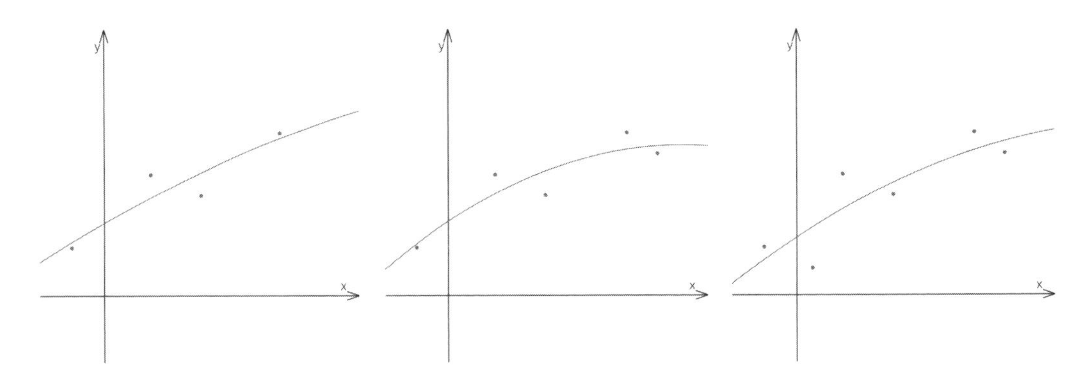

Figure 6.15 A 2-degree approximating polynomial fits data sets of 4 points and more with minimal error.

The formula for quadratic approximation generalizes nicely onto polynomials of any degree. Because we already use *N* for the number of points, let's use the letter *M* to denote the polynomial degree. For a quadratic polynomial, for example, M would be 2. But the general formula for any M is

$$\begin{pmatrix} N & \sum_{i=1}^{N} x_i & \cdots & \sum_{i=1}^{N} {x_i}^M \\ \sum_{i=1}^{N} x_i & \sum_{i=1}^{N} {x_i}^2 & \cdots & \sum_{i=1}^{N} {x_i}^{M+1} \\ \vdots & \vdots & \vdots & \vdots \\ \sum_{i=1}^{N} {x_i}^M & \sum_{i=1}^{N} {x_i}^{M+1} & \cdots & \sum_{i=1}^{N} {x_i}^{M+M} \end{pmatrix} \begin{pmatrix} a_0 \\ a_1 \\ \vdots \\ a_M \end{pmatrix} = \begin{pmatrix} \sum_{i=1}^{N} y_i \\ \sum_{i=1}^{N} y_i x_i \\ \vdots \\ \sum_{i=1}^{N} y_i {x_i}^M \end{pmatrix}$$

You can do linear, quadratic, cubic, or any other polynomial approximation with this single formula. You can even compute a mean value for an array of numbers with a special case of it when $M = 0$. Having an instrument this versatile is handy, because you don't have to duplicate your code to do all these tasks separately.

SEE ALSO The Visualize It website has an interactive plot for the least-squares approximation called Polynomial Regression at http://mng.bz/Jl4V. You can experiment there if you don't feel like writing your own implementation.

6.2.3 Practical example: Showing a trend with approximation

Let's say you have some data points. They could be sales data in units per quarter, the average temperature for your hometown in degrees per year, or your personal best for 2000 meters on a rowing machine per workout session. The idea is that you have something per something, so you have a set of y values for every x. Those are your data points.

If you want to show this data as a smooth, continuous function, the polynomial approximation might help. One caveat, though: there's a huge difference between showing data as a function and seeing a function behind the data. You can *show* a nice plot by using a polynomial graph, but if you want to *see* the forces that govern your progress in sports, weather changes, or sales rise, you need another book. Understanding data is the domain of data science, not geometry.

Showing data, however, or at least making sets of data points into curves you can interpret is a common geometrical task. The difference may appear to be subtle, but it will become visual soon.

Let's take some real-world data to see how the approximation works with it. Table 6.1 shows my 2000-meter rowing times per workout from when I started rowing.

Table 6.1 2000-meter rowing times per workout session

Session number	2000-meter time
1	8'20"
2	8'04"
3	7'56"
4	7'52"
5	8'00"

Here's the same data in Python:

```
xs = [1, 2, 3, 4, 5]
ys = [8 + 20/60., 8 + 4/60., 7 + 56/60., 7 + 52/60., 8 + 0/60.]
N = len(xs)
M = 1
```

Here, `N` is the size of the data set, and `M` is the degree of the target polynomial. Let's show these data points as a trend and presume that the trend is expected to be linear. This makes the degree of our target polynomial 1, so `M = 1`.

To compute the coefficients of the approximation polynomial with the method of least squares, we need to compose the matrix, compose the vector of free coefficients, and make and solve the equation. As a reminder, the equation for this task is

$$\begin{pmatrix} N & \sum_{i=1}^{N} x_i & \cdots & \sum_{i=1}^{N} x_i{}^M \\ \sum_{i=1}^{N} x_i & \sum_{i=1}^{N} x_i{}^2 & \cdots & \sum_{i=1}^{N} x_i{}^{M+1} \\ \vdots & \vdots & \vdots & \vdots \\ \sum_{i=1}^{N} x_i{}^M & \sum_{i=1}^{N} x_i{}^{M+1} & \cdots & \sum_{i=1}^{N} x_i{}^{M+M} \end{pmatrix} \begin{pmatrix} a_0 \\ a_1 \\ \vdots \\ a_M \end{pmatrix} = \begin{pmatrix} \sum_{i=1}^{N} y_i \\ \sum_{i=1}^{N} y_i x_i \\ \vdots \\ \sum_{i=1}^{N} y_i x_i{}^M \end{pmatrix}$$

All the steps together translate into the Python code in the following listing (ch_06/approximate.py in the book's source code).

Listing 6.1 Computing the approximating polynomial

```
LS = Matrix([[sum([xs[i]**(row + column)
    for i in range(N)])
        for column in range(M+1)]
            for row in range(M+1)])

B = Matrix([sum([ys[i]*(xs[i]**row)
    for i in range(N)])
        for row in range(M+1)])

a0, a1 = symbols('a0 a1')
Pol = Matrix([a0, a1])

solution = solve(LS*Pol - B, (a0, a1))
a = [value for key, value in solution.items()]
```

The least-squares matrix (annotates `LS`)

The least-squares column vector (annotates `B`)

Polynomial in its symbolic form (annotates `Pol`)

Solving the system for the polynomial's coefficients (annotates `solution`)

Storing the coefficients in a list for convenient reuse (annotates `a`)

Let's review the listing step by step. The matrix for the least-squares method is the matrix where every element $[i, j]$ is the sum of all *xs* to the power of $i + j$. Using Python list comprehensions, this formula could be written as a single-liner:

```
LS = Matrix([[sum([xs[i]**(row + column)
    for i in range(N)])
        for column in range(M+1)]
            for row in range(M+1)])
```

The column vector on the right side of the equation consists of M elements, where every element is the sum of *ys* multiplied by *xs* to the power of i:

```
B = Matrix([sum([ys[i]*(xs[i]**row)
    for i in range(N)])
        for row in range(M+1)])
```

The following two lines are the overhead of using SymPy instead of some numeric solver. SymPy is a symbolic solver, and even if we don't need any particular symbols, such as `a0` and `a1`, we still have to introduce them to get some numbers as a result. We can live with that fact. All we have to do is to define a polynomial as a column vector of coefficients:

```
a0, a1 = symbols('a0 a1')
Pol = Matrix([a0, a1])
```

Now we have the matrix `LS`, the column vector `B`, and the symbolic representation of the column vector with the polynomial coefficients `Pol`. Next, we put it all into a single equation and solve it with SymPy:

```
solution = solve(LS*Pol - B, (a0, a1))
```

When `solve` is done, our symbols `a0` and `a1` will have some numbers assigned to them. We can use this solution to build a polynomial directly, but it would be easier to turn symbolic representation into a plain list of numbers that we could use in a more general polynomial formula. A solution comes as a dictionary with the sorted key, so we can iterate through its items and gather the values for the coefficients:

```
a = [value for key, value in solution.items()]
```

Running the code on our data set gives us coefficients for a predictable linear approximation (figure 6.16).

This plot shows the progress nicely, but it kind of suggests that the nature of the progress is linear, which isn't true. Normally, the better you get, the harder it gets to improve even further. So even if a line is a good approximation in the short term, it doesn't represent the true function behind the data. To show the absurdity of linear progress, let's show the same plot on a larger interval (figure 6.17).

The other problem with the approximation is that by definition, it is approximate. Sometimes, however, you need to fit the data point precisely. For problems like these, we need another approach, and we'll look at one in the following section.

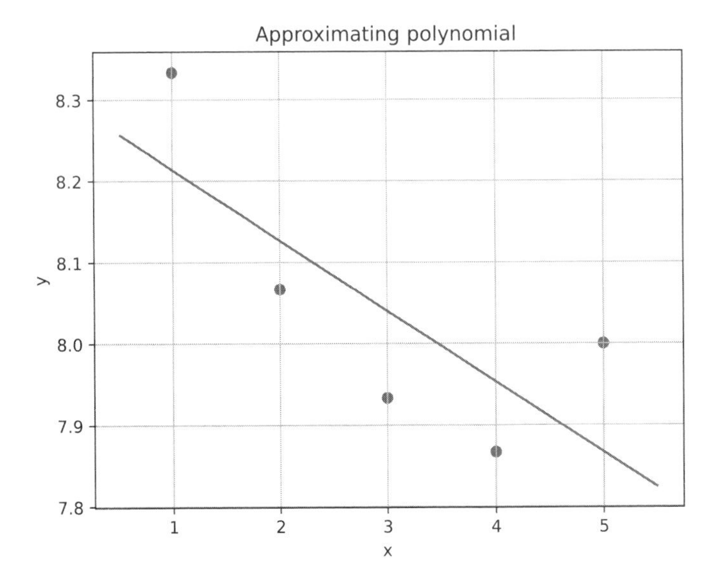

Figure 6.16 Linear approximation is a special case of polynomial approximation.

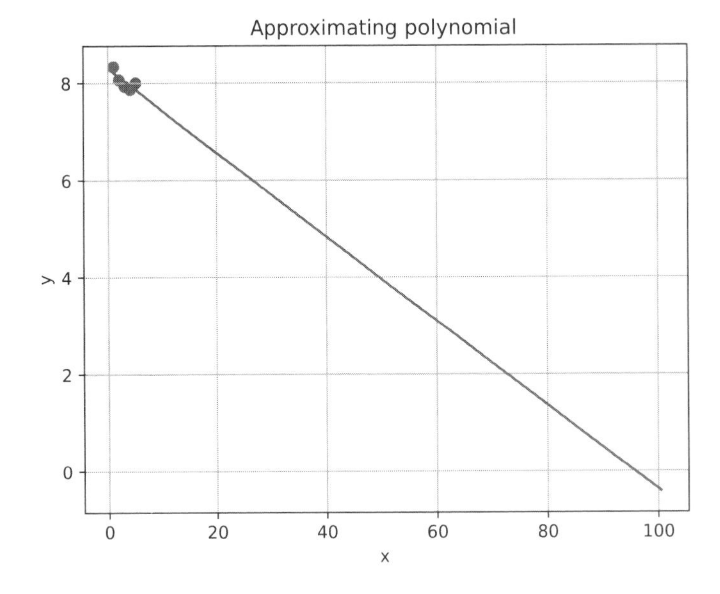

Figure 6.17 This plot suggests that in less than two years, I started paddling myself back in time.

Making plots in Python

This book isn't a Python book, and I promised you wouldn't have to learn much of the language. SymPy alone should be enough to turn formulas into code in the language of your choice.

Starting in this chapter, however, we'll see more code examples that aren't pure SymPy snippets. There, examples won't produce code for the applications but illustrate the applications themselves. For these types of examples, drawing a plot or two in Python may come in handy. The library you usually use for that purpose is called Matplotlib. You can find documentation and examples at https://matplotlib.org.

By the way, all the book's source code examples for this chapter and later chapters are supplied with plotting extensions, which I added for your convenience. Still, because this book isn't about Python, these convenience additions aren't shown in the book itself.

6.2.4 Section 6.2 summary

You can approximate a differentiable function with a polynomial by using the Taylor or Maclaurin series, which allows you to have a cheaper model that works well enough in some limited range.

You can approximate a set of data points with a polynomial by using the least-squares method. This method includes linear, quadratic, cubic approximation, and technically even computing the mean value.

6.3 Polynomial interpolation

Another word for approximating a set of points with a function is *fitting*. A function fits a set of points. When the squared error is 0, the function fits the data perfectly, which means that its graph runs through all the points exactly (figure 6.18).

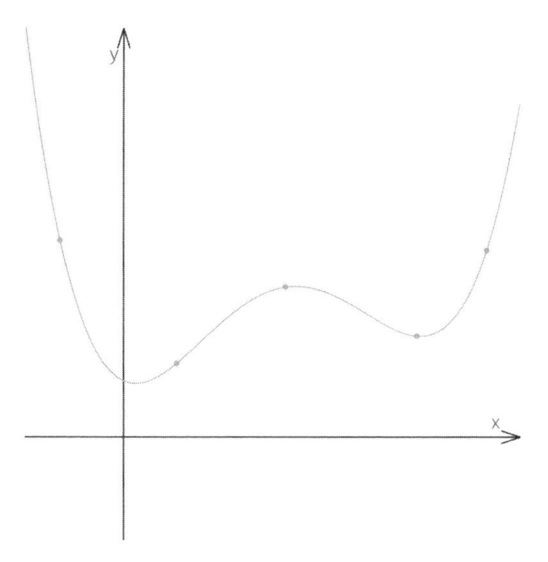

Figure 6.18 When a polynomial approximates the data set with 0 error, every data point belongs to its graph.

In data science, this effect is usually unwanted, but in geometry, when we use approximation to build curves and surfaces, this perfect fitting is a blessing. Without the exact match, we wouldn't be able to conjoin two curves or surfaces perfectly. Our geometric models would have had holes in every place two separate pieces should have touched each other.

The special error-free case of approximation is called *interpolation.* In this section, we'll learn how to interpolate a set of points efficiently. We'll explore the limitations of the polynomial interpolation and see the parallels with what we learned about linear systems in chapter 3.

6.3.1 Using Vandermonde matrix to get the interpolating polynomial

Technically, you don't need to learn yet another formula to compute the coefficients of an interpolating polynomial. The one with the least-squares will do. Remember, when you approximate a set of three points with a quadratic polynomial, the squared error of such approximation is zero, and the approximating polynomial also becomes interpolating. This effect scales nicely to other polynomial degrees and point sets.

A set of four points is being interpolated nicely by a cubic approximating polynomial. A set of five points is being interpolated by a quadric polynomial. For six points, you need a fifth-degree polynomial; for seven points, a sixth, and so on.

Generally, a set of N points may be interpolated by a polynomial of a degree no less than N–1. And because the least-squares method from the previous section would love to minimize the error to 0, given the occasion, you can use it to compute the interpolating polynomial coefficients. But don't, for two reasons.

First, the alternative method we're about to meet is much more effective computationwise. Second, the method also explains why we need (N–1)-degree polynomial to interpolate N points and explains it by using what we've already learned from linear systems. Once you see this connection, you wouldn't be able to unsee it.

So if an interpolating polynomial $P(x)$ goes through a point (x_i, y_i), this means that $P(x_i) = y_i$. That's an equation. We can make an equation per every point, gather them all in a system, and solve this system to get the polynomial coefficients. For a quadratic polynomial, this system looks like this:

$$a_2x_1^2 + a_1x_1 + a_0 = y_1$$

$$a_2x_2^2 + a_1x_2 + a_0 = y_2$$

$$a_2x_3^2 + a_1x_3 + a_0 = y_3$$

Here's the same system in matrix form:

$$\begin{pmatrix} 1 & x_1 & x_1^2 \\ 1 & x_2 & x_2^2 \\ 1 & x_3 & x_3^2 \end{pmatrix} \begin{pmatrix} a_0 \\ a_1 \\ a_2 \end{pmatrix} = \begin{pmatrix} y_1 \\ y_2 \\ y_3 \end{pmatrix}$$

This matrix equation, of course, scales onto polynomials of any degree M where $M = N$ –1 and N is the number of data points:

$$\begin{pmatrix} 1 & x_1 & \cdots & {x_1}^M \\ 1 & x_2 & \cdots & {x_2}^M \\ \vdots & \vdots & \vdots & \vdots \\ 1 & x_N & \cdots & {x_N}^M \end{pmatrix} \begin{pmatrix} a_0 \\ a_1 \\ \vdots \\ a_M \end{pmatrix} = \begin{pmatrix} y_1 \\ y_2 \\ \vdots \\ y_N \end{pmatrix}$$

The matrix on the left is called the *Vandermonde matrix*. It's a matrix made of "the polynomial goes through a point" equations. Here's a specific example:

$$(x_1, y_1) = (-1, 0)$$

$$(x_2, y_2) = (0, 1)$$

$$(x_3, y_3) = (1, 0)$$

The matrix equation for this data set is

$$\begin{pmatrix} 1 & -1 & 1 \\ 1 & 0 & 0 \\ 1 & 1 & 1 \end{pmatrix} \begin{pmatrix} a_0 \\ a_1 \\ a_2 \end{pmatrix} = \begin{pmatrix} 0 \\ 1 \\ 0 \end{pmatrix}$$

You can solve this system numerically or solve it symbolically for a general case in SymPy and put the numbers from a specific problem in the formulas you obtain. For this system, the solution is

$$a_0 = 1,\ a_1 = 0,\ a_2 = -1$$

The polynomial for this set of points looks like figure 6.19.

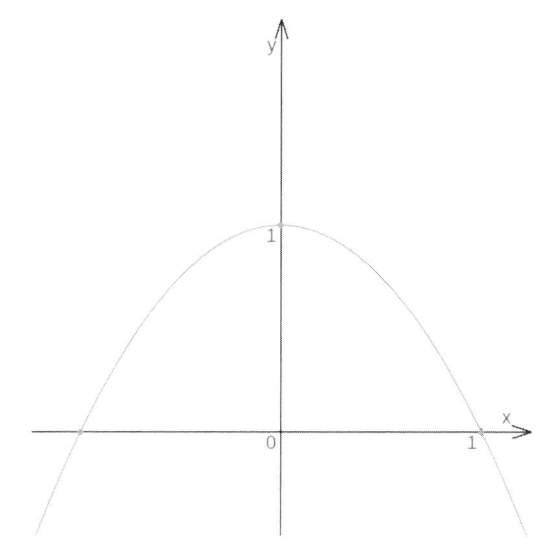

Figure 6.19 The interpolating polynomial for points (-1, 0), (0, 1), and (1, 0)

The same approach generalizes nicely to a set of N points. You collect the equations, put them in a system, and solve the system to get the polynomial's coefficient.

Now it's time to remember what makes a system solvable. The best-case scenario—a system that can easily be solved numerically—is a well-defined or well-specified system in which the number of equations equals the number of variables. An N–1 degree polynomial has N coefficients, which means that you need to gather N equations to make a system well-defined. Every equation comes from a data point, and that's why to interpolate a set of N points, we need a polynomial of exactly N–1 degree.

Or do we? What if a system is underspecified? *Underspecified* means fewer equations than variables or fewer data points than polynomial coefficients. Well, this system is harder to solve numerically because there isn't one solution, but a continuum of solutions. But we can still pick a solution—any solution—from this continuum, and get a nice interpolating polynomial. Let's look at our earlier example with the third point removed:

$$(x_1, y_1) = (-1, 0)$$

$$(x_2, y_2) = (0, 1)$$

Now we have only two equations, so we can't make a well-specified system out of them. But what if we add a third one?

$$a_2 = 0$$

Now the system is well-defined:

$$\begin{pmatrix} 1 & -1 & 1 \\ 1 & 0 & 0 \\ 0 & 0 & 1 \end{pmatrix} \begin{pmatrix} a_0 \\ a_1 \\ a_2 \end{pmatrix} = \begin{pmatrix} 0 \\ 1 \\ 0 \end{pmatrix}$$

The solution would be

$$a_0 = 1,\ a_1 = 1,\ a_2 = 0$$

This is a straight line, not exactly a quadratic polynomial. But, technically, no one prevents us from picking any other coefficient for a_2. Computationwise, the straight line is the most economical way to interpolate a set of two points, but it's not the only possible way. We can interpolate this set with an infinite amount of quadratic parabolas (figure 6.20).

Just as an underspecified system has an infinite continuum of solutions, the polynomial of N degree can interpolate a set of M points in an infinite amount of ways if $N \geq M$.

What about underspecified systems, then? Well, all that we know about the underspecified system translates for polynomial interpolation, too. In the general case, an

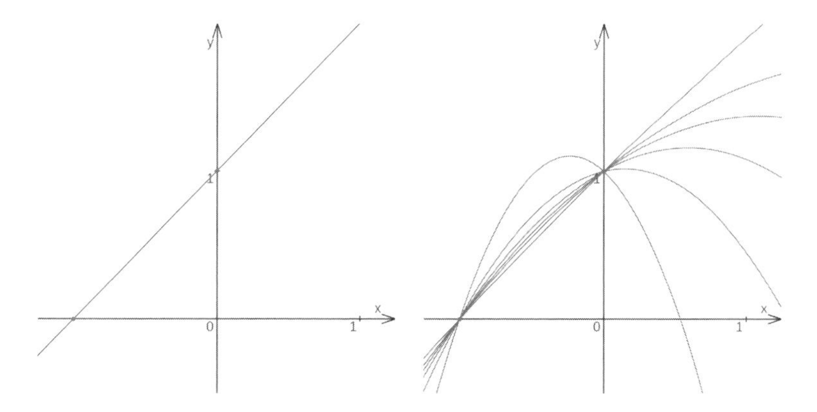

Figure 6.20 A two-point data set may be interpolated with a single straight line or with an infinite amount of quadratic parabolas.

underspecified system is unsolvable, and there's no interpolating polynomial of N degree for a set of M points if $N < M–1$. You can't draw a straight line through a set of three points—unless, of course, the points are already in a straight line.

In terms of linear systems, this happens when the third point makes an equation that is linearly dependent on the other two. So the point on a line is a weighted sum of any other different two points on the same line.

This effect scales to other degrees and set sizes. Normally, you can't interpolate N points with $(N – 1 – k)$ degree polynomial, but if for some reason, at least k points already fit your interpolating function, you can (figure 6.21). In practice, this condition is rare, so it's best to rely on well-defined systems from the start.

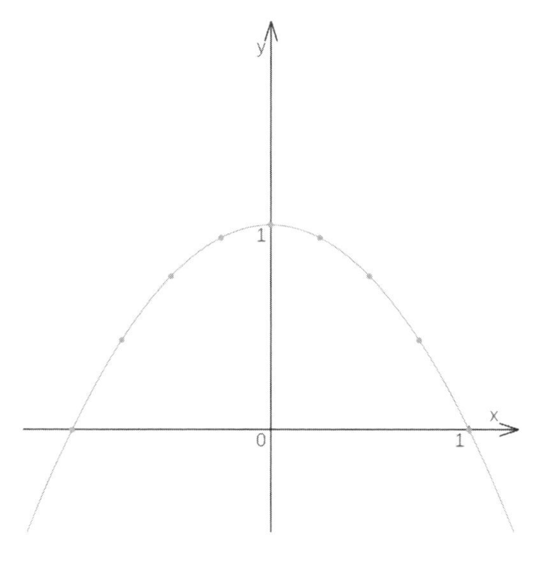

Figure 6.21 A quadratic polynomial may technically interpolate a set of more than three points, but not in general. In this figure, points are generated from the polynomial, not vice versa.

But even if the system seems to be well-defined, if the number of variables matches the number of equations, the system still may not have a solution. Geometrically speaking, this happens when some of the hyperplanes that constitute the system's equations are parallel, and it translates to the polynomial interpolation as well. A pair of parallel hyperplanes will have the left parts of their equations exactly the same (up to scaling), but a free coefficient on the right is different. Given that the Vandermonde matrix consists solely of *x*s and that all the free coefficients are *y*s, having a pair of parallel hyperplanes' equations in a system would imply that

$$P(x_a) = y_a$$

$$P(x_b) = y_b$$

while $x_a = x_b$ but $y_a \neq y_b$. In other words, we want a function to go through two different points at the same time (figure 6.22).

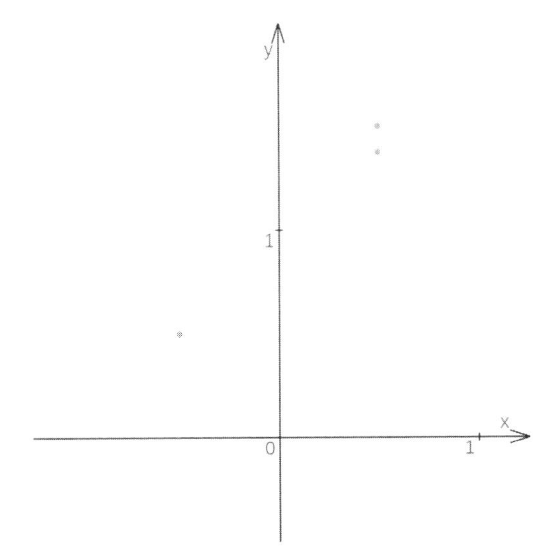

Figure 6.22 An interpolating polynomial can't go through the points with the same *x* but a different *y*.

This contradicts the definition of a function. A polynomial can't do that.

To sum things up, unless there's a contradiction in our data set, a set of *N* points can be successfully interpolated with *N*–1 degree polynomial. Just as there is one and only one solution point for a well-defined linear system with no parallel hyperplanes' equations in it, there is one and only one interpolating polynomial of *N*–1 degree. Technically, we can raise the degree of the interpolating polynomial, but then we'd have to deal with an underspecified system somehow, and this nice unambiguity will be lost.

NOTE By the way, because the *N*–1 degree interpolating polynomial is unambiguous, the coefficients you get by minimizing the squared error are exactly the same as those you get from solving the Vandermonde system.

To keep things simple, from now on, when I talk about an interpolating polynomial for a set of N points, assume that it's of N–1 degree.

6.3.2 What limits polynomial interpolation application to small data only?

Let's remember a few facts about polynomials. A derivative of an M-degree polynomial is an (M–1)-degree polynomial. A polynomial of M-degree may have up to M roots; that's where the function intersects the x-axis. Every time, a derivative of a function crosses the x-axis, the function itself makes a bump: it reaches an extremum point: either a local minimum or a maximum.

Now, why is this suddenly important for interpolation? You see, we can make a polynomial go through N points, but we can't control the exact way it does that. An (N–1)-degree polynomial has an (N–2)-degree derivative, and this derivative may have up to N–2 roots. Then the interpolating polynomial has up to N–2 bumps (figure 6.23).

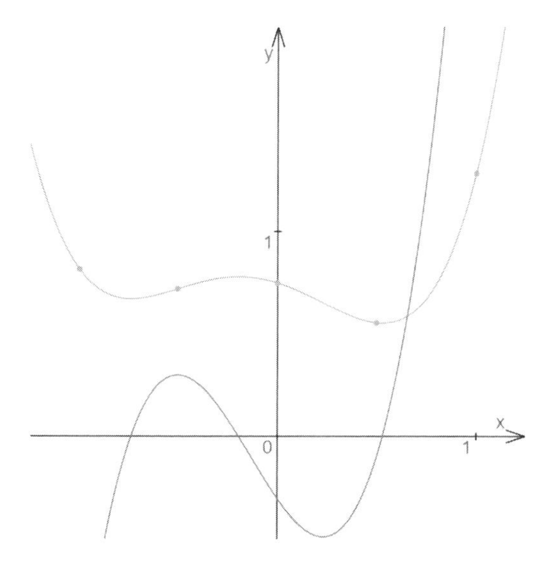

Figure 6.23 The (N–1)-degree interpolating polynomial may have up to N–2 extremums.

For small degrees, this situation is rarely a problem. But as we add more points and raise the degree respectively, we also raise the potential number of bumps. And we have no control over whether or where these bumps will occur. The function starts to wave (or *oscillate*) more, especially toward the ends of the interpolation interval (figure 6.24).

This situation is called *Runge's phenomenon*, which is one thing that restricts the applicability of polynomial interpolation in real-world applications. Normally, when we interpolate between a pair of points (x_i, y_i) and (x_{i+1}, y_{i+1}), we expect the interpolating function on x in the range (x_i, x_{i+1}) to be somewhere close to (y_i, y_{i+1}). With uncontrollable oscillations, this may not be the case.

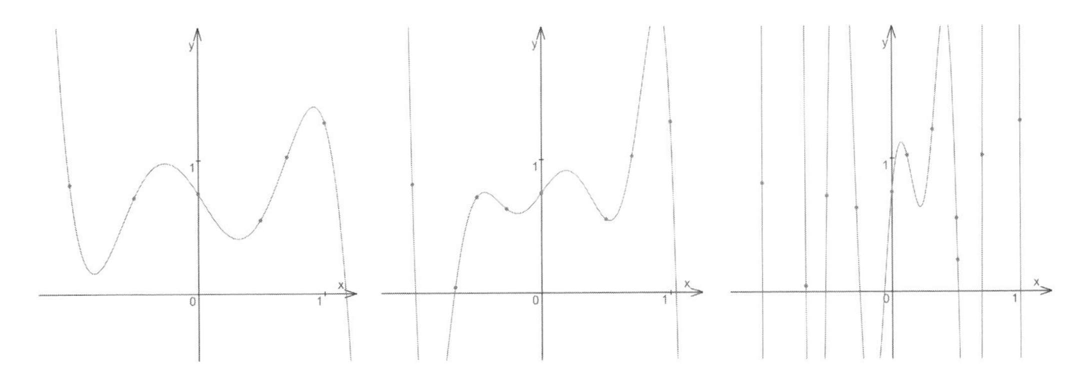

Figure 6.24 Polynomial oscillations usually get worse when more points are added to the data set.

For large sets of points to interpolate, polynomial interpolation may not be the best solution, considering Runge's phenomenon. We may want to try something else. In chapter 7, we'll look at polynomial splines, rational interpolants, and rational splines. But let's not give up on simple polynomials yet.

6.3.3 *How to lessen unwanted oscillations*

In general, we have no control over waving. All we have is the data set, and the data set governs how the interpolating function goes. But what if we're allowed to choose how to compose this set? What if we have the *x*s set for us beforehand and are allowed to choose only *y*s to set the function?

As it turns out, we have a way to place *x*s on the *x*-axis to minimize the oscillations. The "good" places for *x*s are called Chebyshev nodes, and you can find them as shown in figure 6.25:

1. Draw a circle with radius 1 and center at (0, 0).
2. Split the circle into 2N equal arcs.
3. A projection of each arc's end from the upper semicircle on the *x*-axis is a Chebyshev node.

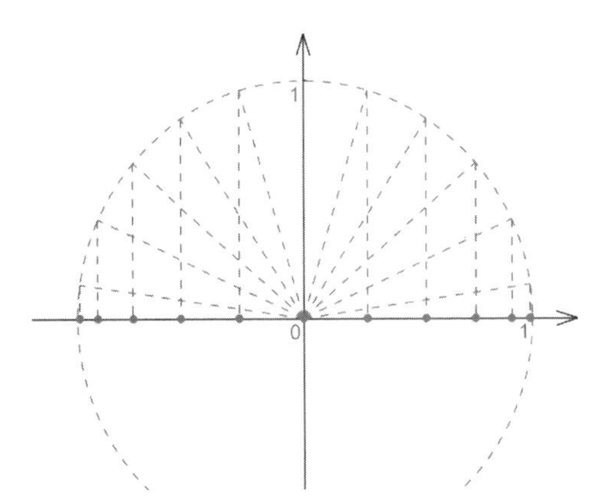

Figure 6.25 Chebyshev nodes and how to find them

Now you can assign every node a y, and the resulting polynomial won't oscillate—well, not much, and not in the (–1, 1) range anyway (figure 6.26).

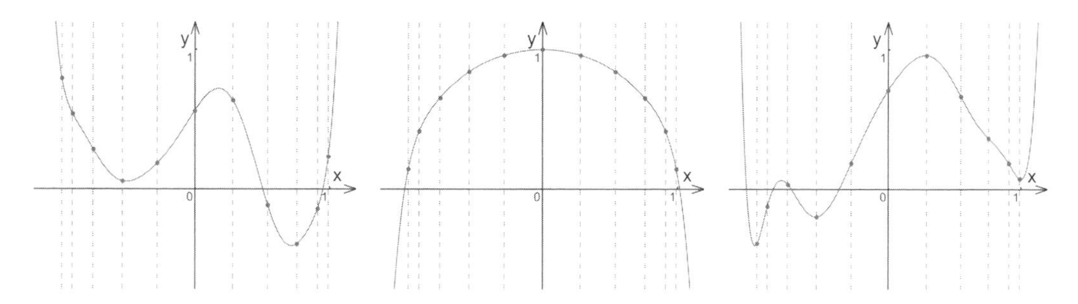

Figure 6.26 Interpolating a data set based on Chebyshev nodes minimizes the function's oscillation.

You can scale or move this polynomial to match the range you want, of course. You're not bound to (–1, 1). If you want to interpolate something in an arbitrary range, you transform your range to (–1, 1); then you interpolate your data and transform the polynomial back. We saw a similar trick earlier with the Taylor series, which is essentially a Maclaurin series with an extra pair of translations.

Although the use of polynomial interpolation in real-world applications is limited by their unpredictable oscillations, it's not completely excluded. There's a way to mitigate these oscillations, even if it restricts the way we chose data points. Also, polynomials are still the fastest smooth and continuous functions we can compute. They imply no branching and no division. Polynomials are economical.

But this isn't the only reason why polynomial interpolation is taught so widely in multiple data science and numeric analysis courses. Understanding polynomial interpolation helps you understand other ways of modeling better.

6.3.4 Lagrange interpolation: Simple, genius, unnecessary

When I learned about Lagrange interpolation, I already knew about the least-squares method (section 6.2.2) and Vandermonde matrix (section 6.3.1). I also knew that the N–1-degree polynomial for N points is unambiguous. Lagrange interpolation is also polynomial and also results in an N–1-degree polynomial for N points, so it brings nothing new to the picture. This new method seemed to me to be redundant.

Later, however, when I learned about basis splines and Bézier curves (and we'll get to them in chapter 7), I learned to appreciate the elegance of the Lagrange interpolation. It simultaneously uses and illustrates one groundbreaking idea: as you can compose any vector in space out of scaled basis vectors, you can compose any polynomial from scaled basis polynomials.

Let's see how this works. A Lagrange polynomial $L(x)$ for (x_i, y_i), $i = 1..N$ is a sum of N scaled basis polynomials $g_i(x)$ where every scale is the corresponding y_i value:

$$L(x) = \sum_{i=1}^{N} y_i g_i(x)$$

Each basis polynomial $g_i(x)$ is a product of all the distances between each data point x_j and the x divided by the distance between x_j and x_i, except when $j = i$:

$$g_i(x) = \prod_{j=1, j \neq i}^{N} \frac{x - x_j}{x_i - x_j}$$

These two formulas don't look like a polynomial, let alone like an interpolating one. So how do they compute? Let's look at the ratio from the basis polynomial's product:

$$\frac{x - x_j}{x_i - x_j}$$

When $x = x_i$, the ratio turns into

$$\frac{x_i - x_j}{x_i - x_j}$$

The nominator becomes the same as the denominator. Unless $i = j$—and in the basis polynomial, it never is by design—the ratio for $x = x_i$ is always 1.

Also, although it looks like a rational function, all the xs in it, apart from the polynomial argument, are already known. x_i and x_j come from the data set, so they turn into numbers in computation, making this fake rational function linear. We can write it slightly differently to make this fact obvious:

$$\frac{1}{x_i - x_j} x - \frac{x_j}{x_i - x_j}$$

When $x = x_j$, the function turns into

$$\frac{x_j}{x_i - x_j} - \frac{x_j}{x_i - x_j}$$

So for any $x = x_j$, the ratio is always 0.

A product of these linear functions masquerading as ratios would be a polynomial. In every basis polynomial, the product has N–1 factors, because we define a ratio for every $j = 1..N$ apart from $i = j$, so the resulting degree of this polynomial would be N–1, too.

In each data point x_j, apart from when $j = i$, the polynomial would be 0 because the corresponding factor will be 0, too. And when $j = i$, the polynomial would be 1 because all the factors will be 1 as well.

Let's see a concrete example. This time, let our data set be

$$(x_1, x_1) = (-3, 2.5)$$

$$(x_2, y_2) = (-1, -2.5)$$

$$(x_3, y_3) = (2, -2)$$

The first basis polynomial for this set is

$$g_1(x) = \frac{x+1}{-3+1}\frac{x-2}{-3-2} = \frac{x+1}{-2}\frac{x-2}{-5} = 0.1x^2 - 0.1x - 0.2$$

Let's evaluate it in x_1, x_2, and x_3:

$$g_1(-3) = 0.9 + 0.3 - 0.2 = 1$$

$$g_1(-1) = 0.1 + 0.1 - 0.2 = 0$$

$$g_1(2) = 0.4 + 0.2 - 0.2 = 0$$

By the way, in SymPy, the evaluation looks like this:

```
>>> (0.1*x**2 - 0.1*x - 0.2).evalf(subs={x:-3})
1.00000000000000
>>> (0.1*x**2 - 0.1*x - 0.2).evalf(subs={x:-1})
0.e-126
>>> (0.1*x**2 - 0.1*x - 0.2).evalf(subs={x:2})
0.e-125
```

evalf evaluates the expression in floating-point numbers, and subs tells SymPy which numeric value to substitute for x.

Ignoring the `e-125` part, which is a computational error, the first basis polynomial is 1 in x_1, and 0 in x_2 and x_3, as it should be. That's what makes it a basis polynomial by definition. All the basis polynomials for our data set look like figure 6.27.

Now if we multiply each polynomial by the corresponding value y_i, each polynomial will go through the respective data points (figure 6.28).

And because in all the other data points, the polynomials will be 0, when we add them up, we get a polynomial that goes exactly through every data point we have (figure 6.29).

That's how the Lagrange polynomial is the interpolating polynomial. We'd get the same polynomial if we used the Vandermonde matrix or the least-squares method to compute it.

Lagrange interpolation may not be the best choice for practical use, but as a learning tool, it's brilliant! It makes the idea of basis functions as simple as counting ones and zeroes.

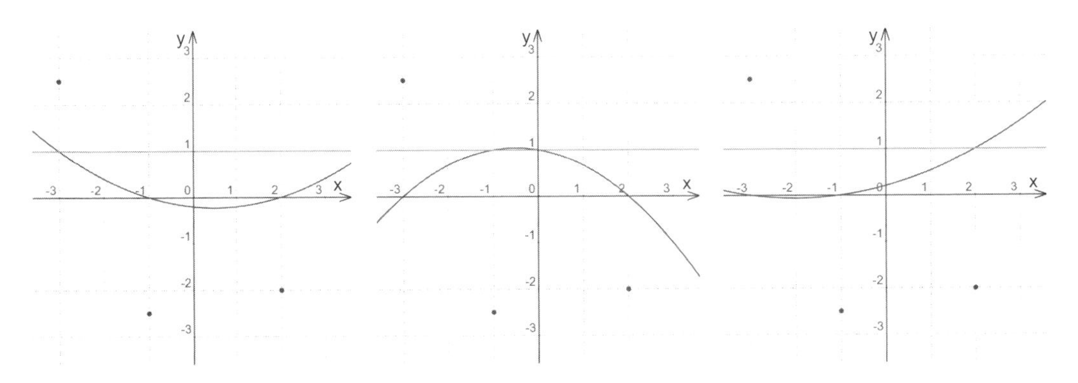

Figure 6.27 Every basis polynomial $g_i(x)$ equals 1 when $x = x_i$, and 0 for all the other points $x = x_j$, $j \neq i$.

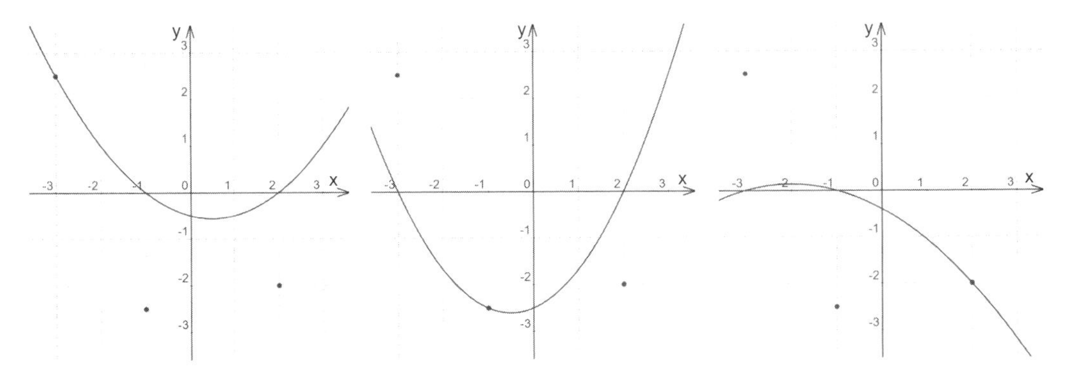

Figure 6.28 A basis polynomial multiplied by the y_i is still 0 for all the points x_j, $j \neq i$, but for x_i, it is now interpolating.

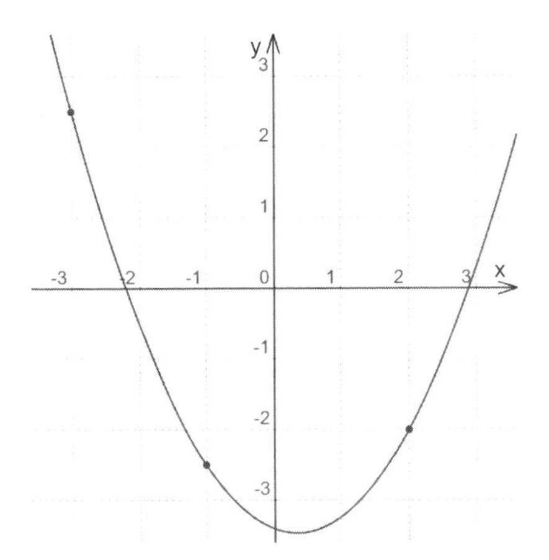

Figure 6.29 A sum of y_i-scaled basis polynomials is the interpolating polynomial.

6.3.5 Practical example: Showing the trend with interpolation

Let's interpolate a set of five data points with a polynomial. At this point, the origin of these data is irrelevant; all we want to do is to join them with a pretty line. Let's say the data we have is

```
xs = [1, 2, 3, 4, 5]
ys = [4, 2, 3, 2, 4]
N = len(xs)
```

The interpolating polynomial is a function. The *xs* values come from its domain or the input in programming terms, and the *ys* values go to its codomain, the output. To find the coefficients of the function, we need to fill the Vandermonde matrix and solve the linear system: `Vandermonde×Pol = ys`. In Python, this solution could look like listing 6.2 (ch_06/interpolate.py in this book's source code).

Listing 6.2 Computing the interpolating polynomial

```
Vandermonde = Matrix([[xs[i]**j for j in range(N)] for i in range(N)])
Ys = Matrix(ys)
a0, a1, a2, a3, a4 = symbols('a0 a1 a2 a3 a4')
Pol = Matrix([a0, a1, a2, a3, a4])

solution = solve(Vandermonde*Pol - Ys, (a0, a1, a2, a3, a4))
a = [value for key, value in solution.items()]
```

The Vandermonde matrix (annotates the `Vandermonde = ...` line)

The target function's values (annotates the `Ys = ...` line)

Polynomial's coefficients in symbolic form (annotates the `a0, a1, ...` and `Pol = ...` lines)

Solving the system for the polynomial's coefficients (annotates the `solution = ...` line)

Putting the coefficients in a list for convenient storage and evaluation (annotates the `a = ...` line)

Let's review the code step by step. The Vandermonde matrix is an $N \times N$ matrix that represents the polynomial in *xs*. As a reminder, in mathematical notation, it looks like this:

$$\begin{pmatrix} 1 & x_1 & \cdots & {x_1}^M \\ 1 & x_2 & \cdots & {x_2}^M \\ \vdots & \vdots & \vdots & \vdots \\ 1 & x_N & \cdots & {x_N}^M \end{pmatrix}$$

Here, we compose the matrix by using Python's native list comprehension feature. The matrix source here is a list of lists, where every nested list carries x_i^j for every j in the corresponding i row. To turn it into a matrix that SymPy can operate on, we pass it as a parameter to the `Matrix` constructor. Let's look at it again, now with every piece on its own line:

```
Vandermonde = Matrix([
                      [xs[i]**j
                               for j in range(N)]
                                                  for i in range(N)])
```

The `Ys` is also a matrix made out of a list `ys`. They're numerically equivalent, but they have different types. We pass the list into the Matrix constructor to convert types:

```
Ys = Matrix(ys)
```

Conventionally, lists are translated into SymPy Matrix as a column vector. If you need a row vector, you should make it a nested list like this: `Matrix([ys])`. But this time, a column vector is exactly what we need.

As in the approximation example, the next line is overhead from using SymPy instead of some numeric solver. We don't need symbols `a0` to `a4`, but we define them so that SymPy can fill them with numeric values:

```
Pol = Matrix([a0, a1, a2, a3, a4])
```

The solution we expect is the dictionary with polynomial coefficients. As a small reminder, they come from solving this equation:

$$\begin{pmatrix} 1 & x_1 & \cdots & {x_1}^M \\ 1 & x_2 & \cdots & {x_2}^M \\ \vdots & \vdots & \vdots & \vdots \\ 1 & x_N & \cdots & {x_N}^M \end{pmatrix} \begin{pmatrix} a_0 \\ a_1 \\ \vdots \\ a_M \end{pmatrix} = \begin{pmatrix} y_1 \\ y_2 \\ \vdots \\ y_N \end{pmatrix}$$

In SymPy form, solving this equation looks like this:

```
solution = solve(Vandermonde*Pol - Ys, (a0, a1, a2, a3, a4))
```

Then we gather the numeric values for the coefficients:

```
a = [value for key, value in solution.items()]
```

These coefficients correspond to the a_0, a_1, ..., a_{N-1} coefficients of the interpolating polynomial: $P(x) = a_{N-1}x^{N-1} + ... + a_2x^2 + a_1x + a_0$. The particular polynomial for our input data looks like figure 6.30.

It looks nice with the made-up data, but can it represent a real-world trend? Let's take the workout data from the approximation example:

```
xs = [1, 2, 3, 4, 5]
ys = [8 + 20/60., 8 + 4/60., 7 + 56/60., 7 + 52/60., 8 + 0/60.]
N = len(xs)
```

The interpolating polynomial for this data set looks like figure 6.31.

Showing data as a continuous function implies that we may not only interpolate, but also extrapolate data. *Extrapolation* is using the functional approximation to imply data beyond the interval our data points cover.

Be warned, though: if it's not backed with solid data science, a polynomial graph isn't a much better model than the mere line we saw in section 6.2.3. This graph suggests

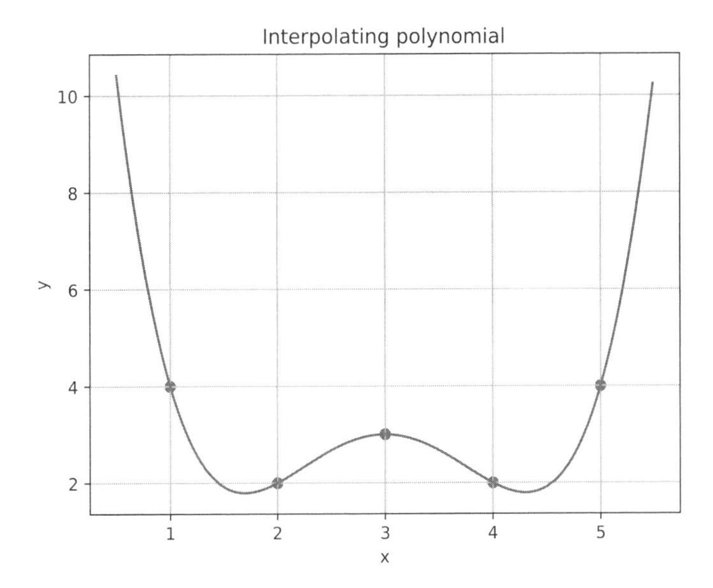

Figure 6.30 The interpolating polynomial for a set of points: (1, 4), (2, 2), (3, 3), (4, 2), (5, 4)

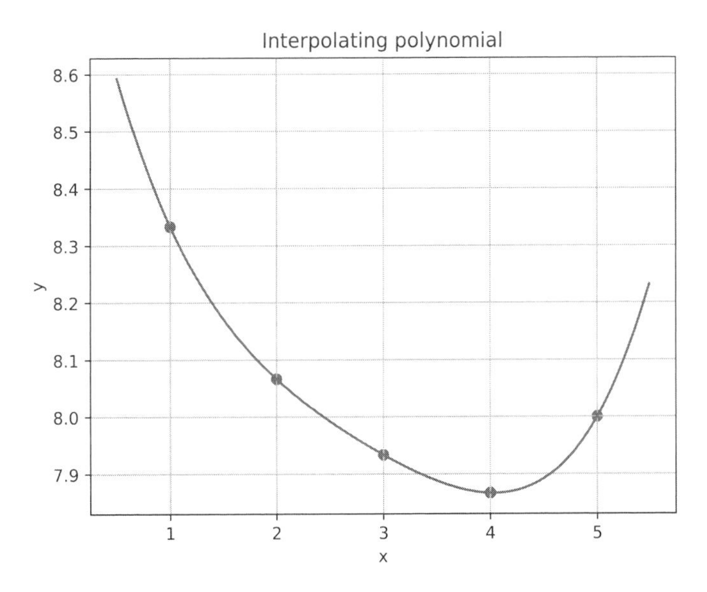

Figure 6.31 Interpolating real-world data doesn't necessarily show a trend, but at least it's precise.

not only that my sixth workout was terrible, but also that my results got worse with every session (figure 6.32).

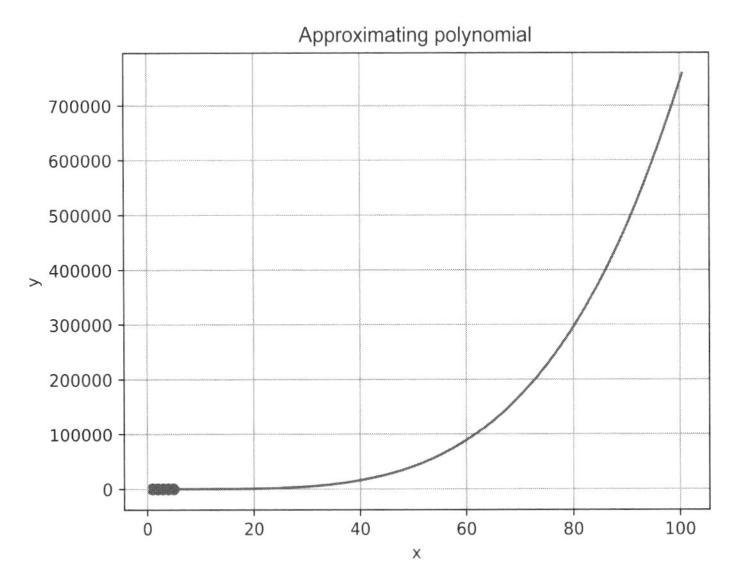

Figure 6.32 This plot suggests that my 100th workout session alone lasted roughly a year and a half.

I must repeat the same statement I made in the approximation exercise: there's a difference between showing data and seeing the driving forces behind the data. Interpolation can help you with only the former task.

Numeric computations in Python

So far, we've used SymPy to do all our math. But SymPy is meant to deal with symbolic computations—hence the *Sym* in the name. In our examples, we compute polynomial coefficients as numbers and don't care which symbols they're assigned to. So normally, we would have chosen NumPy for the job instead. The *Num* part of the name means "numeric," and it's a library for efficient numeric computations. Its website (https://numpy.org) has extensive documentation.

As an example, this is how you declare a matrix in NumPy:

```
a = np.array([[1, 2], [3, 5]])
```

This is how you declare a column vector:

```
b = np.array([1, 2])
```

And this is how you solve a system of linear equations:

```
x = np.linalg.solve(a, b)
```

Yes, it's that simple. Try it!

Or don't. Like Matplotlib, NumPy is only a suggestion. Even though the code examples in this book are in Python, you're free to use any other tools, languages, and libraries you know and love.

6.3.6 Section 6.3 summary

Interpolation is the special case of approximation in which the graph of the interpolating function goes strictly through all the interpolating points. We can use the method of least squares to find the interpolating polynomial, but using the Vandermonde matrix is more pragmatic. Lagrange interpolation is another alternative, but it's valuable mostly as a tool for understanding the idea of basis polynomials.

6.4 Practical example: Showing a trend with both approximation and interpolation

We've already seen that approximation and interpolation together with extrapolation have their downsides for representing data as a graph of a continuous function. Approximation may work better as a way of modeling, but it's inherently imprecise. Interpolation fits all the data properly, but the model it suggests may not reflect the process you want to illustrate.

Ideally, we may want to have something that works as an approximation globally but remains precise in a specific point or a few points.

6.4.1 The problem

Let's say that you have temperature data from the past five hours, and now you want to show how the weather might change in the following half an hour. Fitting an interpolating polynomial would be misleading, but at the same time, your approximating polynomial should start extrapolating from exactly the last data point because it's the temperature right now. It should start with zero error to be both helpful and precise.

Weather forecasting is a complicated task. Normally, you can't predict the weather by using a simple polynomial model; it's not against the law to try, though. If your predicting interval is short and the precision expectations are unassuming, why not?

So you have the temperature data for the past five hours, and you want to predict the temperature half an hour after the last data point. It's a sunny winter day, so a rapid change of weather is unlikely. Still, some midday oscillation occurs in the temperature graph, and that's what we'll try to catch.

We don't want to make separate columns for Fahrenheit and Celsius numbers, so we'll choose a temperature of around –40, which is where the scales meet: –40 Celsius is also –40 Fahrenheit. Table 6.2 shows the complete data set.

Table 6.2 Temperature per hour in a hypothetical winter midday

Hour	Temperature (pretend that it's in your favorite units)
10:00	–42
11:00	–41
12:00	–38
13:00	–37
14:00	–37

Because we're trying to show a midday temperature bump, the linear approximation wouldn't work. We need at least a quadratic polynomial. For our data, the approximating quadratic polynomial looks like figure 6.33.

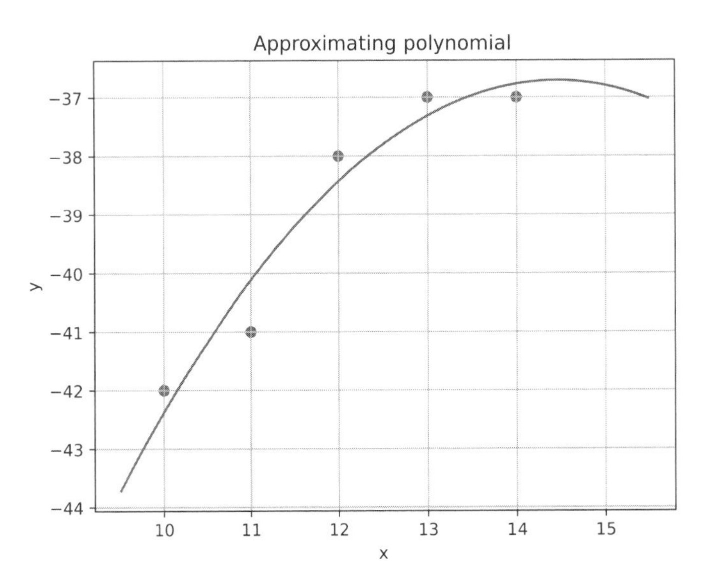

Figure 6.33 It gets warmer in the middle of the day, but then it gets cold again.

There's a problem with this picture: it shows that by 2 p.m., the temperature is approximately –36.7 degrees, but we know for certain that it's exactly –37. The picture shows the trend right, but it misses the most important data point, so it doesn't give us a reliable prediction.

6.4.2 The solution

No single formula makes approximation and interpolation work together. Once again, the problem stops being mathematical and starts being engineering. We have to devise our own way to make an approximating polynomial hit the last data point with good enough precision.

One way to do so would be to use the least-squares method but reenter the last point multiple times. The least-squares method minimizes the total squared error of the model, and each reentrance of the last point will make its error more pronounced compared with the others. The polynomial will tend to come closer to the last data point, as now the polynomial deviation at the last point has more weight than at other points.

Let's add the last point ten more times to see what will happen. The code is almost the same as in listing 6.1, but with the addition of matrices for extra reentrances:

```
from sympy import *

xs = [10, 11, 12, 13, 14]
ys = [-42, -41, -38, -37, -37]
N = len(xs)
M = 2

LS = Matrix([[sum([xs[k]**(i+j) for k in range(N)])
                                    for j in range(M+1)]
                                        for i in range(M+1)])
B = Matrix([sum([ys[k]*(xs[k]**i) for k in range(N)])
                                    for i in range(M+1)])

last_point_reentrance = 10                              <-- The last point is added to the equation ten more times.
LS_more = Matrix([[last_point_reentrance*xs[N-1]**(i+j)
                                    for j in range(M+1)]
                                        for i in range(M+1)])
B_more = Matrix([last_point_reentrance*ys[N-1]*xs[N-1]**i
                                    for i in range(M+1)])

a0, a1, a2 = symbols('a0 a1 a2')
Pol = Matrix([a0, a1, a2])

solution = solve((LS+LS_more)*Pol - (B+B_more), (a0, a1, a2))
a = [value for key, value in solution.items()]
```

The `LS_more` and `B_more` resemble the `LS` and `B`, but instead of summing all the points' powers, they sum the same for the last point in the data set. Now when we add them to the respective counterparts, we get the polynomial shown in figure 6.34.

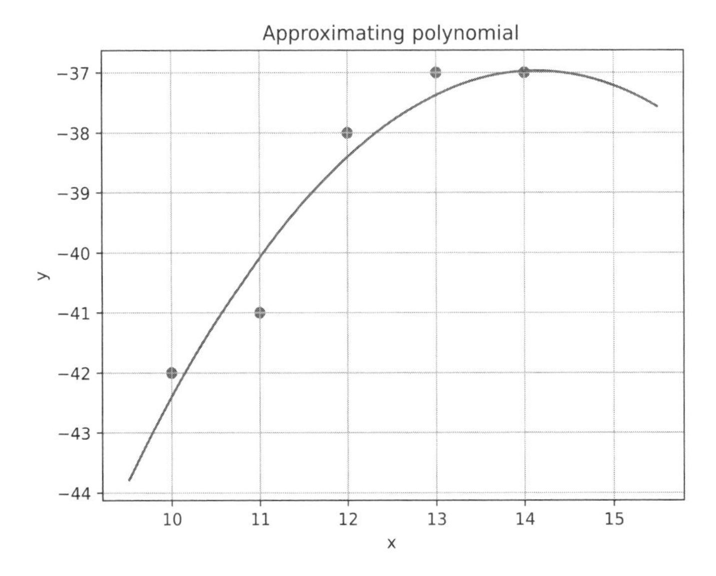

Figure 6.34 This polynomial fits the last point better but still misses it a bit.

We can't make the polynomial go exactly through the point, but if we settle on the precision we want to achieve beforehand, we might keep adding the last point until the polynomial reaches the target precision fitting that point. This algorithm translates to the Python code in listing 6.3 (ch_06/both_approximate_and_interpolate.py in the book's source code).

Listing 6.3 Computing an approximating polynomial that interpolates one point

```
from sympy import *

xs = [10, 11, 12, 13, 14]
ys = [-42, -41, -38, -37, -37]
N = len(xs)
M = 2

LS = Matrix([[sum([xs[k]**(i+j) for k in range(N)])
                                            for j in range(M+1)]
                                                for i in range(M+1)])
B = Matrix([sum([ys[k]*(xs[k]**i) for k in range(N)])
                                                for i in range(M+1)])

last_point_reentrance = 0
a = [0, 0, 0]

def P(x, a):
    return sum([a[i]*x**i for i in range(len(a))])

precision = 0.01
while abs(P(xs[N-1], a) - ys[N-1]) > precision:
```

The algorithm tries adding the last point again and again until the precision expectations are met.

```
LS_more = Matrix([[last_point_reentrance*xs[N-1]**(i+j)
                                              for j in range(M+1)]
                                                  for i in range(M+1)])
B_more = Matrix([last_point_reentrance*ys[N-1]*xs[N-1]**i
                                                  for i in range(M+1)])

a0, a1, a2 = symbols('a0 a1 a2')
Pol = Matrix([a0, a1, a2])

solution = solve((LS+LS_more)*Pol - (B+B_more), (a0, a1, a2))
a = [value for key, value in solution.items()]
last_point_reentrance += 1
```

On each iteration, exactly one new last point is added. But you can add more if you want the algorithm to converge faster.

This iterative approach results in a graph that almost hits the point. There's some error, but it's less than the naked eye can notice, and because we're talking about data visualization, the plot kind of works (figure 6.35).

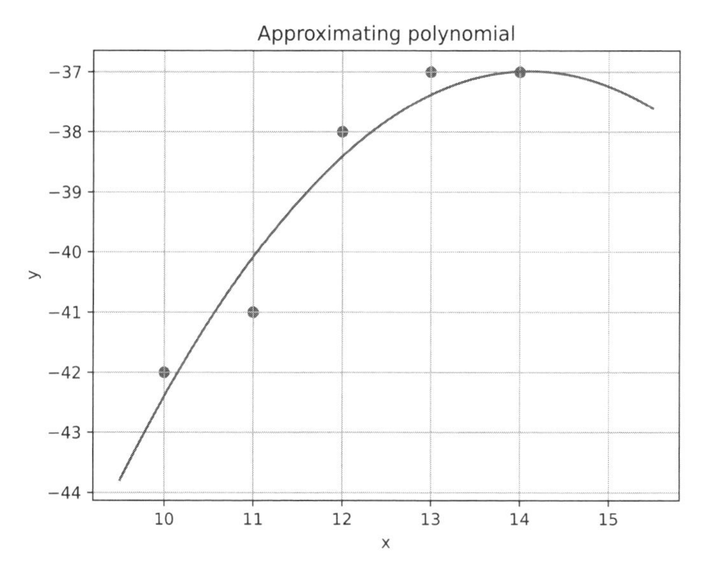

Figure 6.35 The approximating polynomial hits the last data point within the predetermined precision.

6.4.3 Section 6.4 summary

You can use polynomial approximation and interpolation to *show* data obtained as a set of points. When you want to show the general trend, you can use approximation. When you want to show data between data points, use interpolation. With the power of engineering, you can make them both work together.

Whether you should use polynomials to *see* the driving force behind this data, however, is not a question that geometry can answer. The choice of a model that represents your data falls in the domain of data science; geometry covers only the properties of a curve you draw.

> **SEE ALSO** *Math and Architectures of Deep Learning*, by Krishnendu Chaudhury (http://mng.bz/eJ2q), has a whole chapter about how approximation works in the context of neural networks. If you want to learn more about how approximation is used in machine learning applications, please see that book.

6.5 Exercises

Exercise 6.1 Which of these functions are polynomials?

1. $y = x^3 + 3x^2 - x - 1$
2. $y = (x + 3)(x + 1)(x - 1)$
3. $y = x(x(x + 3) - 1) - 1$
4. $y = x^2 + 3x - 1 - \frac{1}{x}$

Exercise 6.2 Which of these functions can be approximated by a Mclaurin series?

1. $y = \frac{1}{x}$
2. $y = \log(x)$
3. $y = e^{2x} + 2e^x$
4. $y = |x|$

Exercise 6.3 A set of four points in the general case can be interpolated by

1. a linear function.
2. a quadratic function.
3. a cubic function.
4. a quadric function.

Exercise 6.4 Please find the first three terms of a Mclaurin series for this function:

$$y = \sin(x) + \cos(x)$$

Exercise 6.5 An open question: How can you mitigate Runge's phenomenon without constraining *x*s to Chebyshev nodes?

6.6 Solutions to exercises

Exercise 6.1 The first three are polynomials. The first is a polynomial in its canonical form, the second one is the same polynomial factorized, and the third one is the same polynomial written by Horner's scheme.

Whether the last one is a polynomial isn't a trivial question. The term *polynomial* is also used in a broader sense to describe any function that consists of a sum of similar parts. There are Fourier polynomials that consist of sines and cosines instead of x^is, for example. In that broader sense, the fourth function from the exercise might also be some kind of a polynomial.

But for us, it's important to distinguish polynomial functions that consist of additions and multiplications from the functions that also include division. The functions that represent a polynomial division are called rational functions:

$$x^2 + 3x - 1 - \frac{1}{x} = \frac{x^3 + 3x^2 - x - 1}{x}$$

Rational functions are also important in geometry, and we'll get back to them in chapter 7. For now, let's agree to consider the fourth function from the example to be a rational function but not a polynomial.

Exercise 6.2 Only the third one. The McLaurin series is a special case of the Taylor series that reassembles the function from its derivatives in x = 0. The first two examples from the exercise aren't defined in 0, and the derivative of the fourth one is discontinuous in 0.

Exercise 6.3 By both cubic and quadric functions. Normally, you should interpolate *N* points with an (*N*–1)-degree polynomial, which leaves out linear and quadratic functions. Sure, in a special case, such as when the points are on the same line, they can be interpolated by a linear function, but the problem states "in the general case," so we can't assume any lucky coincidences.

As for the quadric, a degree of interpolating polynomial can technically exceed *N*–1 for *N* points, although it makes the interpolation ambiguous, and nobody likes that.

Exercise 6.4 1, *x*, and $\frac{-x^2}{2}$. You can find them with pen and paper or by using a SymPy function `series` like this one: `series(sin(x) + cos(x), x0 = 0, n = 3)`.

Exercise 6.5 There's no single correct answer; all the solutions you can come up with are good. I can give you a hint, though: with (*N*–1)-degree polynomial for *N* points, the interpolation is unambiguous. You can't make this polynomial less oscillating. But if you raise the polynomial degree, you lose the unambiguity. Use this effect to your advantage.

Summary

- A polynomial is a function that consists only of additions and multiplications. It's economical for storage and computation.
- An approximation is about finding a function that approximately fits the set of points or another function.
- You can approximate a differentiable function with a polynomial by using the Taylor or Maclaurin series.

- You can approximate a set of data points by using the least-squares method.
- Interpolation is the special case of approximation in which the interpolating function goes strictly through all the interpolating points.
- Normally, a set of N data points is interpolated with $(N-1)$-degree polynomial, but using higher degrees is also possible.
- You can use the least-squares method to interpolate a data set, but using the Vandermonde approach is more practical computationwise.
- Approximation (and interpolation in particular) helps us build nice graphs, but their applicability for real-world modeling is a matter of data science, not geometry.

7 Splines

This chapter covers

- Making interpolating functions from points and from differential and integral conditions
- Understanding and composing Bézier splines
- Understanding and composing NURBS

In mathematics, a *spline* is a compound function, often continuous and smooth, made out of other simpler functions. You can stick two parabolas together and make yourself a quadratic polynomial spline. It's as simple as that (figure 7.1).

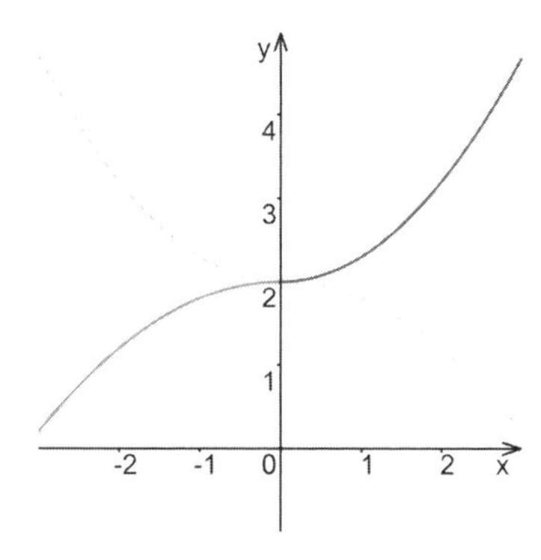

Figure 7.1 For $x < 0$, $y = -0.2x^2 + 2$; for $x \geq 0$, $y = 0.3x^2 + 2$. This is already a spline function.

In the real world, you can see splines everywhere—not only in graphic design tools, virtual reality, computer games, geoinformation systems, and other software, but, interestingly, in almost every physical piece ever produced. Because before the software was even invented, engineers used metal splines to make smooth lines on technical drawings. Almost every human-made smooth surface you see around you first appeared as a spline.

What we call a *Bézier curve* is a mathematical model of a metal spline. But splines are much more than that. *NURBS* is an acronym that stands for "nonuniform rational basis spline," and this kind of spline generalizes Bézier curves, giving us many more options to control the properties and shape of the modeled curve.

This chapter explains Bézier curves and splines based on what you've learned in the previous chapters. It also explains NURBS and how to get to them from Bézier. Moreover, this chapter proposes thinking tools you can use to create your own custom functions and your own custom splines when Bézier and even NURBS aren't enough.

7.1 Going beyond the interpolation

Let's start with a recap. With polynomial interpolation, we find a polynomial that runs through the set of points. If a polynomial $P(x)$ runs through a point (x_i, y_i), this means $P(x_i) = y_i$ which gives us a nice equation:

$$a^n x_i^n + a_{n-1} x_i^{n-1} \dots + a_2 x_i^2 + a_1 x_i + a_0 = y_i$$

The equation looks polynomial because of all the x^ns, but remember, we're hunting for the coefficients a_0, a_1, a_2, and so on, not x. For every known x_i, all the x_i^ns are concrete numbers, not variables. If we replace *x*s and *y*s with corresponding numbers, the equation becomes obviously linear. The following example is a quadratic polynomial going through the point (2, 3):

$$4a_2 + 2a_1 + a_0 = 3$$

Generally, we need N equations to find N coefficients, and vice versa. If a polynomial runs through five points, for example, it has to have five coefficients. A polynomial of the fourth degree has exactly five coefficients, so except for exceptions, we can make a fourth-degree polynomial run through five points.

The interpolating polynomial of (N–1) degree for the set of N points is unambiguous. Consequently, this means that we can't choose how it looks or how bumpy it is between the interpolating points. On the other hand, we can raise the degree, forfeit the unambiguity, and have more freedom. And we can trade this freedom for more restrictions that don't have to come from points alone.

7.1.1 Making polynomial graphs with tangent constraints

To put two parabolas back to back smoothly (as in figure 7.1, which opens the chapter), we need to make sure that the tangent line for the first parabola at a conjunction point has the same slope as the tangent line of the second parabola. Luckily, this slope could be measured and computed numerically. The tangent of the line's angle toward the *x*-axis is the derivative of the polynomial at a point.

Let's say the numerical value of the slope of the tangent at the point x_i is t_i. We've already been differentiating polynomials, and a derivative of a polynomial is also a polynomial; it's simply of a lesser degree. So we can differentiate our target polynomial and make this "polynomial has the slope of the tangent t_i in x_i" statement into an equation:

$$na_n x_i^{n-1} + \ldots + 2a_2 x_i + a_1 = t_i$$

This equation is also linear, of course. Remember, our variables are coefficients a_i. If a quadratic polynomial has the tangent value 1 in $x = 3$, the equation to reflect that is

$$(a_2 x^2 + a_1 x + a_0)' = (2a_2 x + a_1) = 6a_2 + a_1 = 3$$

We can put this equation, along with "the polynomial runs through (x_i, y_i)," in a single system. Solving this system gives us coefficients for a polynomial that is both interpolating and has its tangent vectors constrained. In figure 7.2, a cubic polynomial runs through a pair of points, and its tangent lines in these points are constrained to –45 degrees to the *x*-axis.

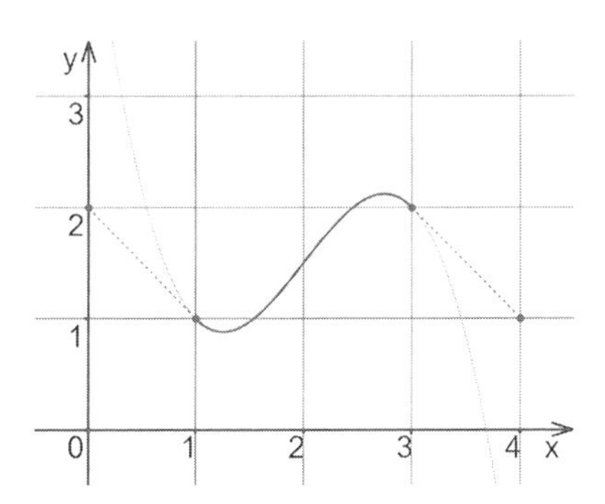

Figure 7.2 A cubic polynomial set by a pair of points and a pair of tangents

The equations for the polynomial in figure 7.1 are as follows:

$$P(x_1) = 1a + 1b + 1c + 1d = 1$$

$$P(x_2) = 27a + 9b + 3c + d = 2$$

$$P(x_1)' = 3a + 2b + 1c = tan(\text{-}45°) = \text{-}1.0$$

$$P(x_2)' = 27a + 6b + 1c = tan(\text{-}45°) = \text{-}1.0$$

The first pair of equations states that the polynomial runs through points (1, 1) and (3, 2). The second pair states that the polynomial derivative in the first point is –1 and that it's also –1 in the second point.

If we want our target polynomial curve to have a specific curvature at a point, we can account for that curvature as well. The curvature is connected to the second derivative of a function, so we can differentiate the target polynomial twice and make a new equation for the curvature:

$$n(n-1)\,a_n x_i^{n-2} + \ldots + 2a_2 = c_i$$

Certainly, this equation is also linear. If a quadratic polynomial has its second derivative in x = 4 as 0, this fact results in the following equation:

$$(a_2x^2 + a_1x + a_0)'' = (2a_2x + a_1)' = 2a_2 = 0$$

You can do the same to the polynomial derivative of any degree or an integral, for that matter. You can turn the "polynomial in x_i has the mth derivative d_{mi}" statement into a linear equation:

$$P^{(m)}(xi) = d_{mi}$$

I've run out of coefficients in my example, but you can imagine that if you want to set the 256th derivative of a polynomial in a point, you can do that by taking the 256th derivative of a polynomial and putting it on the left side of the equation. Your polynomial has to be of 256 degrees minimum, though; otherwise, you'll run out of coefficients, too.

But is it that important to have some predetermined derivatives in a point or two? We have to spend polynomial coefficients for every new constraint, so the polynomial becomes larger and heavier to compute. Adding new equations to the system isn't completely free either. So what are we paying for?

Having predetermined derivatives is important when we put functions back to back and make them look like a single continuous function, which is the main idea behind splines in general. But before we go there, let's talk about a practical example in slightly another area.

7.1.2 Practical example: Approximating the sine function for a space simulator game

In 2006, I was working on a 3D engine for a space simulator game. The funny thing about game engines is that they never have enough performance. If by pure miracle, you manage to make your engine twice as fast by Wednesday, by Thursday, a game designer will bring in a new graphic effect that makes the engine run as slow as it was on Tuesday. Game companies don't sell benchmarks; they sell games. So good game designers use all the means they have to make games sellable.

We were constantly optimizing every bit of our engine, and at some point, we got to trigonometry. Functions such as sine and cosine are used heavily to describe circular motion, and as you can imagine, space simulation is all about circular motion. Planets, asteroids, satellites, and even plasma particles all dance the same trigonometric waltz. Luckily, in a game and not a real simulator, they're allowed to be a little sloppy. We can trade some precision for speed. One way would be to replace sine and cosine, which use relatively slow processor instructions, with cheaper polynomial models. We can do that with interpolation, for example.

We can compute several points of a real sine function and make a polynomial run through them. We already did this in chapter 6, but let's reiterate. The script that gets us polynomial coefficients for a set of points looks like the following listing (ch_07/sine_ipol.py in the book's source code).

Listing 7.1 Modeling the sine function with polynomial interpolation

```
from math import pi
from math import sin
from sympy import *

xs = [(i / 4.) * pi  - pi / 2. for i in range(5)]
ys = [sin(xi) for xi in xs]
N = len(xs)

Vandermonde = Matrix([[xs[i]**j for j in range(N)] for i in range(N)])
Ys = Matrix(ys)
a0, a1, a2, a3, a4 = symbols('a0 a1 a2 a3 a4')
Pol = Matrix([a0, a1, a2, a3, a4])

solution = solve(Vandermonde*Pol - Ys, (a0, a1, a2, a3, a4))
ais = [value for key, value in solution.items()]
```

Let's spread xs evenly over the range from –pi/2 to +pi/2. (annotation on the `xs` line)

ys will be the values of sin(x). (annotation on the `ys` line)

The coefficients of the interpolating polynomial will come from solving the Vandermonde equation (chapter 6). (annotation on the `Vandermonde` line)

This approach works, although not brilliantly (figure 7.3).

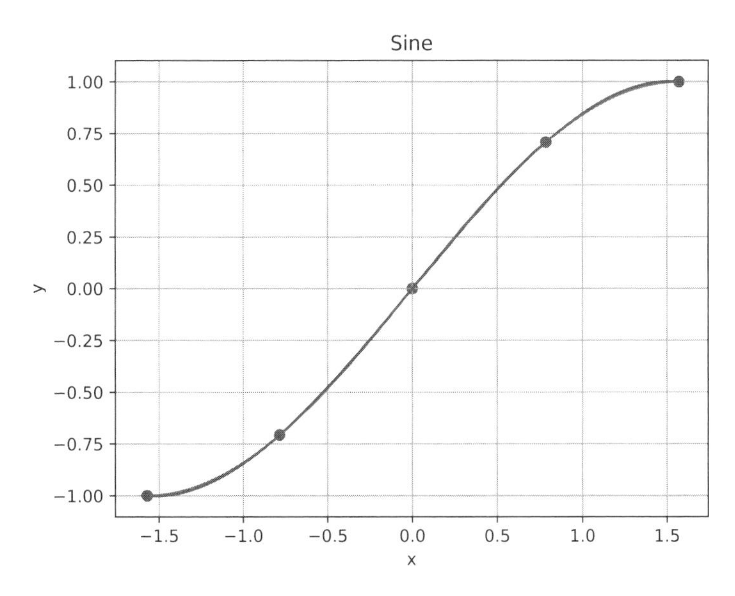

Figure 7.3 Modeling the sine function with polynomial interpolation seems fine to the naked eye.

The mean error of such modeling is more than 0.003. The error isn't spread evenly, of course, and in the five points we used to build the polynomial, the error is exactly 0. But this precision doesn't take into account the derivatives or the tangent lines. This means, among other things, that when we use this sine model to draw a circle, the circle doesn't appear to be perfectly smooth (figure 7.4).

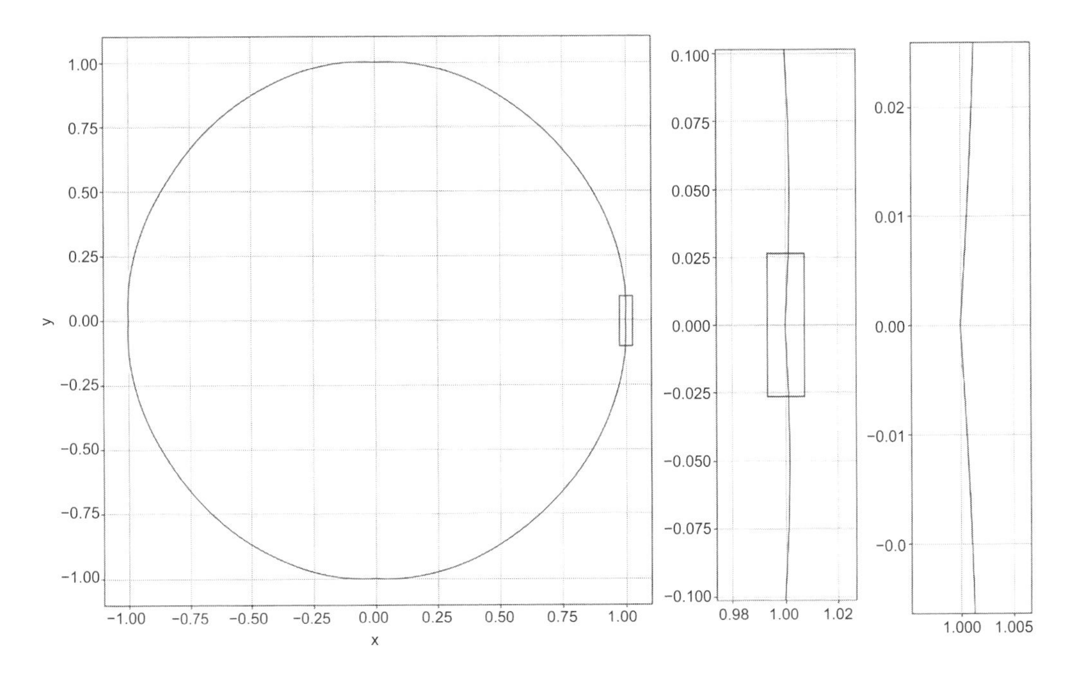

Figure 7.4 An imperfect circle made with the polynomial model of sine. As we zoom in to (1, 0) point on the right, we see the imperfection more clearly.

We could add precision to our model by adding more points to interpolate from, of course, but then the computational cost of the model would rise, and we'd do the polynomial modeling because of how cheap it is. So we need to make a more precise model with roughly the same number of coefficients.

Perhaps we can try the power series. Chapter 6 explained how to build a Maclaurin series for any function you want by computing its derivatives at 0 and dividing them by the factorial of a corresponding degree. This chapter, however, proposes an alternative way of finding the formula for a sine's power series: Googling it. The formula is

$$\sin(x) \approx x - \frac{x^3}{3!} + \frac{x^5}{5!} - \frac{x^7}{7!} + \ldots$$

Because we already have the formula and don't have to compute it symbolically with SymPy, we can program it in plain Python directly, as in the following listing (ch_07/sine_series in the book's source code).

Listing 7.2 Modeling the sine function with a Maclaurin series

```
from math import factorial

ais = []
for i in range(4):
    n = i*2+1
    ais += [0, (1 if i % 2 == 0 else -1) / factorial(n)]

print(ais)
```

We need 0 coefficients for even degrees, too, so we add coefficients as pairs of [0, 1/n!].

This model is more precise. Its relative error is as little as 0.000018. Compared with the 0.003 from before, this precision is a huge improvement, and it results in a much nicer circle, too (figure 7.5).

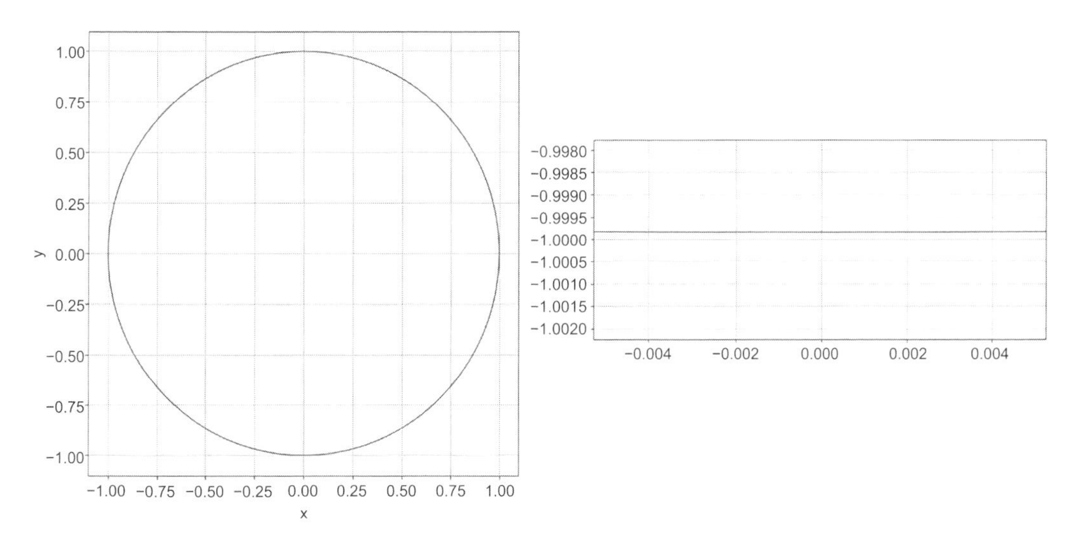

Figure 7.5 A circle made with a Maclaurin series instead of a real sine. Zooming into the circle's bottom shows only a minuscule imperfection.

There's still a small imprecision at the bottom, though. It comes from the fact that the series is being computed for the zero point, and as the model goes away from the 0, it becomes less and less precise.

This bottom-most point is important, however. If you put several circles in a vertical line according to their mathematical formulas, you wouldn't expect a gap between them. And although the gap is negligible for small circles, in a space simulator, where orbits of moons and satellites could easily be in thousands of kilometers, this imprecision may become noticeable. So how do we make a lightweight sine model that

- guarantees a smooth circle;
- is precise at angles of $\frac{-\pi}{2}$, 0, and $\frac{\pi}{2}$; and
- has better overall precision than the piece of power series?

The answer, of course, is to turn the requirements into equations and feed them to SymPy. Observe!

First, let's introduce a bunch of coefficients for the polynomial and the variable x:

```
x, a, b, c, d, e, f = symbols('x a b c d e f')
```

A small reminder, as with polynomial interpolation from chapter 6, for our system, a, b, c, and so on are going to be variables, and x will be a numeric parameter. By solving the system, we'll get the coefficients for the polynomial that models the sine such as this:

```
sine = a*x**5 + b*x**4 + c*x**3 + d*x**2 + e*x + f
```

Now let's introduce the model's derivative. You can do the formula by hand, of course; it's not too hard. But to lessen the chance of a typo, let's use SymPy here as well. Obtaining the derivative formula takes a single function call:

```
sine_d = diff(sine, x)
```

Now for the equations. We know that the sine(x) is 0 in $x = 0$, 1 in $x = \frac{\pi}{2}$, and –1 in $x = \frac{-\pi}{2}$. Usually, SymPy is used for symbolic computations, so normally, it doesn't need any specific numbers to do its job. But you can still specify the numbers for the expressions you want by using the `subs` method:

```
sine.subs(x, 0),
sine.subs(x, pi / 2) - 1,
sine.subs(x, -pi / 2) + 1,
```

The `subs` here stands for *substitution:* it substitutes a specific numeric value for the parameter x. This substitution turns a parametric symbolic equation into a numeric one.

The first three equations guarantee the model's precision in its most important points: $\frac{-\pi}{2}$, 0, and $\frac{\pi}{2}$. Now let's add some smoothness. Smoothness comes from the continuity of the first derivative, which in our case means that the derivative should be exactly 0 at the $\frac{-\pi}{2}$ and $\frac{\pi}{2}$. We also know that the derivative of sine(x) is 1 in $x = 0$, so we'll throw this fact in, too:

```
sine_d.subs(x, pi / 2),
sine_d.subs(x, -pi / 2),
sine_d.subs(x, 0) - 1,
```

Now we have six equations for six coefficients. Let's make them into a system and solve it. The whole script looks like the following listing (ch_07/sine_naive.py in the book's source code).

Listing 7.3 Modeling the sine with a system made from values and derivatives in points

```
from sympy import *
from math import pi

x, a, b, c, d, e, f = symbols('x a b c d e f')

sine = a*x**5 + b*x**4 + c*x**3 + d*x**2 + e*x + f
sine_d = diff(sine, x)

the_system = [
    sine.subs(x, 0),
    sine.subs(x, pi / 2) - 1,
    sine.subs(x, -pi / 2) + 1,
    sine_d.subs(x, 0) - 1,
    sine_d.subs(x, pi / 2),
    sine_d.subs(x, -pi / 2),
]

res = solve(the_system, (a, b, c, d, e, f))

for var, value in res.items():
    print(var, value)
```

- The symbols are the variable x and the polynomial coefficients.
- The sine is modeled by this polynomial.
- The derivative of the model
- The system consists of three "the function runs through a point" equations and three "the function has this derivative in a point" equations.

The program prints out these coefficients:

```
a 0.00740306120838731
b 0.0
c -0.165538780474714
d 0.0
e 1.00000000000000
f 0.0
```

You might notice that this model has all the even coefficients as 0, which is completely normal for odd or antisymmetric polynomials—the ones for which $P(x) = -P(-x)$. The property works the other way too: a polynomial with all nonzero coefficients being odd is antisymmetric.

Sine is antisymmetric, so it only makes sense to model it with an antisymmetric polynomial. Let's do that, restating our model to not include the even coefficients. And because our model is going to be odd or antisymmetric anyway, we won't bother supplying it the equations for $\frac{-\pi}{2}$ at all. $P\left(\frac{-\pi}{2}\right) = -P\left(\frac{\pi}{2}\right)$. Our model will become simpler to compute, and the system to compute this model will have fewer equations. It's a win-win situation, as shown in listing 7.4 (ch_07/sine_odd.py in the book's source code)!

Listing 7.4 Modeling the odd sine without specifying anything for x < 0

```
from sympy import *
from math import pi

x, a, b, c = symbols('x a b c')          <-- Now we need fewer coefficients.

sine = a*x**5 + b*x**3 + c*x             <-- But the polynomial is still
sine_d = diff(sine, x)                       of the fifth degree . . .

the_system = [                           <-- . . . and the system
    sine.subs(x, pi / 2) - 1,                consists of only
    sine_d.subs(x, 0) - 1,                   three equations.
    sine_d.subs(x, pi / 2),
]

res = solve(the_system, (a, b, c))

for var, value in res.items():
    print(var, value)
```

This model is odd, tiny, and fast, but also not especially accurate. Its mean error is 0.00017, which is better than 0.003 for the model we got by interpolation but almost ten times worse than 0.000018 from the model we got with the power series.

How can we make our model more precise? Can we make it more like sine without adding more points or taking more derivatives? Once again, it turns out that we can.

One secret ingredient never pops up with interpolation or series: integration. Remember, we briefly touched on that topic in chapter 5. Integration is an operation that kind of undoes differentiation. You can differentiate the sine function infinitely because its differentiation goes in a loop:

$$\sin(x) \rightarrow \text{cosine}(x) \rightarrow -\text{sine}(x) \rightarrow -\text{cosine}(x) \rightarrow \text{sine}(x) \ldots$$

For the integration, the loop simply goes the other way. Then the integral of sine is minus cosine, minus cosine integrates into minus sine, minus sine integrates into cosine, and the cosine integrates into sine:

$$\sin(x) \rightarrow -\text{cosine}(x) \rightarrow -\text{sine}(x) \rightarrow \text{cosine}(x) \rightarrow \text{sine}(x) \ldots$$

I won't get deep into the geometrical meaning of integration, but note one interesting thing. The difference between the integral of a function between point b and point a is the area between the function graph and the x-axis on the $[a..b]$ interval. In our sine model, the area under the sine on $\left[0..\frac{\pi}{2}\right]$ is $-\cos\left(\frac{\pi}{2}\right) - (-\cos(0)) = 0 - (-1) = 1$ (figure 7.6).

This "integral is area" notion is so cool that it probably deserves to be put into an equation and thrown into the system along with the model's values and derivatives at key points. Let's do that, as in listing 7.5 (ch_07/sine_integral_odd.py in the book's source code).

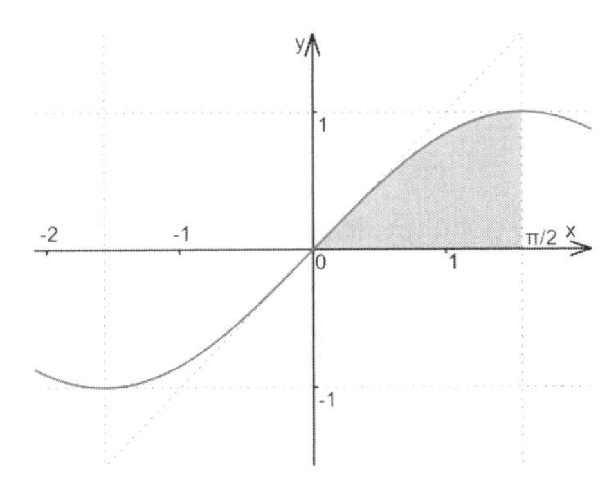

Figure 7.6 The area under the sine on [0, π/2] equals 1.

Listing 7.5 Modeling the sine with derivatives and integrals

```
from sympy import *
from math import pi

x, a, b, c, d = symbols('x a b c d')

sine = a*x**7 + b*x**5 + c*x**3 + d*x
sine_d = diff(sine, x)
sine_i = integrate(sine, x)

the_system = [
    sine.subs(x, pi / 2) - 1,
    sine_d.subs(x, 0) - 1,
    sine_d.subs(x, pi / 2),
    sine_i.subs(x, pi / 2) - sine_i.subs(x, 0) - 1
]

res = solve(the_system, (a, b, c, d))
```

As SymPy can differentiate your expression with diff, it can integrate your expression with integrate.

This line states that the area under the model's graph on [0..p/2] should be 1.

Now, this model is fully comparable to the one we got from the series in terms of complexity. It has four coefficients, like the one from the series. And the coefficients apply only to even degrees, like with the one from the series. The models are almost exactly alike, apart from some numeric differences in the coefficients (see table 7.1).

Table 7.1 Coefficients for the series and the new model with the integral equation added

Member	Coefficient from the series	Coefficient from the new model
x	1.00000000000000	1.00000000000000
x3	–0.16666666666666666	–0.166651012143690
x5	0.008333333333333333	0.00830460224186793
x7	–0.0001984126984126984	–0.000182690409228785

The new model, however, is more precise than the power series. Its mean error is 0.000001, which is more than ten times smaller than the 0.000018 from before.

Does this new model guarantee a smooth circle? Yes. The derivatives on the ends of our model are exactly 0. This means that you can replicate the model through the whole $\mathbb{R}$, and in every joint, the derivative on the right will be exactly as the derivative on the left: 0.

Is the model precise at $\frac{-\pi}{2}$, 0, and $\frac{\pi}{2}$? Yes. We set its values in 0 and $\frac{\pi}{2}$ explicitly, and the value in $\frac{-\pi}{2}$ comes automatically from its oddness.

Does the model have better precision than the series? Yes, as we've established.

Mission accomplished. Congratulations!

Measure your optimizations

In 2006, central processing units (CPUs) were simpler, and graphics processing unit (GPU) programming was in its infancy. Approximating a heavy instruction with a handful of simple ones made sense both theoretically and empirically. Now, however, you can't rely on a thought experiment alone; you have to measure everything. Using a simple benchmark won't cut it, either.

Branch prediction, caching, and superscalarization all play a larger role than ever before, and it's almost impossible to predict the cost of branch misprediction in a real-world scenario. So whenever you attempt an optimization such as the one described in this section, please measure its effect in vivo.

When rewritten with Horner's scheme (chapter 6), the polynomial model we got works almost five times faster than the sine CPU instruction on my Intel Core i7-9700F. Awesome! But the model is defined only on $\left[\frac{-\pi}{2}, \frac{\pi}{2}\right]$. If you glue this model back to back into an infinite strip, the branching eats up almost all the performance benefits. The resulting model works only about 25% faster, and that's for only one concrete CPU. To be confident that the optimization makes sense for most of your user base, you have to measure it on a plethora of devices.

And not only CPUs. Nowadays, a lot of general-purpose computing, not to mention visual effects, is done on GPUs. I've measured our model on GeForce GTX 1050 Ti Mobile, and although it does work twice as fast with default compiler parameters, with `--use_fast_math`, the polynomial outperforms the standard sine only marginally.

The good news: on mobile devices, the optimization still seems to work. On ARMv7, the full model runs almost three times faster than the sine from the standard C++ library.

My general advice is this: when in doubt, measure; when not in doubt, measure and be surprised.

7.1.3 *Unexpected fact: Explicit polynomial modeling generalizes power series*

As you saw in the preceding section, by using the whole arsenal of polynomial modeling— points, tangents, and integral properties—you can make yourself a model that both reflects exactly the traits you want and outperforms the easily searchable solutions. Power series is the common approach for polynomial modeling, but our custom model tops it in precision.

No wonder. If you think about it from the modeler's perspective, power series is only a special case of generalized polynomial modeling. You add one point to your model and then add a few derivatives at this point. If this point is a zero, you get a diagonal matrix system that has all the factorials in the matrix part and the derivatives' values on the right. In our case, the system for the sine Mclaurin series looks like this:

$$\begin{pmatrix} 1 & 0 & 0 & 0 & 0 & 0 & 0 & 0 \\ 0 & \frac{1}{1!} & 0 & 0 & 0 & 0 & 0 & 0 \\ 0 & 0 & \frac{1}{2!} & 0 & 0 & 0 & 0 & 0 \\ 0 & 0 & 0 & \frac{1}{3!} & 0 & 0 & 0 & 0 \\ 0 & 0 & 0 & 0 & \frac{1}{4!} & 0 & 0 & 0 \\ 0 & 0 & 0 & 0 & 0 & \frac{1}{5!} & 0 & 0 \\ 0 & 0 & 0 & 0 & 0 & 0 & \frac{1}{6!} & 0 \\ 0 & 0 & 0 & 0 & 0 & 0 & 0 & \frac{1}{7!} \end{pmatrix} \begin{pmatrix} a_0 \\ a_1 \\ a_2 \\ a_3 \\ a_4 \\ a_5 \\ a_6 \\ a_7 \end{pmatrix} = \begin{pmatrix} 0 \\ 1 \\ 0 \\ -1 \\ 0 \\ 1 \\ 0 \\ -1 \end{pmatrix}$$

The resulting polynomial, of course, is

$$P(x) = x - \frac{x^3}{3!} + \frac{x^5}{5!} - \frac{x^7}{7!}$$

This follows the series formula from the internet:

$$\sin(x) \approx x - \frac{x^3}{3!} + \frac{x^5}{5!} - \frac{x^7}{7!} + \dots$$

So the general polynomial modeling generalizes not only polynomial interpolation, but power series as well.

7.1.4 Section 7.1 summary

General polynomial modeling is about mixing conditions like "the polynomial runs through a point (x, y)," "the polynomial tangent in x is t," and "the polynomial integral on a range (a, b) is s" in a single Vandermonde-like system (chapter 6). Solving this system gives you the polynomial coefficients.

This modeling generalizes polynomial interpolation and, perhaps less obviously, power series as well. Using this technique, you can make fast and reasonably precise models with predetermined differential and integral characteristics.

7.2 Understanding polynomial splines and Bézier curves

A reminder: to define a parametric curve, you should define a separate function for each of its coordinates. An argument shared in the per-coordinate functions is your parameter. You get a pair of coordinates representing a 2D point or a triplet of coordinates representing a point in 3D for every parameter's value. In this section, let's look into a pair of slightly different approaches to building parametric curves.

7.2.1 Explicit polynomial parametric curves

A polynomial parametric curve, as you can guess, is a parametric curve with polynomial per-coordinate functions. This is a 2D cubic curve:

$$P_x(t) = a_x t^3 + b_x t^2 + c_x t + d_x$$

$$P_y(t) = a_y t^3 + b_y t^2 + c_y t + d_y$$

The curve requires a set of eight coefficients: four per coordinate. These coefficients could be obtained from four guiding points with interpolation. For that purpose, you need to compose two Vandermonde systems, also one per coordinate.

Let's say we have these points: (x_1, y_1), (x_2, y_2), (x_3, y_3), (x_4, y_4). For a $Y(x)$ graph, the Vandermonde equations would be $P(x_i) = y_i$, but this isn't what we want; we want $P_x(t)$ and $P_y(t)$. A mere set of points doesn't give us any hints about how these points correspond to the parameter. No worries; we can set this correspondence ourselves. We can assign parameter values evenly from the [0, 1] interval, as in this example:

$$t = 0,\ x = x_1,\ y = y_1$$

$$t = \frac{1}{3},\ x = x_2,\ y = y_2$$

$$t = \frac{2}{3},\ x = x_3,\ y = y_3$$

$$t = 1,\ x = x_4,\ y = y_4$$

Now we can compose the equations, first for the $P_x(t)$,

$$0a_x + 0b_x + 0c_x + d_x = d_x = x_1$$

$$\frac{1}{27}a_x + \frac{4}{9}b_x + \frac{1}{3}c_x + d_x = x_2$$

$$\frac{8}{27}a_x + \frac{4}{9}b_x + \frac{2}{3}c_x + d_x = x_3$$

$$a_x + b_x + c_x + d_x = x_4$$

and then use a similar set of equations for the $P_y(t)$:

$$d_y = y_1$$

$$\frac{1}{27}a_y + \frac{4}{9}b_y + \frac{1}{3}c_y + d_y = y_2$$

$$\frac{8}{27}a_y + \frac{4}{9}b_y + \frac{2}{3}c_y + d_y = y_3$$

$$a_y + b_y + c_y + d_y = y_4$$

By solving these equations, we get the coefficients for the polynomials. You don't have to solve the equations every time you want a polynomial curve, of course. With SymPy, you can solve them once symbolically and reuse the formula, as in the following listing (ch_07/parametric_polynomial.py in the book's source code).

Listing 7.6 Getting the formula for the coefficients of a parametric polynomial curve

```
from sympy import *

ax, bx, cx, dx = symbols('ax bx cx dx')
ay, by, cy, dy = symbols('ay by cy dy')
x1, x2, x3, x4 = symbols('x1 x2 x3 x4')
y1, y2, y3, y4 = symbols('y1 y2 y3 y4')

x_system = [
    dx - x1,
    (1/27)*ax + (1/9)*bx + (1/3)*cx + dx - x2,
    (8/27)*ax + (4/9)*bx + (2/3)*cx + dx - x3,
    ax + bx + cx + dx - x4
]

y_system = [
    dy - y1,
    (1/27)*ay + (1/9)*by + (1/3)*cy + dy - y2,
    (8/27)*ay + (4/9)*by + (2/3)*cy + dy - y3,
    ay + by + cy + dy - y4
]

axs = solve(x_system, (ax, bx, cx, dx))
ays = solve(y_system, (ay, by, cy, dy))

print(axs)
print(ays)
```

Coefficients for the x(t) polynomial

Coefficients for the y(t) polynomial

x-values in points 1, 2, 3, and 4

y-values in points 1, 2, 3, and 4

The system to compute the coefficients for the x(t) polynomial

The system for the y(t) polynomial coefficients. As you can see, the systems are identical, but x becomes y in the latter. In cases such as this, you can save yourself some time on symbolic computation by solving one and using its solution for every coordinate.

The solutions to these systems are

```
{ax: -4.5*x1 + 13.5*x2 - 13.5*x3 + 4.5*x4,
 bx: 9.0*x1 - 22.5*x2 + 18.0*x3 - 4.5*x4,
 cx: -5.5*x1 + 9.0*x2 - 4.5*x3 + x4,
 dx: x1}
{ay: -4.5*y1 + 13.5*y2 - 13.5*y3 + 4.5*y4,
 by: 9.0*y1 - 22.5*y2 + 18.0*y3 - 4.5*y4,
 cy: -5.5*y1 + 9.0*y2 - 4.5*y3 + y4,
 dy: y1}
```

You can use the formulas as a code to get coefficients for every set of four points and then use the coefficients to define per-coordinate polynomials and polynomials to build an actual curve (figure 7.7).

This result is great by itself, but it's not something we want if we want to move toward splines. We want to conjoin a lot of curves together smoothly, and with interpolation, we have little control of the tangent vectors of those curves. Usually, this motivation is enough to cause us to jump from explicit polynomials to Bézier curves, but we won't go there yet. Instead, we'll use what we learned in chapter 6 to

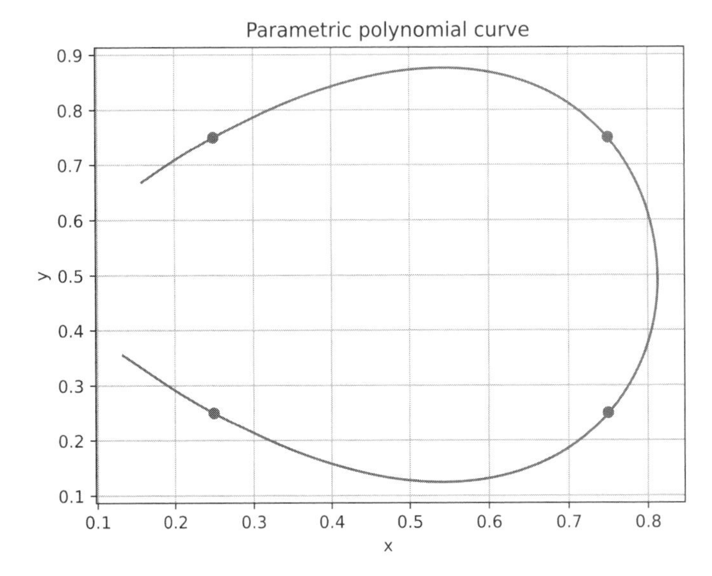

Figure 7.7 A parametric polynomial curve that goes through four points

build a curve that gives us full control of the tangent vectors on its ends. So our 2D cubic curve is still

$$P_x(t) = a_x t^3 + b_x t^2 + c_x t + d_x$$

$$P_y(t) = a_y t^3 + b_y t^2 + c_y t + d_y$$

We still have to fill in eight coefficients. Now, instead of computing them all from the "curve goes through the point" equations, let's get half of them from the "curve has a given tangent vector at a point" equations.

We need two tangent vectors complementing two interpolation points. The points will be (x_1, y_1), (x_2, y_2), and the tangent vectors are (d_{x1}, d_{y1}), (d_{x2}, d_{y2}).

The parameter-to-points problem, as before, isn't a problem but a matter of choice. Because I don't like messy numbers in my system, and because ones and zeroes help keep equations tidy, I propose this assignment:

$$t = 0: x = x_1, y = y_1, d_x = d_{x1}, d_y = d_{y1}$$

$$t = 1: x = x_2, y = y_2, d_x = d_{x2}, d_y = d_{y2}$$

A small reminder: a derivative of a cubic polynomial is $3ax^2 + 2bx + c$. So the systems that reflect both the points and tangent values are as follows. One system is for the x,

$$d_x = x_1$$

$$c_x = d_{x1}$$

$$a_x + b_x + c_x + d_x = x_2$$

$$3a_x + 2b_x + c_x = d_{x2}$$

and the other is for the y:

$$d_y = y_1$$
$$c_y = d_{y1}$$
$$a_y + b_y + c_y + d_y = y_2$$
$$3a_y + 2b_y + c_y = d_{y2}$$

Let's feed these systems to SymPy, as in the following listing (ch_07/parametric_polynomial_tangents.py in the book's source code).

Listing 7.7 Parametric polynomial curve with given tangent vectors

```
from sympy import *

ax, bx, cx, dx = symbols('ax bx cx dx')
ay, by, cy, dy = symbols('ay by cy dy')
x1, x2, dx1, dx2 = symbols('x1 x2 dx1 dx2')
y1, y2, dy1, dy2 = symbols('y1 y2 dy1 dy2')

x_system = [
    dx - x1,
    cx - dx1,
    ax + bx + cx + dx - x2,
    3*ax + 2*bx + cx - dx2
]

y_system = [
    dy - y1,
    cy - dy1,
    ay + by + cy + dy - y2,
    3*ay + 2*by + cy - dy2
]

axs = solve(x_system, (ax, bx, cx, dx))
ays = solve(y_system, (ay, by, cy, dy))

print(axs)
print(ays)
```

Annotations:
- Coefficients for the x(t) polynomial
- Coefficients for the y(t) polynomial
- x-values in points 1 and 2 complemented with tangent x-components in points 1 and 2
- y-values in points 1 and 2 complemented with tangent y-components in points 1 and 2
- The system to compute the coefficients for the x(t) polynomial
- The system to compute the coefficients for the y(t) polynomial

The solutions are even simpler than when we didn't have tangent vectors:

```
{ax: dx1 + dx2 + 2*x1 - 2*x2,
 bx: -2*dx1 - dx2 - 3*x1 + 3*x2,
 cx: dx1,
 dx: x1}
{ay: dy1 + dy2 + 2*y1 - 2*y2,
 by: -2*dy1 - dy2 - 3*y1 + 3*y2,
 cy: dy1,
 dy: y1}
```

These coefficients set a pair of polynomial functions that form a curve going through a pair of points while aligning with the given tangent vectors at these points (figure 7.8).

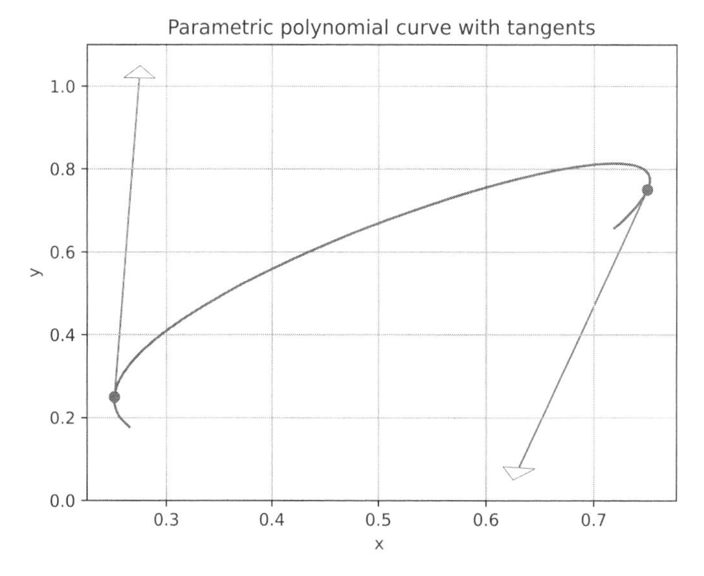

Figure 7.8 A polynomial parametric curve with given tangent vectors

Because the curve is based on a cubic polynomial, we can't run it through more than two points and have two tangent vectors at the same time. We have no control of its curvature, either. With polynomial modeling, however, we can raise our degree and have that, too. A pair of 5-degree polynomials can make a curve that has its curvatures constrained at the ends. And with another degree, we can afford another guiding point in the middle. We can add degrees and have the properties we want until the polynomial becomes too large and unpredictable due to the Runge phenomenon.

What is the Runge phenomenon?

The *Runge phenomenon* concerns polynomial interpolation (section 6.3.2). As you add more points to the data set, the polynomial tends to oscillate more as well, making the interpolating function impractical, as shown in this figure:

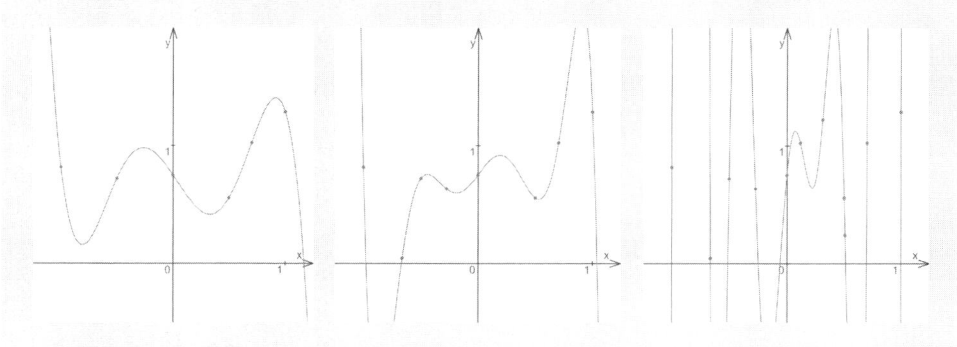

Polynomial oscillations usually get worse when more points are added to the data set.

Considering this phenomenon, you have to use other means of interpolation for large data sets, such as splines.

To avoid that unpredictability, we should split our target curve into a set of smaller ones and conjoin them back to back concerning their tangent vectors based on the first derivative (figure 7.9).

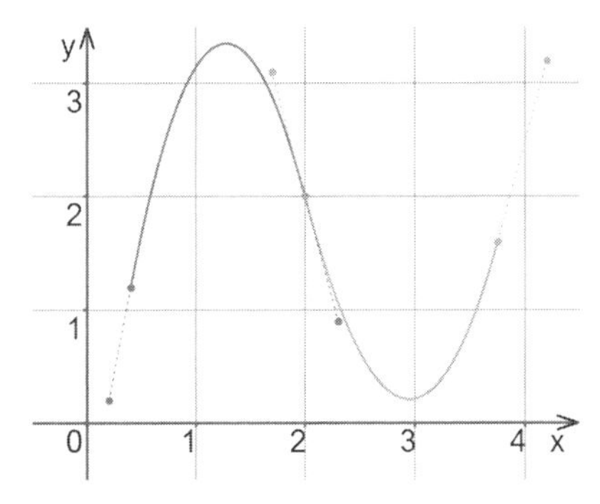

Figure 7.9 Two cubic polynomial curves conjoined smoothly by keeping the tangent vectors in the joint symmetrical

If a problem we're solving requires a higher degree of continuity, we can also exploit the second derivative or the third, but such problems are rare. Usually, keeping lines smooth is enough.

7.2.2 Practical example: Crafting a spline for points that aren't there yet

This particular example also comes from game development. We had a mini-game in which a player was supposed to lead a motorboat through an island maze by putting a set of waypoints on the map. Because a motorboat can't turn around on a spot, the path had to be both continuous and smooth. And because the game was a mini-game, we couldn't afford to spend a month setting up a real physics engine.

We already knew how to make continuous and smooth paths. All we needed was a set of points and a set of tangent vectors for each point. Then we could make a piece of a path between any of the points by putting a cubic polynomial there, based on the "the curve goes through a point" and "the tangent vector in a point is" equations. Because adjacent pieces were sharing tangents, as in figure 7.9, the whole path would be smooth.

But here comes the catch: the boat was supposed to start moving while a player was still adding the points, not after they put them all on the map and clicked Go. And of course, we wouldn't ask players to draw tangents for every point explicitly. So at the beginning, we had no waypoints and no tangents, but as soon as a player started clicking, we wanted our boat to start sailing along a smooth path. How could we do that?

Let's tackle the problem ourselves, step by step. As soon as a user puts down the first waypoint, a boat should start moving from wherever it was drifting before to this new point. We don't have any additional data yet, so a pair of points is all we got. We can put a parametric straight line through the points, which already gives us four equations, two per coordinate:

$$a_x t_0 + b_x = x_0$$

$$a_y t_0 + b_y = y_0$$

$$a_x t_1 + b_x = x_1$$

$$a_y t_1 + b_y = y_1$$

Here, (x_0, y_0) are the coordinates of the boat's position at the beginning. We can keep them (0, 0) to simplify the equations and compute the whole path in relative coordinates. Then (x_1, y_1) are the coordinates of the first waypoint, and t_0 and t_1 are parameters that set the interval for the first piece of the path. I advise you to keep them 0 and 1, respectively, but the system would work fine with any other numbers.

Feed the system to SymPy, and you'll get two pairs of coefficient formulas to form a parametric line, or two functions that when put together return an (x, y) point for every t:

$$x = a_x t + b_x$$

$$y = a_y t + b_y$$

When t changes from t_0 to t_1, the boat follows the line segment (figure 7.10).

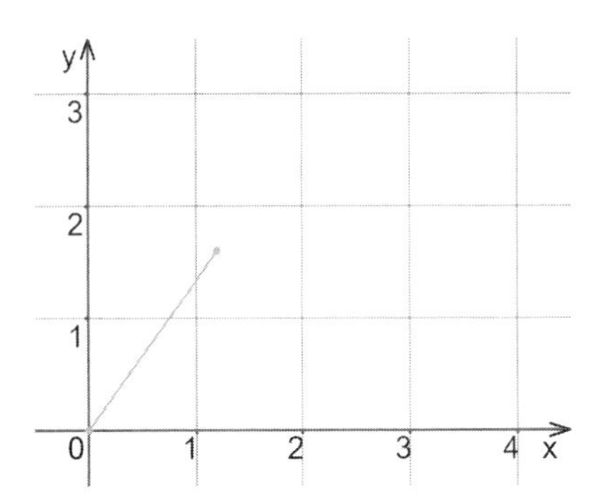

Figure 7.10 The starting point and the first waypoint are enough to build a straight segment.

Then the user adds another point. The second piece should start from where the first ended, and it should share its tangent vector. So for the second piece, we can compose six equations, three per coordinate or two per each of the three conditions:

- "The polynomial goes through the first waypoint."

$$a_x t_1^2 + b_x t_1 + c_x = x_1$$

$$a_y t_1^2 + b_y t_1 + c_y = y_1$$

- "The tangent vector in this point is whatever we get from the first piece."

$$2a_x t_1 + b_x = d_{x1}$$

$$2a_y t_1 + b_y = d_{y1}$$

- "The polynomial goes through the second waypoint."

$$a_x t_2^2 + b_x t_2 + c_x = x_2$$

$$a_y t_2^2 + b_y t_2 + c_y = y_2$$

The point conditions are fairly trivial—we've seen them before—but what about the tangent condition? Where do the d_{x1} and d_{x2} come from?

Well, from the first piece's equations, of course! We can differentiate them symbolically and compute the derivatives in the t_1 point. In fact, at this step, we wouldn't have to compute anything at all. Because the first piece is a line segment, its equations are linear, and their derivatives are simply constants:

$$d_{x1} = (a_x t + b_x)' = a_x$$

$$d_{y1} = (a_y t + b_y)' = a_y$$

Now we have all the coefficients for the input equations. Put the equations in SymPy, and get the polynomial coefficients for the second piece, which will be a quadratic curve (figure 7.11):

$$x = a_x t^2 + b_x t + c_x$$

$$y = a_y t^2 + b_y t + c_y$$

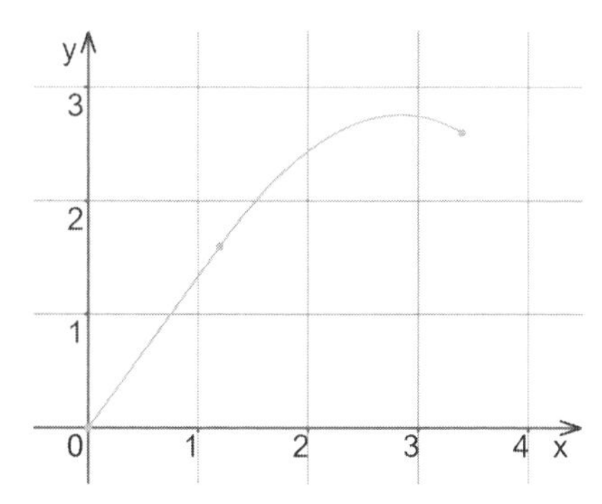

Figure 7.11 The piece between the first waypoint and the second one is a quadratic curve if we borrow the tangent vector from the first segment.

Let's add the third piece. As before, we have the six equations from two points and one tangent vector:

$$a_x t_2^2 + b_x t_2 + c_x = x_2$$

$$a_y t_2^2 + b_y t_2 + c_y = y_2$$

$$2a_x t_2 + b_x = d_{x2}$$

$$2a_y t_2 + b_y = d_{y2}$$

$$a_x t_3^2 + b_x t_3 + c_x = x_3$$

$$a_y t_3^2 + b_y t_3 + c_y = y_3$$

This time, however, d_{x2} and d_{y2} come from the previous quadratic polynomials' differentiation, not linear functions:

$$d_{x2} = (a_x t_2^2 + b_x t_2 + c_x)' = 2a_x t_2 + b_x$$

$$d_{y2} = (a_y t_2^2 + b_y t_2 + c_y)' = 2a_y t_2 + b_y$$

When we solve the equations, we get the coefficients for the third piece. The third segment, of course, is also a quadratic curve (figure 7.12).

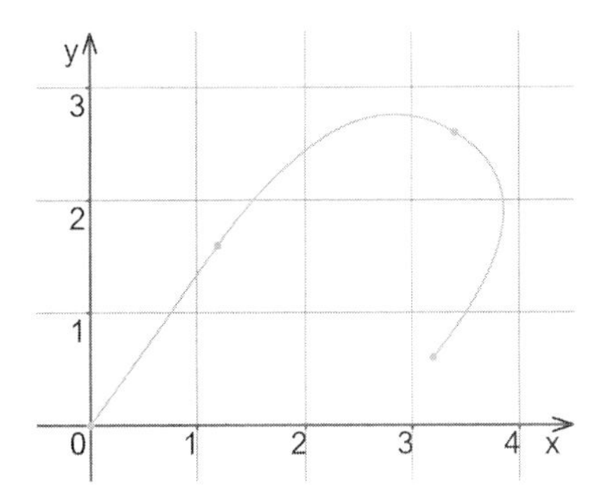

Figure 7.12 The third piece and all the pieces afterward are quadratic because we can always take a tangent vector from the last point of the previous segment.

At this point, we can start generalizing our formula. The piece that goes through the i-th and $(i+1)$-th points is defined by these equations:

$$a_{xi} t_i^2 + b_{xi} t_i + c_{xi} = x_i$$

$$a_{yi} t_i^2 + b_{yi} t_i + c_{yi} = y_i$$

$$2a_{xi} t_i + b_{xi} = 2a_{x(i-1)} t_i + b_{x(i-1)}$$

$$2a_{yi} t_i + b_{yi} = 2a_{y(i-1)} t_i + b_{y(i-1)}$$

$$a_{xi}t_{(i+1)}^2 + b_{xi}t_{(i+1)} + c_{xi} = x_{(i+1)}$$

$$a_{yi}t_{(i+1)}^2 + b_{yi}t_{(i+1)} + c_{yi} = y_{(i+1)}$$

The formula is recursive. To get the coefficients for the i-th piece, you have to know the coefficients for the $(i–1)$-th.

Now for some tips and tricks. In real life, a stationary boat still has a movement direction, or rather an orientation, so there's a tangent vector for the zero point after all. This makes the first linear equation quadratic, and now all the pieces follow the same recursive formula.

We don't have to ask a player for this vector. Because it's part of the initial game state, we can get an initial orientation from a game designer. If a game designer also refuses to specify the initial boat direction, we can set the 0-tangent vector like this:

$$d_{x0} = 0$$

$$d_{y0} = 0$$

The equations for the first piece will automatically turn out linear because

$$a_{x0} = d_{x0} = 0$$

$$a_{y0} = d_{y0} = 0$$

The next trick is about the boat's speed. The speed is regulated by the length of the tangent vector, whereas the direction of the boat is regulated by the tangent vector direction. If you divide or multiply the tangent vector by a scalar, this operation changes the speed and the appearance of the path, respectively, but not the path's smoothness. To make the tangent vector shorter or longer at each waypoint, modify your tangent equations like this:

$$2a_{xi}t_i + b_{xi} = k(2a_{x(i-1)}t_i + b_{x(i-1)})$$

$$2a_{yi}t_i + b_{yi} = k(2a_{y(i-1)}t_i + b_{y(i-1)})$$

Here, k is the speeding/slowing coefficient. It isn't computed anywhere; you pick it yourself. The coefficients affect the appearance of the curve. In figure 7.13, two more paths are going through the same points as before, but one path has $k = 0.8$, and the other path has $k = -1.5$.

So we have a path that is smooth, continuous, and governable, and it continues to be so as a player adds more points. This isn't some classic polynomial spline you can Google up; it's a quadratic spline that we reinvented to cover the task at hand. Knowing the mathematics behind splines, you could just as well craft your own custom splines for your tasks.

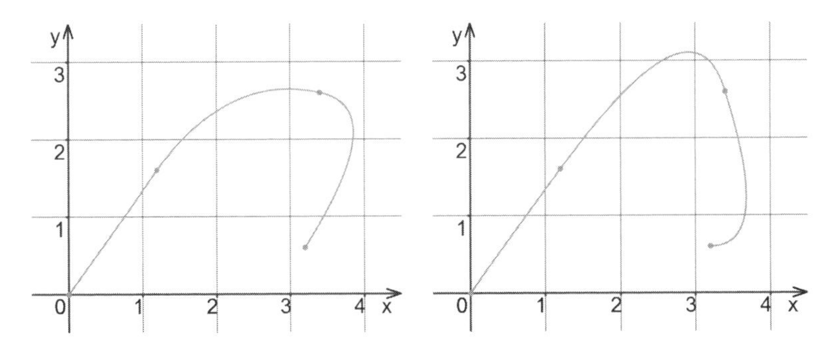

Figure 7.13 The path goes through the same points, but with *k* = 0.8, the boat tends to turn early, and with *k* = 1.5, it tends to turn late.

7.2.3 Bézier curves

Explicit polynomial modeling gives us all the control we want to make long, smooth, predictable spline curves. In practice, however, curves with polynomials often go by the name Bézier. What are Bézier curves, and how do they differ from what we've already seen? Let's find out.

Bézier curves are parametric curves. They're also produced by a pair (in 2D) or a triplet (in 3D) of per-coordinate functions. Moreover, in classical Bézier curves, the per-coordinate functions are also polynomial. But instead of finding the polynomials' coefficients explicitly, we build up the polynomials from basis functions, as like we did with Lagrange interpolation (chapter 6).

Like polynomials, Bézier functions have degrees. Let's start with the simplest possible case: the first-degree Bézier function.

Let's say we have a parameter t running from 0 to 1, and we have a pair of values, y_1 and y_2. We want our function to be y_1 when $t = 0$ and y_2 when $t = 1$. The function we're looking for is a line equation like this:

$$B_1(t) = (1 - t)y_1 + ty_2$$

When $t = 0$, $(1 - t)$ is 1, the function turns into $1y_1 + 0y_2$, so y_1. When $t = 1$, the function is $(1 - 1)y_1 + 1y_2 = 0y_1 + 1y_2 = y_2$.

The function is also linear. If you expand the brackets, you'll get $y_1 - ty_1 + ty_2$. And when you collect the parameter t, $(y_2 - y_1)t + y_1$. This is a line equation in its explicit form.

The function consists of two *basis polynomials*. The first is $1 - t$, and the second is t. We make a formula for a specific first-degree Bézier function by multiplying the first basis polynomial by the first value, y_1, and the second basis polynomial by multiplying by the second value, y_2.

Take these values: $y_1 = 0.2$ and $y_2 = 0.7$. The Bézier function for these values is

$$B_1(t) = (1 - t)*0.2 + t*0.7$$

- When $t = 0$, the $B_1(t) = 1 * 0.2 + 0 * 0.7 = 0.2$.
- When $t = 0.2$, the $B_1(t) = 0.8 * 0.2 + 0.2 * 0.7 = 0.16 + 0.14 = 0.3$.
- When $t = 0.4$, the $B_1(t) = 0.6 * 0.2 + 0.4 * 0.7 = 0.12 + 0.28 = 0.4$.
- When $t = 0.6$, the $B_1(t) = 0.4 * 0.2 + 0.6 * 0.7 = 0.08 + 0.42 = 0.5$.
- When $t = 0.8$, the $B_1(t) = 0.2 * 0.2 + 0.8 * 0.7 = 0.04 + 0.56 = 0.6$.
- When $t = 0.4$, the $B_1(t) = 0.0 * 0.2 + 1.0 * 0.7 = 0.7$.

Figure 7.14 shows the function's graph.

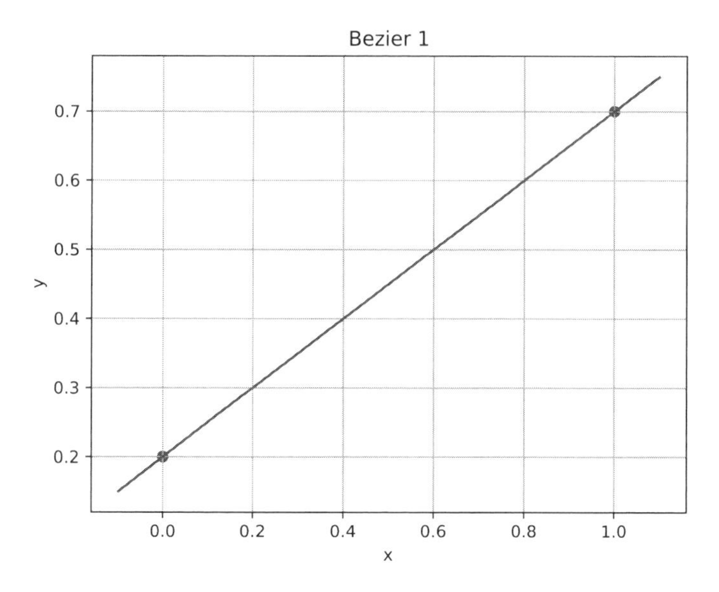

Figure 7.14 The first-degree Bézier function is linear and interpolating.

This function has a lot in common with the Lagrange interpolation. If you remember the basis polynomials, there were products of ratios:

$$g_i(x) = \prod_{j=1, j \neq i}^{N} \frac{x - x_j}{x_i - x_j}$$

These basis polynomials are defined for every i-th point, and they take all the ratios for the points that aren't i-th.

In our case, we use t instead of x. And because t runs from 0 to 1, $t_1 = 0$, and $t_2 = 1$. Because we have only two points, the basis polynomial for the first point is

$$g_1(t) = \frac{t - t_2}{t_1 - t_2} = \frac{t - 1}{0 - 1} = 1 - t$$

and the basis polynomial for the second point is

$$g_2(t) = \frac{t - t_1}{t_2 - t_1} = \frac{t - 0}{1 - 0} = t$$

So the basis polynomials for the first-degree Bézier function are exactly the same as for the Lagrange interpolation! And no wonder. You can't put more than one straight line through a pair of points, so all the polynomials of the first degree, no matter how they're obtained, are deemed to be the same.

The difference starts when we raise the degree. We don't want yet another formula for an interpolating polynomial; we already have more than enough. We want a function that still starts in y_1 and ends in y_n but uses only the intermediate values to guide the shape of the function.

On the other hand, we want to preserve one property of a Bézier formula because we'll use it later. The property is "the sum of all the basis functions in a Bézier formula is always 1." This property is trivial for the first degree; indeed, $1 - t + t = 1$. But we want to keep this property while raising the degree. We want the sum of our basis polynomials to be 1 for the quadratic Bézier, for the cubic one, for the quadric one, and so on. Luckily, this property is surprisingly easy to keep.

Let's call 1-t an a and t a b. We know that whatever t really is, $a + b = 1$.

Now let's multiply $(a + b)$ by . . . $(a + b)$. The product will still be 1 because we essentially multiply 1 by 1. The result of this multiplication, then, is $a^2 + ab + ba + b^2$, or because $ab = ba$, $a^2 + 2ab + b^2$.

Putting the t back into the formula, we'll get the basis polynomials for the second-degree Bézier formula:

$$(1 - t)^2, 2(1 - t)t, t^2$$

Put the values in, and you have the formula itself:

$$B_2(t) = (1 - t)^2 y_1 + 2(1 - t)t y_2 + t^2 y_3$$

It's easy to see that when $t = 0$, the formula still boils down to y_1 because the only member that doesn't have a nullifying t in it is the first one. Just as well, when $t = 1$, the only member without the (1-t) is the last one, and the formula is reduced to y_3.

You can check it out yourself. Listing 7.8 is a code snippet that implements a second-degree Bézier curve out of two second-degree Bézier formulas (ch_07/bezier_2. py in the book's source code).

Listing 7.8 Quadratic Bézier curve

```
xs = [0, 0.3, 1]          # Points
ys = [0.2, 0.8, 0.5]
```

```
def b2x(t, xs):
    return (1-t)*(1-t)*xs[0] + 2*(1-t)*t*xs[1] + t*t*xs[2]

def b2y(t, ys):
    return b2x(t, ys)
```

Bézier function

Because both functions share a basis, there's no point in copying and pasting the formula.

The curve from the example looks like figure 7.15.

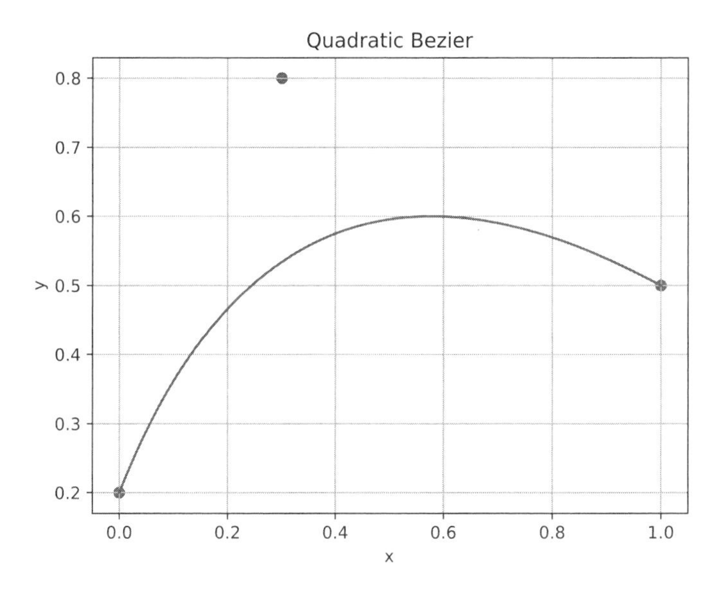

Figure 7.15 The second-degree or quadratic Bézier curve starts and ends in control points, but the middle point only shapes the curve.

You can obtain the basis for the third-degree or cubic Bézier formula:

$$(a + b)(a + b)(a + b) = a^3 + 3a^2b + 3ab^2 + b^3$$

The formula is

$$B_3(t) = (1\text{-}t)^3y_1 + 3(1\text{-}t)^2ty_2 + 3(1\text{-}t)t^2y_3 + t^3y_4$$

NOTE You can calculate the basis for the fourth degree (or any degree) Bézier formula with a simple one-liner in SymPy: `expand((a + b)**4)`.

The following listing shows the code for the cubic Bézier curve (ch_07/bezier_3.py in the book's source code).

Listing 7.9 An example of a cubic Bézier curve

```
xs = [0, 0.3, 0.7, 1]
ys = [0.2, 0.8, 1., 0.5]
```

Points

```
def b3x(t, xs):
    return ((1-t)*(1-t)*(1-t)*xs[0]
            + 3*(1-t)*(1-t)*t*xs[1]
            + 3*(1-t)*t*t*xs[2]
            + t*t*t*xs[3])

def b3y(t, ys):
    return b3x(t, ys)
```

The cubic Bézier formula

Python famously restricts the indentation, but in an expression, it allows you everything as long as you keep it in parentheses.

Figure 7.16 shows the curve for this example.

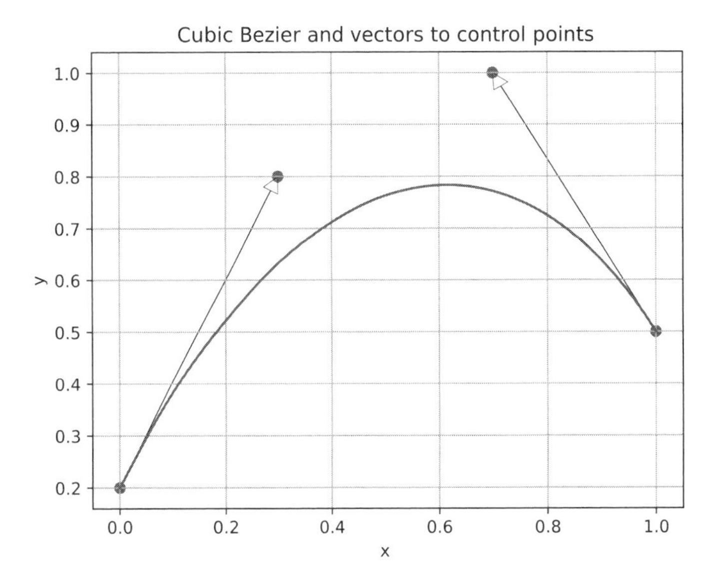

Figure 7.16 The third-degree or cubic Bézier curve has two points to control the curve's shape.

Note that the vectors from point 1 to point 2, and from point 3 to point 4 are collinear to the tangent vectors. Also, a simple proportion makes Bézier points into tangent vectors for explicit polynomial modeling, and vice versa. The proportion is three. Yes, it's that simple. The tangent vectors are three times the point differences (figure 7.17).

This correspondence is nice, of course, because it allows us to build smooth complex curves out of cubic Béziers as we can already do with explicit polynomials. But why do we need Bézier curves if we already have polynomials?

The answer is: we're not done with Bézier yet. Remember the property "the sum of all basis polynomials is always 1"? This property allows us to rewrite the cubic Bézier function like this:

$$B_3(t) = \frac{(1-t)^3 y_1 + (1-t)^2 t y_2 + (1-t)t^2 y_3 + t^3 y_4}{(1-t)^3 + (1-t)^2 t + (1-t)t^2 + t^3}$$

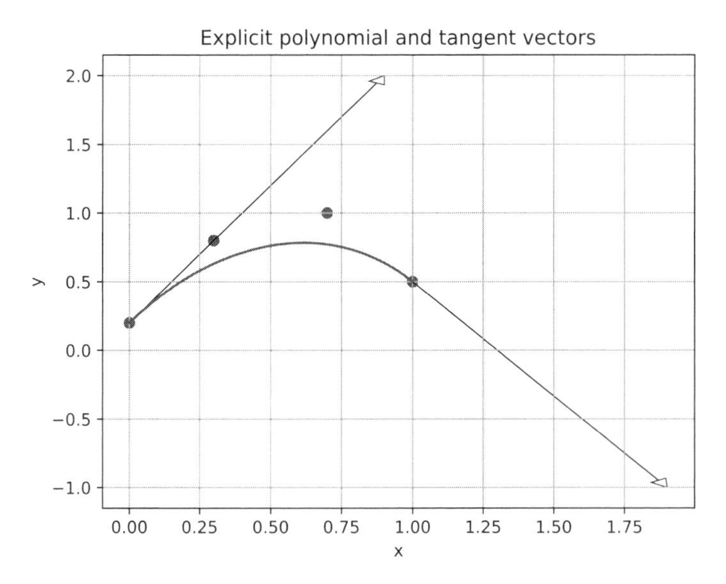

Figure 7.17 The same curve as in figure 7.16, now built with explicit polynomial modeling based on points and tangent vectors

This is still the same formula, as the denominator is always 1 anyway. But now we can add weights to every basis polynomial, both in the nominator and the denominator:

$$B_3(t) = \frac{(1-t)^3 w_1 y_1 + (1-t)^2 t w_2 y_2 + (1-t) t^2 w_3 y_3 + t^3 w_4 y_4}{(1-t)^3 w_1 + (1-t)^2 t w_2 + (1-t) t^2 w_3 + t^3 w_4}$$

A Bézier function with weights is called a *rational Bézier function*. These weights give us additional control of the function and, consequently, the shape of the curve. The following listing shows the sample implementation (ch_07/bezier_rational.py in the book's source code).

Listing 7.10 Rational Bézier curve example

```
xs = [0, 0.3, 0.7, 1]
ys = [0.2, 0.8, 1., 0.5]
ws = [1., 2., 2., 1.]
```

Points and weights

```
def b3x(t, xs):
    n = ((1-t)*(1-t)*(1-t)*xs[0]*ws[0]
        + 3*(1-t)*(1-t)*t*xs[1]*ws[1]
        + 3*(1-t)*t*t*xs[2]*ws[2]
        + t*t*t*xs[3]*ws[3])
    d = ((1-t)*(1-t)*(1-t)*ws[0]
        + 3*(1-t)*(1-t)*t*ws[1]
        + 3*(1-t)*t*t*ws[2]
        + t*t*t*ws[3])
    return n / d
```

Rational Bezier function

```
def b3y(t, ys):
    return b3x(t, ys)
```

The curve with these weights looks like the first one in figure 7.18. The figure includes several other weight sets for comparison.

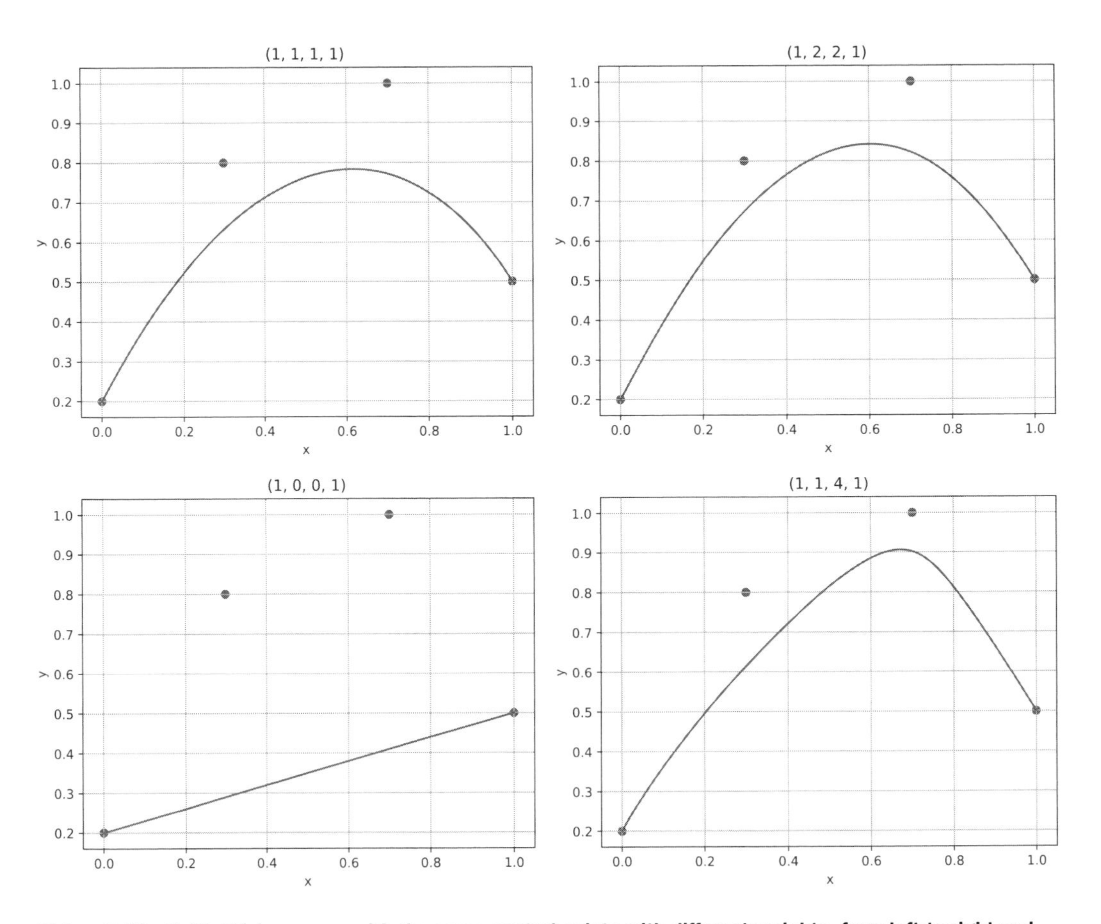

Figure 7.18 Cubic Bézier curves with the same control points with different weights, from left to right and top to bottom: (1, 1, 1, 1), (1, 2, 2, 1), (1, 0, 0, 1), (1, 1, 4, 1)

One important thing: with rational Bézier curves, we can describe conic sections such as circles, ellipses, and hyperbolas. At the same time, rational functions are more computationally heavy because of all the extra multiplications and one extra division. That's why in 2D graphics or typography, regular polynomial Bézier curves are still preferable, whereas rational Bézier curves are indispensable in computer-aided design (CAD) applications, in which you need precision more than speed.

SEE ALSO Check out a free online book when you really need to know how to do Bézier things: https://pomax.github.io/bezierinfo. It not only explains

Bézier curves, but also showcases typical problems such as fitting, point projection, and intersections. It features source code and interactive widgets, too, so it's both educational and fun.

7.2.4 Section 7.2 summary

Polynomial splines and Bézier functions allow you to build curves with full control of their points and tangents. With polynomials, you set the tangents explicitly; with Bézier curves, you set them by using a set of control points. The main advantage of Bézier functions over polynomials is that you can give the Bézier functions weights by making them rational functions.

7.3 Understanding NURBS

NURBS stands for "nonuniform rational basis spline." If you understand every word of the acronym, you can safely skip this section.

NURBS is a powerful instrument for building curves and surfaces out of arrays of control points. Most of the human-made objects were at some point designed in a CAD or computer-aided engineering (CAE) application as a set of NURBS.

The only problem with NURBS is that it simply works. In this chapter alone, we had to add an integral condition to a polynomial model to outperform the power series, and we added a new level of control to polynomial splines by doing the "scale the derivative" trick. NURBS doesn't need any of that; it already has all the controls to guide the spline's shape and position. As a result, in your work, you're likely to use some variant of a ready-made library without having to modify it in any way. You won't have to get deep into the subject.

In this section, we'll build our understanding of NURBS by looking into each of its aspects one by one, as well as by building parallels with the polynomial splines and Bézier curves we used in previous sections.

7.3.1 BS stands for "basis spline"

Like a Bézier function, a NURBS consists of scaled basis polynomial functions. Also, as in a Bézier function, these bases have a common degree. As with polynomial splines, this degree corresponds to the degree of smoothness of the whole function. When the degree is 1, the function is piecewise linear, and the curve formed by a pair of such functions is a continuous but not smooth polyline (figure 7.19). N control points form N-1 segments.

A second-degree NURBS preserves the smoothness or the C^1 continuity, but now N control points produce only (N–2) segments, although these segments are curved and conjoined nicely (figure 7.20).

Third- and fourth-degree NURBS follow the same pattern. The continuity rises: the third-degree spline preserves a C^2 continuity, and the fourth-degree preserves a C^3 continuity. But the number of segments you can get from the set of control points goes down (figure 7.21).

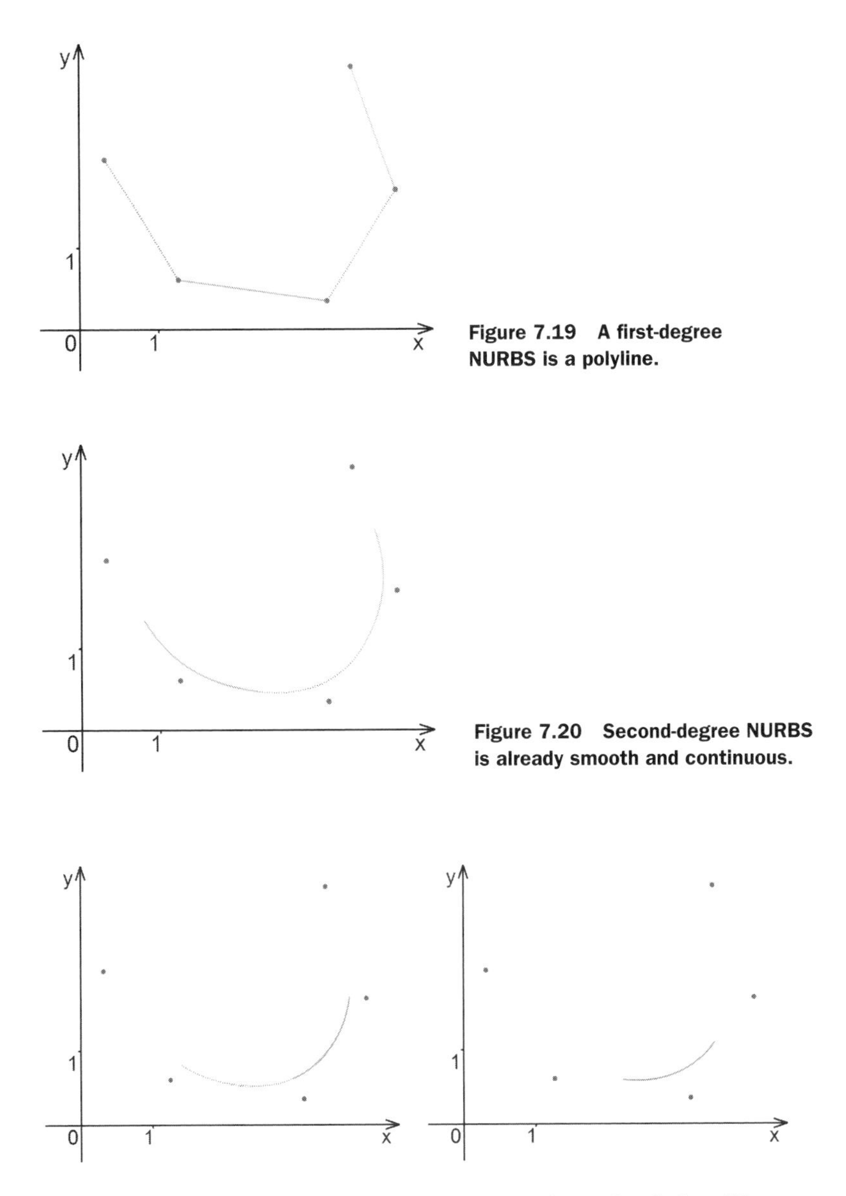

Figure 7.19 A first-degree NURBS is a polyline.

Figure 7.20 Second-degree NURBS is already smooth and continuous.

Figure 7.21 NURBS of degrees 3 and 4 on the same data set as before. When raising the degree, we trade the number of segments for continuity.

So far, this "more continuity = fewer segments" rule is in accord with the polynomial splines. With polynomials, we can also build four linear segments from a set of five points. With the algorithm we saw in section 7.2.2, we can start turning all the segments but the first one into quadratic pieces and make the whole thing smooth. Just like that, we can turn all the segments but the first two into cubics. Or we can turn all the segments except the first three into quadrics. I hope you see the pattern.

THE PATTERN With both explicit polynomial splines and NURBS, we can turn N points into (*N-M*) *M*-degree curve segments. Then the continuity of both splines is $C^{(M-1)}$.

7.3.2 NU stands for "nonuniform"

A Bézier function is always defined on a [0..1] interval, which helps keep its sum of basis polynomials 1. This is what makes the function tick. It also makes a Bézier spline (a couple of Bézier functions put back to back) *uniform,* which means that the parametric distance between points in this spline is always the same: 1.

With NURBS, parametric distances between points are set by a special knot vector, an array of arbitrary parameter values in which every pair of adjacent values stands for a segment or, if the values are equal, for skipping a segment. Skipping a segment is a valuable technique in NURBS because it allows us to make some of the control points interpolating.

In figure 7.22, you can see three NURBS curves. The first has a uniform knot vector: [0, 1, 2, 3, 4, 5, 6, 7]. The second has a skipped quadratic segment. The first three values of the knot vector are the same: [0, 0, 0, 1, 2, 3, 4, 5]. This makes the first point lie on the curve. The third curve has the three last values the same, so the last point also belongs to the target curve, but the knot value distribution is also nonuniform even in other segments: [0, 0.5, 2, 3, 4.5, 5, 5, 5]. This shapes the curve differently from the first two cases (figure 7.22).

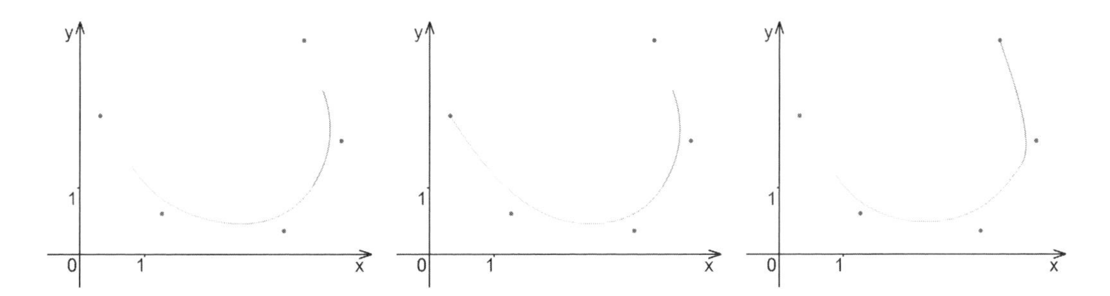

Figure 7.22 Same control points, different knot vectors

Nonuniformity is a valuable trait. With Bézier curves, your first and last points always lie on a curve, and all the points in between work as control points. With NURBS, you choose which points are interpolating and which only help shape the curve.

7.3.3 R stands for "rational"

Essentially, *rational* here means the same thing as in Bézier. Every basis polynomial is equipped with a weight. Bases with higher values of weights have more influence on the way the curve behaves. The basis of the i-th point also affects the surrounding segments. The higher the degree, the more segments are affected by a single weight.

In figure 7.23, three NURBS curves share control points and the knot vector. The only thing that differs is the weight assigned to each point. The first curve has the first weight set to 0.0001. The second curve has the third weight set to 0.1. The third curve has the fifth weight set to 100.

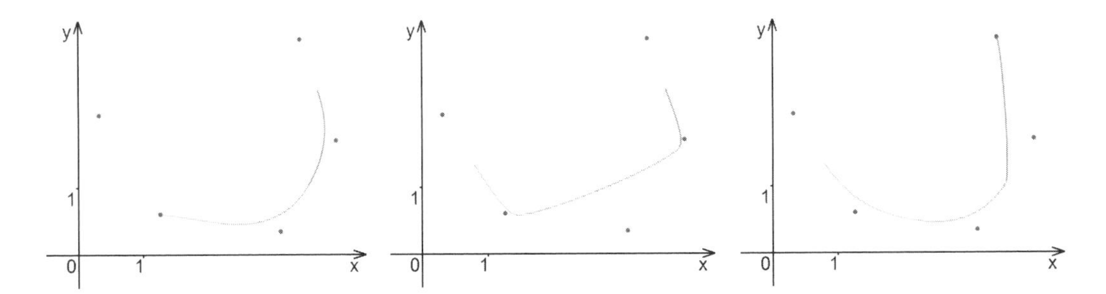

Figure 7.23 Same control points, same knot vectors, different weights

Rationality is the superpower of NURBS. So far, everything we can do with NURBS—trading points for continuity, setting the points on a curve, and shaping the curve with control points and intervals—we could do with explicit polynomial splines. But with rationality, NURBS are capable of doing the same things as rational Bézier functions, describing mathematical curves and surfaces with mathematical precision—curves such as conic sections, including hyperbolas and ellipses; quadric surfaces such as paraboloids; and less exotic spheres and cylinders. Because ellipses, spheres, and cylinders are staples of engineering, NURBS, not generic polynomial splines, took the leading role in CAD.

7.3.4 *Not-so-practical example: Building a circle with NURBS*

To be honest, this example isn't so much practical as it is anecdotal. Yes, we can build a circle with NURBS. But there are plenty of other ways to build a circle. Euclid didn't need NURBS to draw circles, so why do we?

Well, we don't. But building a hyperboloid instead would complicate the equations and make understanding the visualization much more difficult while illustrating essentially the same concept. So because the didactic value is the same, let's do the easier thing, shall we?

One way to build a NURBS surface is to use de Boor's algorithm. This algorithm doesn't compute the basis polynomials explicitly; instead, it evaluates them by using a recursive relation.

SEE ALSO This algorithm is explained well in several open source documents. One is "Generating a B-Spline Curve by the Cox-De Boor Algorithm" on the WOLFRAM Demonstrations Project website (http://mng.bz/Q8e6). For more information on NURBS with explicit basis functions and other splines, see

"An Interactive Introduction to Splines," by Evgeny Demidov, at https://www.ibiblio.org/e-notes/Splines/Intro.htm.

The following listing is an implementation of the algorithm in Python (ch_07/nurbs_circle.py in the book's source code).

Listing 7.11 De Boor's algorithm

```
def nurbs_in_t(t, control_xs, control_ys, control_ws, knot, p):
    k = index_in_knots(t, knot)

    d_x = []
    d_y = []
    d_w = []
    for j in range(p + 1):
        d_x += [control_xs[j + k - p]]
        d_y += [control_ys[j + k - p]]
        d_w += [control_ws[j + k - p]]

    for r in range(1, p+1):
        for j in reversed(range(r, p+1)):
            a = (t - knot[j+k-p]) / (knot[j+1+k-r] - knot[j+k-p])
            d_x[j] = (1.0 - a) * d_x[j-1] + a * d_x[j]
            d_y[j] = (1.0 - a) * d_y[j-1] + a * d_y[j]
            d_w[j] = (1.0 - a) * d_w[j-1] + a * d_w[j]

    return [d_x[p] / d_w[p], d_y[p] / d_w[p]]
```

t is the parameter in which we want to evaluate the curve; control_xs is the list of control points' x coordinates; control_ys is the list of y coordinates; control_ws is the list of point weights; knot is the knot vector; and p is the degree.

The function returns a pair of coordinates (x, y) for every t.

The algorithm evaluates the NURBS surface given by arrays of control points, weights, the knot vector, and the degree value in *t*. But to do this, it also needs to know what segment the *t* corresponds to. For this purpose, it uses another function that finds out the knot index for the interval containing the *t* value:

```
index_in_knots(t, knot).

def index_in_knots(t, nurbs_knot):
    for i in range(len(nurbs_knot) - 1):
        if t >= nurbs_knot[i] and t < nurbs_knot[i+1]:
            return i
    return -1
```

The function accepts the parameter value and the knot vector.

If a parameter is found between two adjacent values of the knot vector, the index of the first value is returned.

Now all we have to do is feed it the data for the circle model, and we get this data:

```
sr2i = 1.0 / (2**0.5)
xs = [1.0,  sr2i,  0.0, -sr2i, -1.0, -sr2i,  0.0,  sr2i,  1.0]
ys = [0.0,  sr2i,  1.0,  sr2i,  0.0, -sr2i, -1.0, -sr2i,  0.0]
ws = [1.0,  sr2i,  1.0,  sr2i,  1.0,  sr2i,  1.0,  sr2i,  1.0]
n = len(xs)
knot = [0, 0, 0, 1, 1, 2, 2, 3, 3, 4, 4, 4];
degree = 2;
```

The magic constant that is the inverse of the square root of 2

There's a catch: unlike with the rational Bézier, the control points coordinates for NURBS, or rather for this algorithm in particular, are put in homogeneous coordinates (chapter 4)! So the second point which is `(sr2i, sr2i, sr2i)` in the code or $\left(\frac{1}{\sqrt{2}}, \frac{1}{\sqrt{2}}, \frac{1}{\sqrt{2}}\right)$ in real numbers, is, in fact, the same as (1, 1, 1).

Putting pieces of code together with the input data and making a parameter t go through the interval [0..4], we get a NURBS circle (figure 7.24).

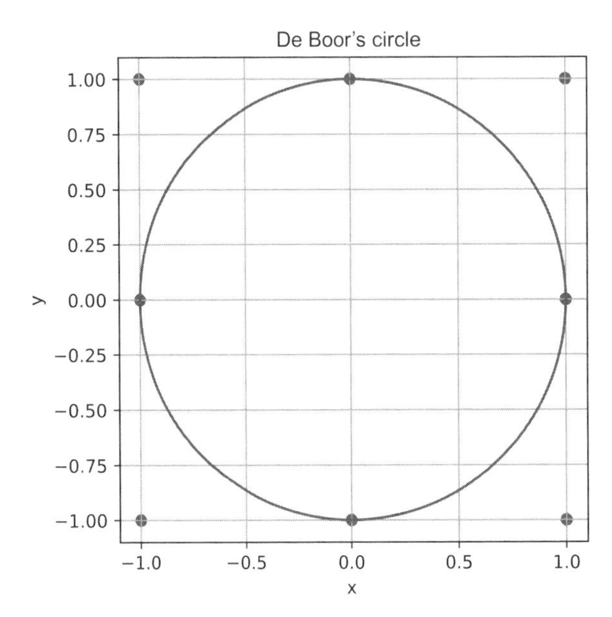

Figure 7.24 This isn't a model like the one we got with polynomial modeling of the sine; it's a real circle.

Different implementations of NURBS differ in speed and computational stability, but conceptually, they're all the same. They give you a spline function made of basis polynomials of a certain degree divided by knots and guided by the set of weighted points.

7.3.5 Section 7.3 summary

NURBS is an acronym. To use NURBS efficiently, you should understand all the acronym's letters one by one. Nonuniformity gives you the knot vector with which you can make your curve go through points. Rationality gives you enough control to describe engineering curves and, in 3D, surfaces. The basis degree lets you trade control points for continuity. And spline means that the function is a compound one; at least, it's immune to the Runge phenomenon.

SEE ALSO If you want to build deep understanding of NURBS, the book you may want to start with is *The NURBS Book,* by Les Piegl and Wayne Tiller (Springer Science and Business Media, 2012).

7.4 Exercises

Exercise 7.1 If we want to build a parametric polynomial curve that goes through three points and has given tangent vectors in two of them, what could the degree of that polynomial be?

Exercise 7.2 Which of these are valid Bézier functions?

1. $(1-t)^3y_1 + (1-t)^2ty_2 + (1-t)t^2y_3 + t^3y_4$
2. $(1-t)^3y_1 + 3(1-t)^2ty_2 + 3(1-t)t^2y_3 + t^3y_4$
3. $(1-t)^3y_1 + 2(1-t)^2ty_2 + t^3y_3$
4. $(1-t)^2y_1 + 2(1-t)ty_2 + t^2y_3$

Exercise 7.3 Which of these properties are alien for NURBS curves?

1. You can set the continuity degree by choosing the degree of the basis.
2. You can model conic sections by setting the weights properly.
3. You can set the tangent vector at any point explicitly.
4. You can set the curve to go through any control point.

Exercise 7.4 In chapter 5, we made a path out of circles and lines. Its most valuable property was that the linear increment in the parameter corresponded to the linear increment in the path length. In other words, a boat, moving along this path would have had its speed constant. For the example in section 7.2.2, this isn't the case.

Can you make it so? Can you make a boat move along the parametric polynomial curve with constant (or at least constant enough to the naked eye) speed? What would your approach to this task be?

7.5 Solutions to exercises

Exercise 7.1 Fourth or more. To build this curve, we need at least three pairs of per-coordinate equations to respect the "curve goes through points" conditions and two pairs more to respect the "curve has a tangent vector specified in a point" condition. That's five equations. Solving five equations gives us five coefficients, which usually corresponds to a fourth-degree polynomial.

But we can certainly add more conditions and raise the degree depending on the points and tangents themselves. If the curve is symmetrical, the same five coefficients may result in a ninth-degree polynomial with all the even coefficients being 0.

Exercise 7.2 Answers 2 and 4. Answer 1 misses the binomial coefficients. For the third-degree Bézier, they are (1, 3, 3, 1).

Answer 3 masquerades as a legit second-degree Bézier and has its binomial coefficients right, but the degree of all its basis polynomials is 3, not 2.

Exercise 7.3 The third one. Actually, you can still govern the tangent vector at a point by putting the neighboring control point in the right position, but this way isn't explicit.

Exercise 7.4 This question is an open question; there could be different approaches. Here's a hint: the speed of the boat is set by the length of the tangent vector, and the tangent vector is the result of the path segment equations' differentiation.

Summary

- Polynomial modeling allows you to build functions from arbitrary sets of differential and integral conditions, which in turn means that you can build curves that not only go through some given points, but also have given tangent vectors or curvature values in them.
- Classic Bézier functions are weighted sums of polynomial bases and, as such, are also polynomials.
- Rational Bézier functions are more than polynomials, allowing us to describe conic sections such as ellipses and hyperbolas exactly.
- NURBS is an acronym for nonuniform rational basis spline. Nonuniformity lets us pick which points lie on a NURBS curve exactly and which help control its shape. Rationality allows us to describe important curves and surfaces such as ellipses, spheres, and cylinders. Basis degree allows us to select the degree of continuity. And the word spline means a compound function. A spline may have as many control points as you want and not suffer from the Runge phenomenon.

8 Nonlinear transformations and surfaces

This chapter covers

- Performing polynomial transformations in multidimensional space
- Understanding how a multidimensional transformation is the same as a surface
- Using spatial interpolation to build a deformation field

Before this chapter, we barely spoke about 3D space. That's because the math behind linear equations and projective transformations doesn't change much when you go from 2D to 3D. And because things are usually easier to understand in 2D, there was no good reason to add a new dimension—until now. This chapter is about building and manipulating 3D objects.

To get there, we'll once again start with a 2D transformation, but this time, it'll be nonlinear. In chapter 4, we used a linear transformation to make a scanning app. Adding nonlinearity enhances the app by allowing bent book pages to appear flat. Next, we'll reuse the math behind nonlinear transformations to build 3D surfaces. In chapter 5, we learned about parametric curves; now we'll see how parametric surfaces are made. Finally, we'll define a 3D deformation field to help us model not one surface but a whole family of 3D surfaces.

8.1 Polynomial transformations in multidimensional space

In chapter 6, we composed an interpolating polynomial from "a polynomial P in x is y" equations. In chapter 7, we added a few variants, such as "a polynomial P in x has its derivative dx" and "the integral of a polynomial P on the interval $[a, b]$ is c." This gave us a lot of control of the polynomial models we wanted to create, and all we had to do to get the polynomials' coefficients explicitly was solve linear systems.

In multidimensional space, we need multivariate polynomials, which are still polynomials, consisting of additions and multiplications, but now they have more than one variable. Examples of these polynomials are

$$P(x, y) = x^2y^2 + 3x + 2y + 1$$

and

$$P(x, y, z) = x^3 + y^2 + xyz$$

and even this fully linear one:

$$P(x_1, x_2, x_3, x_4) = 3x_1 + 2x_2 + x_3 + 5x_4$$

Now we have more variables in the polynomials, but when we put these polynomials in a kind of Vandermonde system (chapter 6), they still turn into linear equations. By solving such systems, we still obtain the polynomials' coefficients. Business as usual.

8.1.1 The straightforward approach to polynomial transformations

Remember the bilinear transformation from chapter 4 (figure 8.1)?

$$x_t = a_{11}\, x_i\, y_i + a_{12}\, x_i + a_{13}\, y_i + a_{14}$$

$$y_t = a_{21}\, x_i\, y_i + a_{22}\, x_i + a_{23}\, y_i + a_{24}$$

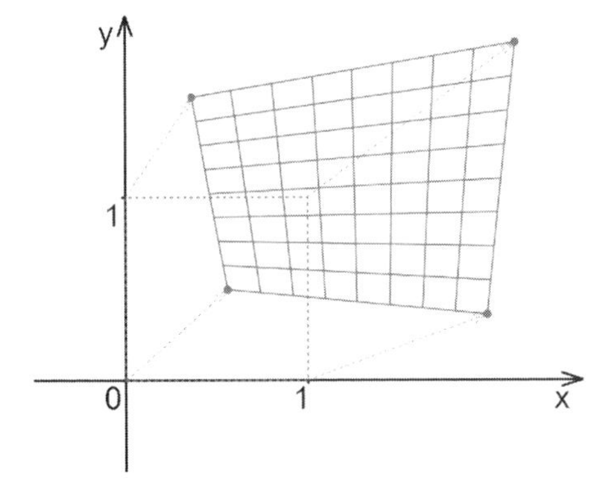

Figure 8.1 An example of a bilinear transformation of the standard square

In bilinear transformation, every coordinate has its own formula, which is a product of two linear polynomials, which is also a polynomial of two variables:

$$(ax + b)(cy + d) = acxy + adx + bcy + bd$$

Now, a polynomial member is not only a scaled power of x, but also a scaled product of a power of x and power of y. Because a bilinear polynomial is a product of two linear single variable polynomials, each member may have its x or y powered only 0 or 1:

$$P_x(x, y) = a_{11}\, x_i\, y_i + a_{12}\, x_i + a_{13}\, y_i + a_{14}$$

$$P_y(x, y) = a_{21}\, x_i\, y_i + a_{22}\, x_i + a_{23}\, y_i + a_{24}$$

In 2D, the bilinear transformation resembles the projective transformation because it's also governable by only four points. Setting how the space transforms in four points sets the whole transformation anywhere, and at this point, you should already see why.

Each bilinear polynomial has four coefficients, and to program the transformation, we need two polynomials: one for x and one for y. That's eight coefficients total. One point transformation gives us the correspondence between a pair of points (x_i, y_i), (x_t, y_t), which leads to a pair of equations. To make a well-defined system that will bring us all eight coefficients, we need eight equations. That's four pairs of points, which makes the transformation a four-point transformation.

With a four-point transformation, we can move four corners of a rectangle, for example, and by this move alone determine the bilinear transformation of the whole space. Let's see how to write the point transformation equations in SymPy. The most straightforward way is to go for the "equation soup" strategy and put all we know about the target transformation into a single system. All the equations are linear, and solving a system of eight linear equations symbolically isn't beyond the realm of possibility. In matrix form, the equation looks like this:

$$\begin{pmatrix} x_1y_1 & x_1 & y_1 & 1 & 0 & 0 & 0 & 0 \\ 0 & 0 & 0 & 0 & x_1y_1 & x_1 & y_1 & 1 \\ x_2y_2 & x_2 & y_2 & 1 & 0 & 0 & 0 & 0 \\ 0 & 0 & 0 & 0 & x_2y_2 & x_2 & y_2 & 1 \\ x_3y_3 & x_3 & y_3 & 1 & 0 & 0 & 0 & 0 \\ 0 & 0 & 0 & 0 & x_3y_3 & x_3 & y_3 & 1 \\ x_4y_4 & x_4 & y_4 & 1 & 0 & 0 & 0 & 0 \\ 0 & 0 & 0 & 0 & x_4y_4 & x_4 & y_4 & 1 \end{pmatrix} \begin{pmatrix} a_{11} \\ a_{12} \\ a_{13} \\ a_{14} \\ a_{21} \\ a_{22} \\ a_{23} \\ a_{24} \end{pmatrix} = \begin{pmatrix} x_{t1} \\ y_{t1} \\ x_{t2} \\ y_{t2} \\ x_{t3} \\ y_{t3} \\ x_{t4} \\ y_{t4} \end{pmatrix}$$

Now let's define the symbols we need, form the system, and solve it in SymPy, as shown in listing 8.1 (ch_08/bilinear_transformation.py in the book's source code).

Listing 8.1 A straightforward approach for the bilinear transformation

```
a11, a12, a13, a14 = symbols('a11 a12 a13 a14')
a21, a22, a23, a24 = symbols('a21 a22 a23 a24')
x1, x2, x3, x4 = symbols('x1 x2 x3 x4')
y1, y2, y3, y4 = symbols('y1 y2 y3 y4')
xt1, xt2, xt3, xt4 = symbols('xt1 xt2 xt3 xt4')
yt1, yt2, yt3, yt4 = symbols('yt1 yt2 yt3 yt4')
```

The symbols are more or less like those in the formula, except that the initial positions of points lose the i index for brevity.

```
system = [
    a11*x1*y1 + a12*x1 + a13*y1 + a14 - xt1,
    a21*x1*y1 + a22*x1 + a23*y1 + a24 - yt1,
    a11*x2*y2 + a12*x2 + a13*y2 + a14 - xt2,
    a21*x2*y2 + a22*x2 + a23*y2 + a24 - yt2,
    a11*x3*y3 + a12*x3 + a13*y3 + a14 - xt3,
    a21*x3*y3 + a22*x3 + a23*y3 + a24 - yt3,
    a11*x4*y4 + a12*x4 + a13*y4 + a14 - xt4,
    a21*x4*y4 + a22*x4 + a23*y4 + a24 - yt4
]
```

The system consists of four pairs of "x transforms to xt" and "y transforms to yt" equations.

```
all_as = solve(system, (a11, a12, a13, a14, a21, a22, a23, a24))
```

On my computer, the solver works for about a quarter of a second.

From solving the system, we get our polynomial coefficients as formulas, where four points before and after the transformation are the formulas' input. The formulas are quite large, so let's peek into only one:

```
a11: (-x1*xt2*y3 + x1*xt2*y4 + x1*xt3*y2 - x1*xt3*y4 - x1*xt4*y2 +
    x1*xt4*y3 + x2*xt1*y3 - x2*xt1*y4 - x2*xt3*y1 + x2*xt3*y4 +
    x2*xt4*y1 - x2*xt4*y3 - x3*xt1*y2 + x3*xt1*y4 + x3*xt2*y1 -
    x3*xt2*y4 - x3*xt4*y1 + x3*xt4*y2 + x4*xt1*y2 - x4*xt1*y3 -
    x4*xt2*y1 + x4*xt2*y3 + x4*xt3*y1 - x4*xt3*y2)/
    (x1*x2*y1*y3 - x1*x2*y1*y4 - x1*x2*y2*y3 + x1*x2*y2*y4 - x1*x3*y1*y2 +
    x1*x3*y1*y4 + x1*x3*y2*y3 - x1*x3*y3*y4 + x1*x4*y1*y2 - x1*x4*y1*y3 -
    x1*x4*y2*y4 + x1*x4*y3*y4 + x2*x3*y1*y2 - x2*x3*y1*y3 - x2*x3*y2*y4 +
    x2*x3*y3*y4 - x2*x4*y1*y2 + x2*x4*y1*y4 + x2*x4*y2*y3 - x2*x4*y3*y4 +
    x3*x4*y1*y3 - x3*x4*y1*y4 - x3*x4*y2*y3 + x3*x4*y2*y4),
```

The other seven formulas are pretty much alike. We can reuse them as code in our own projects. To do so, we should put the guiding points in the expressions either in precomputation or at run time, get all the coefficients, and use the coefficients to build polynomials for both x and y. This approach works—so far.

Let's raise the degree of our transformation to see whether we can keep our approach. A product of two quadratic polynomials would be a biquadratic polynomial, a pair of which in turn describes a corresponding biquadratic polynomial transformation (figure 8.2):

$$x_t = a_{11}\, x_i^2\, y_i^2 + a_{12}\, x_i^2\, y_i + a_{13}\, x_i^2 + a_{14}\, x_i\, y_i^2 + a_{15}\, x_i\, y_i + a_{16}\, x_i + a_{17}\, y_i^2 + a_{18}\, y_i + a_{19}$$

$$y_t = a_{21}\, x_i^2\, y_i^2 + a_{22}\, x_i^2\, y_i + a^{23}\, x_i^2 + a_{24}\, x_i\, y_i^2 + a_{25}\, x_i\, y_i + a_{26}\, x_i + a_{27}\, y_i^2 + a_{28}\, y_i + a_{29}$$

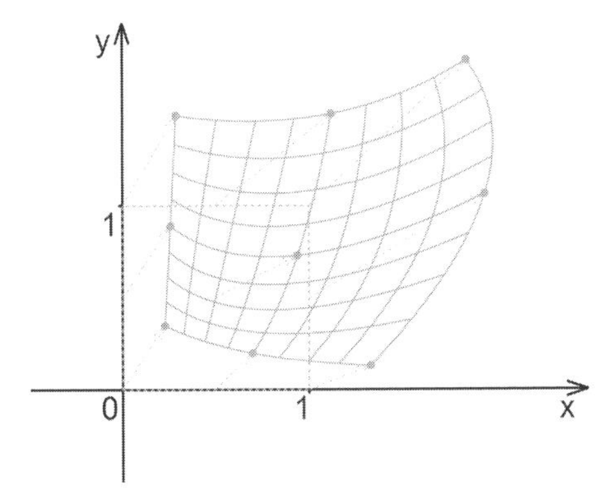

Figure 8.2 An example of a biquadratic transformation of the standard square

Now the polynomials are of the second degree, which means that both *x* and *y* are now represented in degrees 0, 1, and 2. We have 3 × 3 = 9 members in each polynomial, which also means that this transformation is a 9-point transformation.

> **NOTE** We all realize that, for example, $2x^2 + 4x + 1$ is the same polynomial as $4x + 1 + 2x^2$, but the conventional way to write it is the former. The degree of each term decreases from left to right. With two-variable polynomials, this convention stops being unambiguous. Variable *x* has no inherent precedence over *y*; both are simply variables. There's no particular reason why x^2y should come before xy^2. We do need some sort of convention, though, so let's agree that the variable with the lowest position in the alphabet should appear with the higher degree first: $x^2y + xy^2$, not $xy^2 + x^2y$. Again, the polynomials are equal; we merely need some certainty in the notation.

We can still compose an "equation soup" system for the 9-point transformation. But it will consist of 18 equations, two per polynomial member because we need both polynomials: one for *x* and one for *y*. Solving a system of 18 equations symbolically may take too much time to be practical.

On the other hand, if we look closely at our equations, we see that the *x* equations are independent of the *y*s. The former has all the coefficients $a_{1\dots}$ and the latter all the coefficients $a_{2\dots}$. The coefficients from the *x* don't diffuse into the *y* formula, or vice versa. In the case of a bilinear transformation, we can rewrite the matrix equation as follows simply by swapping rows:

$$\begin{pmatrix} x_1y_1 & x_1 & y_1 & 1 & 0 & 0 & 0 & 0 \\ x_2y_2 & x_2 & y_2 & 1 & 0 & 0 & 0 & 0 \\ x_3y_3 & x_3 & y_3 & 1 & 0 & 0 & 0 & 0 \\ x_4y_4 & x_4 & y_4 & 1 & 0 & 0 & 0 & 0 \\ 0 & 0 & 0 & 0 & x_1y_1 & x_1 & y_1 & 1 \\ 0 & 0 & 0 & 0 & x_2y_2 & x_2 & y_2 & 1 \\ 0 & 0 & 0 & 0 & x_3y_3 & x_3 & y_3 & 1 \\ 0 & 0 & 0 & 0 & x_4y_4 & x_4 & y_4 & 1 \end{pmatrix} \begin{pmatrix} a_{11} \\ a_{12} \\ a_{13} \\ a_{14} \\ a_{21} \\ a_{22} \\ a_{23} \\ a_{24} \end{pmatrix} = \begin{pmatrix} x_{t1} \\ x_{t2} \\ x_{t3} \\ x_{t4} \\ y_{t1} \\ y_{t2} \\ y_{t3} \\ y_{t4} \end{pmatrix}$$

Then we can solve it as a pair of smaller independent systems:

$$\begin{pmatrix} x_1y_1 & x_1 & y_1 & 1 \\ x_2y_2 & x_2 & y_2 & 1 \\ x_3y_3 & x_3 & y_3 & 1 \\ x_4y_4 & x_4 & y_4 & 1 \end{pmatrix} \begin{pmatrix} a_{11} \\ a_{12} \\ a_{13} \\ a_{14} \end{pmatrix} = \begin{pmatrix} x_{t1} \\ x_{t2} \\ x_{t3} \\ x_{t4} \end{pmatrix}$$

$$\begin{pmatrix} x_1y_1 & x_1 & y_1 & 1 \\ x_2y_2 & x_2 & y_2 & 1 \\ x_3y_3 & x_3 & y_3 & 1 \\ x_4y_4 & x_4 & y_4 & 1 \end{pmatrix} \begin{pmatrix} a_{21} \\ a_{22} \\ a_{23} \\ a_{24} \end{pmatrix} = \begin{pmatrix} y_{t1} \\ y_{t2} \\ y_{t3} \\ y_{t4} \end{pmatrix}$$

For the biquadratic equations, instead of solving a system of 18 equations, we can also solve 2 systems of 9. Let's take a look at one of them:

```
x_system = [
    a11*x1*x1*y1*y1 + a12*x1*x1*y1 + a13*x1*x1 + a14*x1*y1*y1 +
➥  a15*x1*y1 + a16*x1 + a17*y1*y1 + a18*y1 + a19 - xt1,
    a11*x2*x2*y2*y2 + a12*x2*x2*y2 + a13*x2*x2 + a14*x2*y2*y2 +
➥  a15*x2*y2 + a16*x2 + a17*y2*y2 + a18*y2 + a19 - xt2,
    a11*x3*x3*y3*y3 + a12*x3*x3*y3 + a13*x3*x3 + a14*x3*y3*y3 +
➥  a15*x3*y3 + a16*x3 + a17*y3*y3 + a18*y3 + a19 - xt3,
    a11*x4*x4*y4*y4 + a12*x4*x4*y4 + a13*x4*x4 + a14*x4*y4*y4 +
➥  a15*x4*y4 + a16*x4 + a17*y4*y4 + a18*y4 + a19 - xt4,
    a11*x5*x5*y5*y5 + a12*x5*x5*y5 + a13*x5*x5 + a14*x5*y5*y5 +
➥  a15*x5*y5 + a16*x5 + a17*y5*y5 + a18*y5 + a19 - xt5,
    a11*x6*x6*y6*y6 + a12*x6*x6*y6 + a13*x6*x6 + a14*x6*y6*y6 +
➥  a15*x6*y6 + a16*x6 + a17*y6*y6 + a18*y6 + a19 - xt6,
    a11*x7*x7*y7*y7 + a12*x7*x7*y7 + a13*x7*x7 + a14*x7*y7*y7 +
➥  a15*x7*y7 + a16*x7 + a17*y7*y7 + a18*y7 + a19 - xt7,
    a11*x8*x8*y8*y8 + a12*x8*x8*y8 + a13*x8*x8 + a14*x8*y8*y8 +
➥  a15*x8*y8 + a16*x8 + a17*y8*y8 + a18*y8 + a19 - xt8,
    a11*x9*x9*y9*y9 + a12*x9*x9*y9 + a13*x9*x9 + a14*x9*y9*y9 +
➥  a15*x9*y9 + a16*x9 + a17*y9*y9 + a18*y9 + a19 - xt9]
```

Solving a system of nine equations symbolically is possible, but it still takes a lot of time. We can simplify equations by selecting the input points from the standard square, as we did in chapter 4; this approach will make the solving process much faster. But it means that we'll have to give up on the idea of a generic nine-point transformation.

The whole approach to putting the point transformation equations in a single system isn't scalable. With the arbitrary point positions, it barely works for biquadratic transformations in which we have to solve systems of nine equations. With bicubic transformations, the systems will be 16 equations each; with biquartic transformations, the systems will be 25 equations each, and so on.

As we raise the polynomial degree, the systems get larger, more difficult to solve, and harder to compose even by hand. At some point, we'll need a Python script to write our Python script for us. We need an alternative approach with simpler formulas and, therefore, simpler code.

What about numeric problems?

One less obvious drawback of this approach is the numeric error of the resulting solution. As we discussed in chapter 2, floating-point operations are inherently imprecise, and as we saw in chapter 3, the error may accumulate as the operations go on. Well, with the straightforward solution for linear systems, the error rises as we increment the system's size.

Optimizing expressions for better accuracy is a complicated task that lies far beyond the scope of this book. Luckily, as we use SymPy to do our math for us, we can use Herbie to do numeric optimization.

Herbie isn't part of the Python infrastructure, it's a standalone tool. But like SymPy, it's available online for you to try it out. You can find it at https://herbie.uwplse.org.

8.1.2 The fast and simple approach to polynomial transformations

The alternative approach would be to turn bi-something equations into a product of two somethings. Observe!

As I've mentioned, a bilinear expression is a product of two linear polynomials. This one, for example,

$$a_1\, x_i\, y_i + a_2\, x_i + a_3\, y_i + a_4$$

is the product of

$$(a_{x1} x + a_{x2})\,(a_{y1} y + a_{y2})$$

where

$$a_1 = a_{x1}\, a_{y1}$$

$$a_2 = a_{x1}\, a_{y2}$$

$$a_3 = a_{x2}\, a_{y1}$$

$$a_4 = a_{x2}\, a_{y2}$$

Technically, we don't have to find all the *aij* coefficients explicitly to build a transformation. We could program any bipolynomial expression—including bilinear, biquadratic, or bicubic—as a product of two single-variable polynomials. Let's see how this process works in practice, once again starting with the bilinear interpolation.

In a single-variable (or 1D) case, bilinear interpolation boils down to linear interpolation. If we define our input points in x_1=0 and x_2=1 as y_1 and y_2 (figure 8.3), the formula for the linear interpolation is

$$y(x) = (1 - x)\, y_1 + xy_2$$

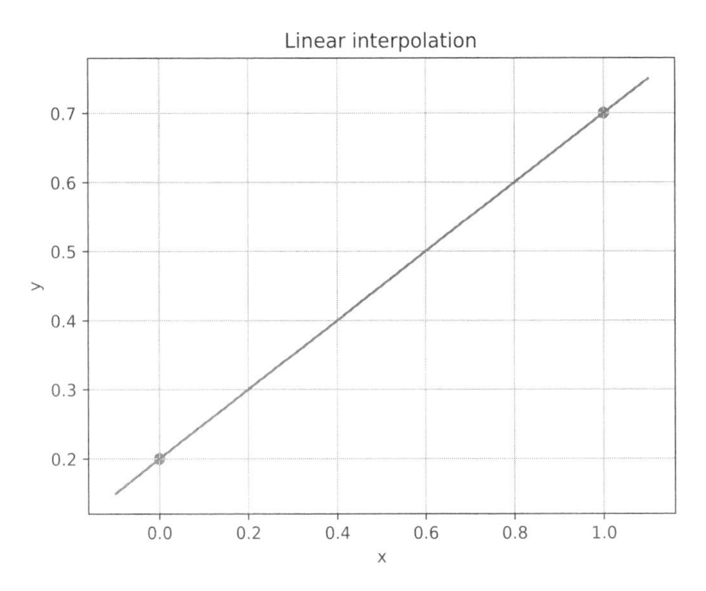

Figure 8.3 Single-variable linear interpolation for points (0, 0), and (1, 1)

In a 2D case, let's say we have four (x, y) points with values z in them: (0, 0) with a value z_{11}, (0, 1) with z_{12}, (1, 0) with z_{21}, and (1, 1) with z_{22}. This example gives us enough fuel for equations, of course, and we can find the bilinear transformation coefficients by solving a system like the one in the preceding section. But let's try something different: interpolate the interpolants. On the line from (0, 0) to (0, 1), the value z changes according to linear, not bilinear, interpolation:

$$z_1(y) = (1 - y)\, z_{11} + yz_{12}$$

On the line from (1, 0) to (1, 1), the value z changes accordingly:

$$z_2(y) = (1 - y)\, z_{21} + yz_{22}$$

Now between these lines, we can use linear interpolation to interpolate between $z_1(y)$ and $z_2(y)$ at any y:

$$z(x, y) = (1 - x)\, z_1(y) + xz_2(y)$$

Using linear interpolation to interpolate the y functions by x makes the resulting interpolant bilinear (figure 8.4).

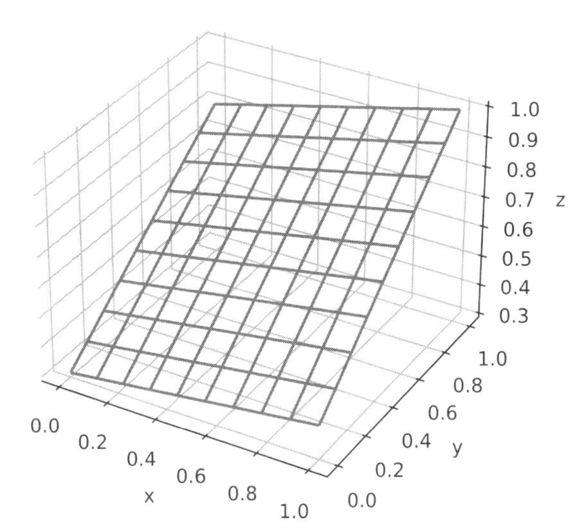

Figure 8.4 Bilinear interpolation for z_{ij} defined in points (x_j, y_i)

With this approach, we don't have to deal with large linear systems and large formulas; it scales nicely to biquadratic, bicubic interpolation, and so forth. But before we go too far, let's see how it works for biquadratic interpolation. Let's define our points in

(0, 0) as z_{11}	(0, 0.5) as z_{12}	(0, 1) as z_{13}
(0.5, 0) as z_{21}	(0.5, 0.5) as z_{22}	(0.5, 1) as z_{23}
(1, 0) as z_{31}	(1, 0.5) as z_{32}	(1, 1) as z_{33}

Now we need a formula for single-variable quadratic interpolation based on points $(0, z_1)$, $(0.5, z_2)$, and $(1, z_3)$ (figure 8.5).

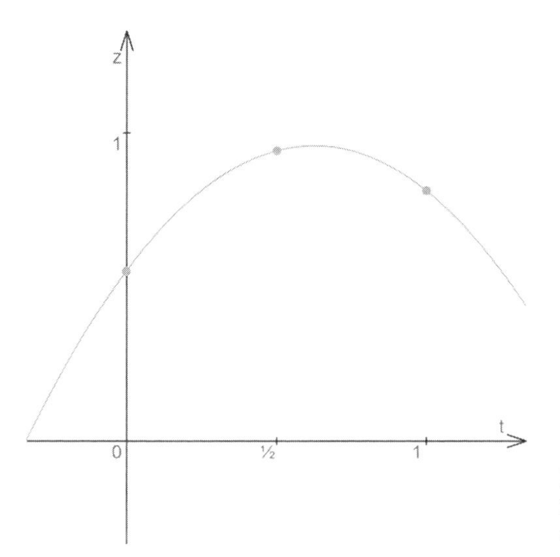

Figure 8.5 A single variable interpolation of three variables defined in 0, 0.5, and 1

We can Google the formula, or we can coin it ourselves with SymPy. We already did that in chapter 6 with polynomial interpolation. The coefficients for the resulting polynomial simply come from solving three "the polynomial graph goes through the point" equations, as shown in the following listing (ch_08/getting_quadratic_interpolation.py in the book's source code).

Listing 8.2 Getting a quadratic interpolant for specific points

```
from sympy import *

z1, z2, z3, a, b, c = symbols('z1 z2 z3 a b c')

print(solve([
    c - z1,                          # A generic (works for both x and y) quadratic polynomial in 0
    a * 0.25 + b * 0.5 + c - z2,     # A quadratic polynomial in 0.5
    a + b + c - z3,                  # A quadratic polynomial in 1
    ], (a, b, c)))
```

The result is

```
{a: 2.0*z1 - 4.0*z2 + 2.0*z3, b: -3.0*z1 + 4.0*z2 - z3, c: z1}
```

or, rewritten as a Python function that does the actual interpolation,

```
def P2(z1, z2, z3, x):
    a = 2.0*z1 - 4.0*z2 + 2.0*z3
    b = -3.0*z1 + 4.0*z2 - z3
    c = z1
    return a*x*x + b*x + c
```

As with bilinear interpolation, we define the interpolants for the *y*-oriented lines first and then interpolate them with a single-*x*-variable interpolation:

```
def z1(y): return P2(z11, z12, z13, y)     # Three
def z2(y): return P2(z21, z22, z23, y)     # y-interpolants
def z3(y): return P2(z31, z32, z33, y)
def z(x, y): return P2(z1(y), z2(y), z3(y), x)    # A single x-interpolant brings them up together.
```

That's it. In figure 8.6, the biquadratic interpolating function is built for the following values:

$$z_{11} = 0.3;\ z_{12} = 0.5;\ z_{13} = 0.9$$

$$z_{21} = 0.4;\ z_{22} = 0.7;\ z_{23} = 1.1$$

$$z_{31} = 0.8;\ z_{32} = 1.4;\ z_{33} = 1.5$$

See the full listing, plotting included, in ch_08/biquadratic_interpolation.py in the book's source code.

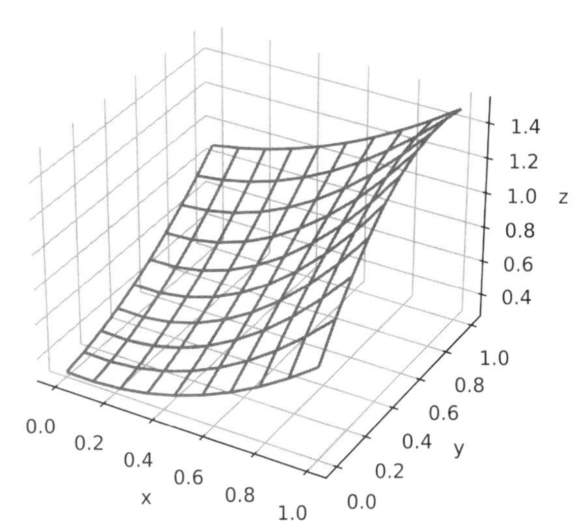

Figure 8.6 A biquadratic interpolation of values z_{ij} in points (x_j, y_i)

The only downside is that if you use this interpolation to define a transformation, your guiding points for the transformation have to be predefined in certain positions (figure 8.7).

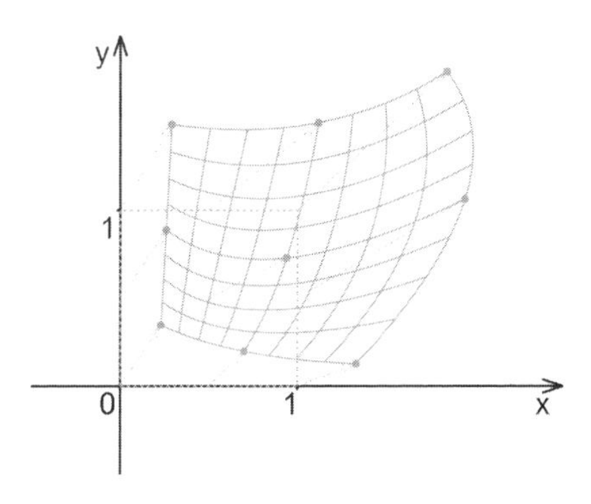

Figure 8.7 A biquadratic transformation as a product of quadratic interpolations

This approach scales nicely to any degree of polynomials and also to any number of dimensions. You can make trilinear or triquadratic transformations in 3D. Moreover, with this approach, the transformation doesn't have to be just bi- or tri-something. It could be linear by one axis and cubic by the other, for example, so it could be linear-cubic. Interestingly, this linear-cubic transformation has a nice practical application.

8.1.3 Practical example: Adding the "unbending" feature to the scanning app

My father is an amateur historian. As a historian, he works a lot with printed sources: books, maps, archive documents, old newspapers, and so on. As an amateur, though, he doesn't spend a lot of time in the archives with good scanners. More often than not, he has to settle for mobile-phone photos of books and documents instead of nice flat scanned pages.

In chapter 4, we discussed a scanning app that helps you turn a photo of a flat page into a scan of that page. But what if a page isn't flat? What if it's bent? This situation means that projective transformation isn't enough for us anymore. We need something more versatile, something that allows curvature.

In 2013, I wrote an app that runs such a transformation to unbend book pages for my father (figure 8.8).

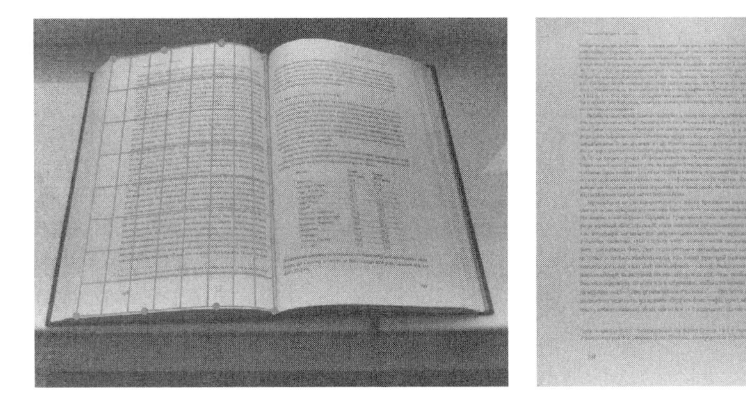

Figure 8.8 The book photo and an example unbent page

The transformation is cubic by the *x*-axis, where the page is bent, and linear by the *y*-axis, where the page is normally straight. We can program this transformation by using the same logic as before: first, we define a single-variable cubic interpolation, and then we interpolate four linear interpolations with it.

Let's say that our points before the transformation are placed evenly along the lines of the standard square and that to set up the points after the transformation, we'll need the user's help. Computer-vision algorithms can do this work for us, but they go way beyond the scope of this book. So let's presume that our user is kind enough to "draw" the page as it appears in figure 8.8 by moving the control points manually. Then the transformation transforms the standard square onto a photo with a bent page in it. As you may remember from chapter 4, this kind of transformation is exactly what we want. We already know how to transform pixel coordinates from a nice flat rectangular page to a standard square, so when we compose the page-to-square and square-to-photo transformations together, we get the way to answer the question "Which color should a pixel on a rectangular page be?" For the single-variable cubic interpolation, the points along the line are located at 0, $\frac{1}{3}$, $\frac{2}{3}$, and 1 (figure 8.9). Let's find the formulas for the cubic polynomial coefficients, as shown in listing 8.3 (ch_08/getting_cubic_interpolation.py in the book's source code).

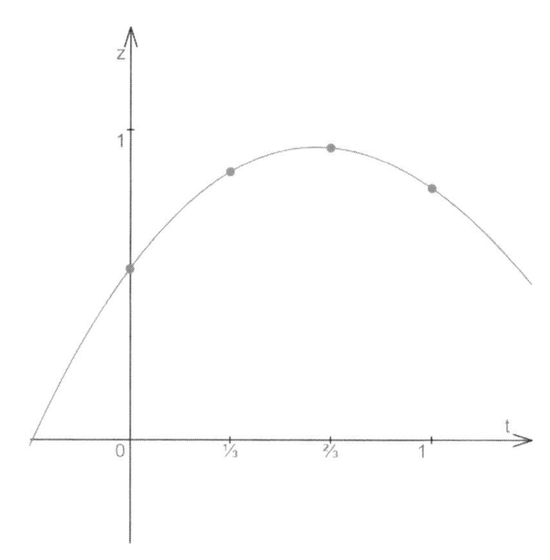

Figure 8.9 A single-variable interpolation of four values that gives us a four-variable system

Listing 8.3 Getting coefficients for a generic cubic interpolation formula

```
from sympy import *

z1, z2, z3, z4, a, b, c, d = symbols('z1 z2 z3 z4 a b c d')

print(solve([
    d - z1,
    a * 1./27. + b * 1./9. + c * 1./3. + d - z2,
    a * 8./27. + b * 4./9. + c * 1./3. + d - z3,
    a + b + c + d - z4,
    ], (a, b, c, d)))
```

The coefficients' formulas are

```
a: 5.4*z1 - 2.7*z2 - 5.4*z3 + 2.7*z4
b: -4.2*z1 - 0.9*z2 + 7.2*z3 - 2.1*z4
c: -2.2*z1 + 3.6*z2 - 1.8*z3 + 0.4*z4
d: z1
```

Once again, let's compute the coefficients and turn them into the interpolating polynomial:

```
def P3(z1, z2, z3, z4, x):
    a = 5.4*z1 - 2.7*z2 - 5.4*z3 + 2.7*z4
    b = -4.2*z1 - 0.9*z2 + 7.2*z3 - 2.1*z4
    c = -2.2*z1 + 3.6*z2 - 1.8*z3 + 0.4*z4
    d = z1
    return a*x*x*x + b*x*x + c*x + d
```

Given that we have the `P1(z1, z2, y)` function for linear interpolation, the whole interpolant looks like this (for the full listing with plotting, see ch_08/linear_cubic_interpolation.py in the book's source code):

```
def z1(y): return P1(z11, z12, y)
def z2(y): return P1(z21, z22, y)
def z3(y): return P1(z31, z32, y)
def z4(y): return P1(z41, z42, y)
def z(x, y): return P3(z1(y), z2(y), z3(y), z4(y), x)
```

Four linear y-interpolants

A single cubic x-interpolant makes them into a two-variable function.

This interpolation does what we want. It's linear by one axis and cubic by the other, so when it's turned into a transformation, it also keeps pages straight by one axis but allows them to bend by the other. In figure 8.10, you see a plot for the linear-cubic interpolation of the following values:

$$z_{11} = 0.5; z_{12} = 0.6$$

$$z_{21} = 0.4; z_{22} = 0.5$$

$$z_{31} = 0.7; z_{32} = 0.9$$

$$z_{41} = 0.6; z_{42} = 0.7$$

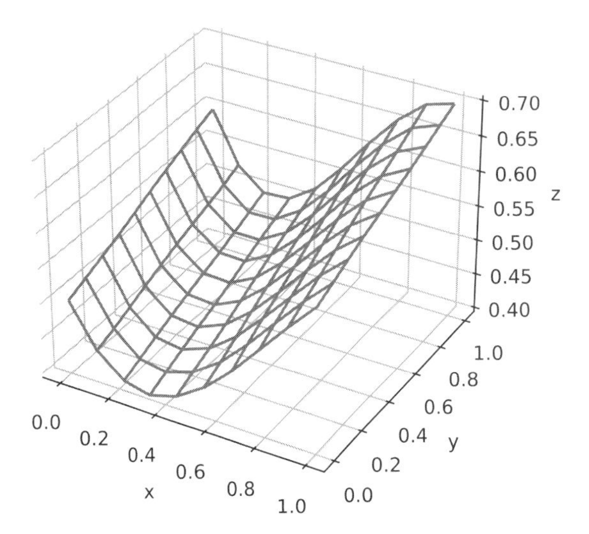

Figure 8.10 Linear-cubic interpolation

Now when we have the interpolation, the transformation is easy to achieve. Instead of interpolating some abstract *z* values, we'll interpolate both *x* and *y* of the guiding points. Instead of a single `z(x, y)`, we have to introduce two functions, `xt(x, y)` and `yt(x, y)`, both of them being linear-cubic interpolation of the grid points of the singular square. Figure 8.11 shows an example of such a transformation.

To do the unbending itself, we'll traverse the resulting rectangular bitmap as we did in section 4.4.1. For every pixel [i, j], the pixel's coordinates on a standard square will be `(x, y) = (j/bitmap.width, i/bitmap.height)`.

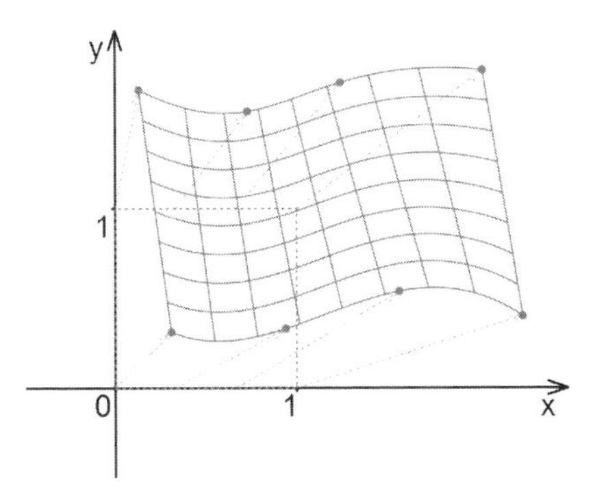

Figure 8.11 An example of a linear-cubic transformation suitable for page unbending

NOTE Do you remember the convention? Normally, *j* is a horizontal index in a bitmap, so it corresponds to the bitmap width and the *x*-axis; *i* is the vertical index, so it corresponds to height and the *y*-axis.

Then the photo's corresponding pixel has coordinates `xt(j/bitmap.width, i/bitmap.height)` and `yt(j/bitmap.width, i/bitmap.height)`. Put its color in the resulting rectangular bitmap, and our unbending is done!

8.1.4 Section 8.1 summary

Polynomial transformations scale nicely to all degrees and dimensions. Using the same approach, you can have a trilinear transformation, a biquadric transformation, or a linear transformation by one axis and a cubic one by another. In the simplest cases, you can compute the polynomial's coefficients explicitly, as we do in single-variable interpolation. But by keeping the data points aligned to the grid, we can decompose our *n*-variable interpolation into *n* single-variable ones with formulas that are much simpler and faster to compute.

8.2 3D surface modeling

A shift from 2D to 3D may seem to be challenging. Indeed, the explanations and examples of 2D things like curves and transformations are generally easier to understand than those of their 3D counterparts. What makes learning hard is that way too often, we're bound to a 2D medium, such as a book page or a screen. Also, the formulas are simply getting larger, with more variables in them. This kind of complexity, however, isn't essential; rather, it's accidental. The mathematics behind objects in both spaces is largely the same. This section shows that all you've learned so far is already enough for you to understand and construct 3D surfaces.

8.2.1 A surface is just a 3D transformation of a plane

As you may have noticed, the functions of two variables have their graphs as surfaces. But hold on—we're not done yet. The graphs are fine, but you can't make a graph into a closed surface, such as a sphere or a cylinder. To have absolute freedom in construction, you have to do one more step and start thinking about surfaces not in terms of graphs, but in terms of transformations.

As a reminder, a *transformation* is a function with both the domain and codomain being points in the same space. A 3D transformation, therefore, is a function like this one:

$$f: (x, y, z) \rightarrow (x_t, y_t, z_t)$$

This function transforms any point in 3D space into some other (or sometimes the same) point of that space. We don't have to transform all the space; we can choose to transform only a plane. We might transform the *x*-*y* plane, for example, which would involve all the points $(x, y, 0)$, where both x and y are real numbers. Then the 3D transformation transforms the *x*-*y* plane into some other surface in 3D. The process is a bit like bending sheet metal, but the sheet is infinite, and instead of a bending machine, we have a transformation function.

Because our initial z will always stay 0, there's no point in mentioning it in the formula at all. Then our surface becomes a two-variable point function. Or we can see it as three separate coordinate functions, all of them two-variable,

$$f: (x, y) \rightarrow (x_t, y_t, z_t)$$

or as

$$f_x: (x, y) \rightarrow x_t$$

$$f_y: (x, y) \rightarrow y_t$$

$$f_z: (x, y) \rightarrow z_t$$

Does the domain have to be in the same space where the range is? Not really. It may be, but the exact source of input variables is irrelevant to the result. So we can "host" the point function in some other 2D space, not necessarily the *x*-*y* plane. Conventionally, the variables of the domain space for surfaces are called u and v. So now the surface is a function like this one:

$$f: (u, v) \rightarrow (x, y, z)$$

This function greatly resembles the one we used to define parametric curves in 2D,

$$f(\mathrm{t}) \rightarrow (x, y)$$

but now we have more variables on both sides. Otherwise, the math for the curves, transformations, and surfaces is exactly the same (table 8.1).

Table 8.1 Geometric objects and conventional functions that represent them

Object	Function
2D curve	$f(t) \rightarrow (x, y)$
2D transformation	$f(x, y) \rightarrow (x_t, y_t)$
3D curve	$f(t) \rightarrow (x, y, z)$
3D surface	$f(u, v) \rightarrow (x, y, z)$
3D transformation	$f(x, y, z) \rightarrow (x_t, y_t, z_t)$

We can still use polynomial interpolation for surface modeling, for example, as we did earlier in this chapter to define transformations. We simply have to pick three two-variable polynomials:

$$f(u, v) = (P_x(u, v), P_y(u, v), P_z(u, v))$$

In modeling, we usually define the surface's borders in UV space as well. If we only want a squarish thing with no complex form or holes in it, the conventional borders are the borders of the standard square:

$$(u, v) = [0..1]\times[0..1]$$

But sometimes, it's more convenient to work in units that closely correspond to the units in *x-y-z* space with the resulting surface so we ignore the convention.

To sum things up, a surface with borders is a transformed piece of a plane. Usually but not necessarily, it's a standard square of UV space. If we talk about polynomial surface modeling, the transforming functions for every coordinate are polynomials.

8.2.2 Practical example: Planting polynomial mushrooms

In 2010, I helped my brother make the Procedure Project game, a 3D shooter in which all the terrain was procedurally generated based on the contents of the user's disk space and processes (figure 8.12). The premise was to rail through the virtual representation of your own PC shooting bugs—which, to a pair of programmers who were spending most of their lives doing real-life debugging, sounded like a great deal of fun. The game didn't sell well, though.

The procedurally generated landscapes differ from stage to stage. My favorite is the mushroom desert (figure 8.12, right) because it incorporates not one but two geometric problems that are applicable to this chapter. The landscape itself—hills and valleys—is a function graph. A function $F(x,y) \rightarrow z$ sets the elevation for every point on the (x, y)-plane, and the landscape is a surface made of $(x, y, F(x, y))$ points. The function isn't too interesting to us right now, but we'll meet the formula behind it later in the chapter.

Generating the mushrooms is another problem that's more relevant to this section. Let's look into this problem. We'll start by building an ugly low-polygonal mushroom

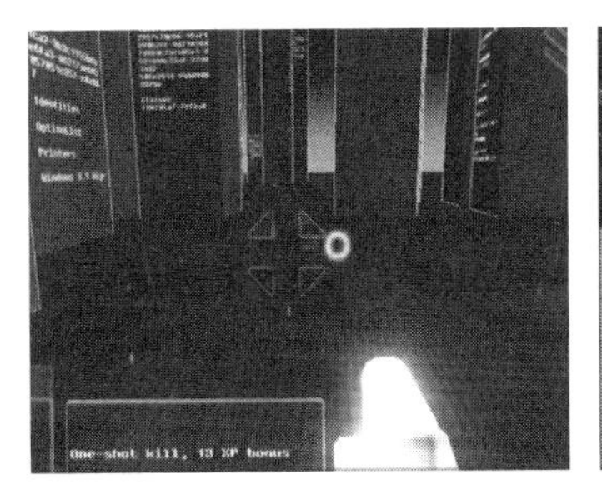  

Figure 8.12 Screenshots from the Procedure Project

(figure 8.13) to see how it works. Then we'll make a randomized generator to plant the whole desert.

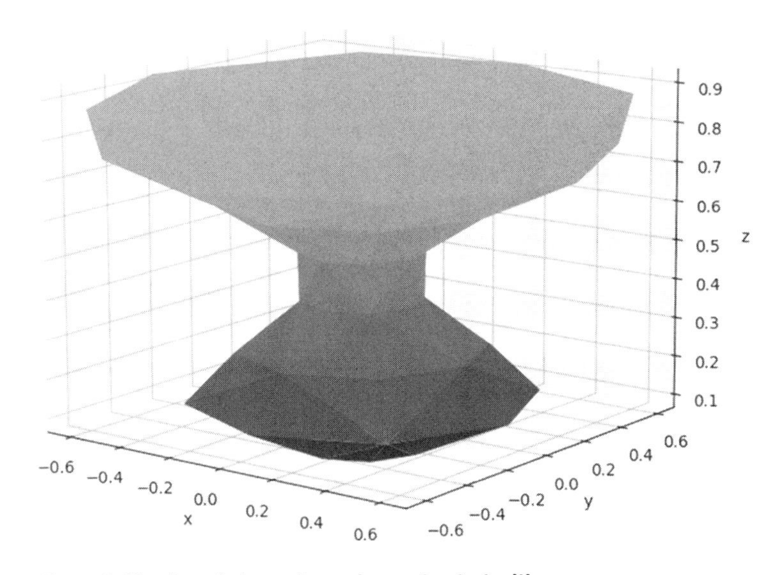

Figure 8.13 An ugly low-poly mushroom to start with

Conceptually, a mushroom here is a *surface of revolution*, meaning that it's a surface built by rotating a curve around some axis. In our case, the axis is the *z*-axis, and the curve is a polynomial curve. Realistically, though, an in-game 3D mushroom is only a bunch of triangles.

A surface is essentially a transformation of a plane. In our case, the surface is closed, so it's not the whole plane, but a closed piece of it. This is the same as a point function of two coordinate variables. We'll follow conventions and define our surface on a standard UV square:

$$S(u, v) \rightarrow (x, y, z), \text{ where u is in range } [0, 1], v \text{ is in range } [0, 1]$$

A bunch of triangles is slightly different; it's a simplification of a surface, a model of a surface, not a real thing. In this model, we fill some point grid with triangles. We don't even have to do the work ourselves; Python does the actual triangulation for us. We'll see how it's done later in the code. So we don't need a continuous function per se; we need only a 2D array of points that come from that function. Let's keep the formula from the surface, but now its domain will be a UV point grid:

$$S(u, v) \rightarrow (x, y, z),$$

$$\text{where } u \text{ is one of } \{0,\ d_u,\ 2d_u,\ ...,\ 3d_j,\ ...,\ 1\},$$

$$\text{and } v \text{ is one of } \{0,\ d_v,\ 2d_v,\ ...,\ 3d_i,\ ...,\ 1\}$$

Our model of the surface is a point grid itself. It keeps a triplet (x_{ij}, y_{ij}, z_{ij}) for every (u_j, v_i). In a more usual representation, it's the same as three 2D arrays that correspond to the UV grid: one for x_{ij}, one for y_{ij}, and one for z_{ij}.

That's what we want. We want three 2D arrays that represent the coordinates of a low-polygonal mushroom's points corresponding to the grid on a UV standard square. And all we have to do to get them is to define the function behind the mathematical, or conceptual, mushroom surface and compute it for every (u_j, v_i).

To build a surface of revolution, let's build a polynomial profile curve first. A small recap: to build a polynomial curve, we'll reuse polynomial interpolation (chapter 6). We'll pick some points, build a Vandermonde matrix for the polynomial going through the points, and use the solution of the Vandermonde equation to find the polynomial coefficients, all in 11 lines of Python code:

```
zs = [0.0, 0.25, 0.5, 0.75, 1.0]
rs = [0.2, 0.4, 0.2, 0.8, 0.0]
N = len(zs)

Vandermonde = Matrix([[zs[i]**j for j in range(N)] for i in range(N)])
Rs = Matrix(rs)
a0, a1, a2, a3, a4 = symbols('a0 a1 a2 a3 a4')
Pol = Matrix([a0, a1, a2, a3, a4])

solution = solve(Vandermonde*Pol - Rs, (a0, a1, a2, a3, a4))
a = [value for key, value in solution.items()]

def P(x, a):
    return sum([a[i]*x**i for i in range(len(a))])
```

The heights for which the mushroom radii are defined (points to `zs = ...`)

The mushroom radii themselves (points to `rs = ...`)

As this is a polynomial interpolation, we'll get the polynomial coefficients from solving a Vandermonde equation. (points to `Vandermonde = ...`)

The radii are the right side of the Vandermonde equation. (points to `Rs = Matrix(rs)`)

The polynomial coefficients themselves (refers to `a0, a1, ...` and `Pol = ...`)

With these points, we have a curvy line that starts small, grows, has a bump, goes small again, gets large again, and finally fades into 0 (figure 8.14).

Let's not go through all the details of the interpolation itself. If you see anything unfamiliar in this code, please refer to chapter 6 for the explanations. The interesting

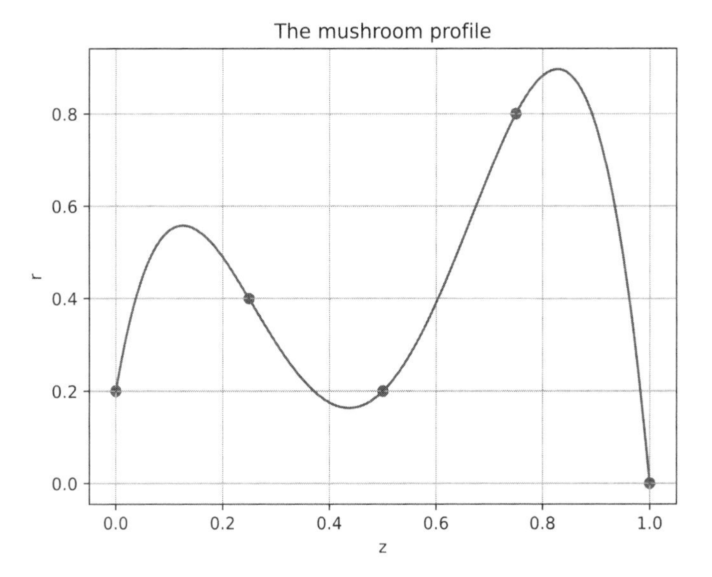

Figure 8.14 The profile curve of a mushroom, showing how the mushroom radius changes as the z-coordinate rises

part here is how to turn this curve into a surface of revolution. To do that, we need to define the x, y, and z functions of (u,v). Let's write a formula for a circle on an x-y plane as a function or some angle θ and radius r:

$$x = r\cos(\theta)$$

$$y = r\sin(\theta)$$

Our profile curve is a graph of a function that represents how radius changes with height, so we already have an $r(z)$ function. With this, we can define z, y, and z as functions of θ and z itself:

$$x = r(z)\ \cos(\theta)$$

$$y = r(z)\ \sin(\theta)$$

$$z = z$$

To turn it into a UV thing, we have to turn u that lies in the range $[0, 1]$ into θ, which will then lie in the range $[0, 2\pi]$, and also turn v from $[0, 1]$ into z that will be in $[0, z_{max}]$. Both transformations are scalings:

$$x = r(z_{max} v)\ \cos(2\pi u)$$

$$y = r(z_{max} v)\ \sin(2\pi u)$$

$$z = z_{max} v$$

It gets even simpler, given that we choose our z to run from 0 to 1 by design:

$$x = r(v)\cos(2\pi u)$$

$$y = r(v)\sin(2\pi u)$$

$$z = v$$

So this is our mushroom surface, or rather its mathematical description. Let's turn the description into Python code and a surface into a bunch of triangles. To do that, we need to go from a continuous representation of a [0, 1] × [0, 1] standard square to point grids, and from continuous segments [0, 1] to arrays with numbers put one after another with a fixed step length. In Python, specifically the NumPy library, the thing that represents arrays pretending to be continuous segments is called a *linear space* or `linspace`.

NOTE We'll use NumPy for this exercise because the array operations that NumPy provides make the code concise. But you can do the same operations in plain Python or in any other programming language, for that matter.

Let's build linear spaces for u and v. Each will consist of 9 points placed on the [0, 1] interval:

```
u = np.linspace(0, 1, 9)
v = np.linspace(0, 1, 9)
```

Their printout is

```
[0.    0.125 0.25  0.375 0.5   0.625 0.75  0.875 1.   ]
[0.    0.125 0.25  0.375 0.5   0.625 0.75  0.875 1.   ]
```

Our linear spaces in Python aren't continuous spaces, but arrays of nine numbers arranged from 0 to 1. So far so good. Now let's scale the u linear space to *theta* (θ in the preceding formulas is a Greek letter), which will run from 0 to 2π:

```
theta = u * 2 * np.pi
```

Now the theta is

```
[0.       0.78539 1.57079 2.35619 3.14159 3.92699 4.71238 5.49778 6.28318]
```

You see how we made the whole array scale with one simple line, which is what we employed NumPy for. But wait—it gets better. With NumPy, we can make a pair of regular arrays into a pair of 2D arrays filled in a per-coordinate manner in one line, too. First, let's compute our polynomial for every value of v and store it in the array that defines the radius of the mushroom for every pair of UV coordinates:

```
r, _ = np.meshgrid(P(v, a), u)
```

We don't care about the second array, so we'll assign it to the _ variable and forget that it ever existed. The first resulting array is the 2D array of radii that we want. Here's how it looks:

```
[[0.2 0.5578125 0.4 0.1953125 0.2 0.4578125 0.8 0.8453125  0]
 [0.2 0.5578125 0.4 0.1953125 0.2 0.4578125 0.8 0.8453125  0]
 [0.2 0.5578125 0.4 0.1953125 0.2 0.4578125 0.8 0.8453125  0]
 [0.2 0.5578125 0.4 0.1953125 0.2 0.4578125 0.8 0.8453125  0]
 [0.2 0.5578125 0.4 0.1953125 0.2 0.4578125 0.8 0.8453125  0]
 [0.2 0.5578125 0.4 0.1953125 0.2 0.4578125 0.8 0.8453125  0]
 [0.2 0.5578125 0.4 0.1953125 0.2 0.4578125 0.8 0.8453125  0]
 [0.2 0.5578125 0.4 0.1953125 0.2 0.4578125 0.8 0.8453125  0]
 [0.2 0.5578125 0.4 0.1953125 0.2 0.4578125 0.8 0.8453125  0]]
```

So every row consists of the nine interpolated values from our computed polynomial, and every row is multiplied nine times. Now if we address it as `r[v*8][u*8]`, we get an approximate radius for a point (u, v). Let's also turn our v and *theta* into such arrays:

```
v, theta = np.meshgrid(v, theta)
```

Now for the revolution part. We have 2D spaces (well, not really spaces, but 2D arrays) for r with all the radii we want, for v containing numbers from 0 to 1, and for *theta* for numbers from 0 to 2π. Let's turn them all into functions that transform (u, v) points to x, y, and z coordinates respectively. (Again, these aren't real functions, but 2D arrays.) With NumPy, this task is a matter of a few lines:

```
x = r * np.cos(theta)
y = r * np.sin(theta)
z = v
```

The x "function" is the radius coming from the r array multiplied by the cosine of the angle coming from *theta*. The y "function" is the radius from r multiplied by the angle from *theta*. And the z "function" is exactly the same as v. This gives us the 2D arrays we can use to build a triangulated surface with special tools from the Matplotlib Python library. As usual, the code for the plotting is left out of the text; it resides only in the code example (ch_08/ugly_mushroom.py in the book's source code).

That's it for the ugly mushroom. Now we know how to make a surface out of 2D arrays. But we're not done yet. Let's add some variability to the input data, enlarge our linear spaces, and make a decent mushroom generator.

To do all that, we need to add only a few lines. Before we compute our Vandermonde matrix, we want to add some variability to the input sample radii from the *rs* array. Even in regular Python, that's a one-liner:

```
rs = [r+random.random()*0.5 for r in rs[:-1]]+[rs[-1]]
```

Adds a random number in the range [0, 0.5] to every radius except the last one

We also want to introduce some height variability:

```
dz = random.random()*0.6
```

Defines the additional height of the mushroom in the range of [0, 0.6]

To apply the *dz* to the mushroom height, we need to tweak only one line:

```
z = v * (1 + dz)
```

Also, we can make our linear spaces denser without losing too much performance, so let's do that, too:

```
u = np.linspace(0, 1, 59)
v = np.linspace(0, 1, 59)
```

These four adjustments are enough to turn our ugly mushroom code into a mushroom generator. Now we can generate random mushrooms of all shapes and sizes (figure 8.15).

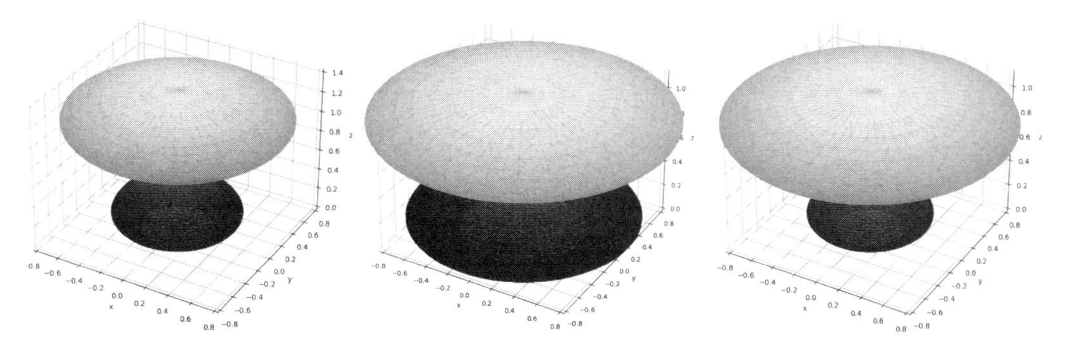

Figure 8.15 A few samples of randomly generated mushrooms

The coefficients 0.5 and 0.6 from the variability lines are totally arbitrary, like the numbers for the input radii. You can pick your own and make yourself an even better fungi forest. (For the full listing see ch_08/mushrooms.py in this book's source code.)

8.2.3 Section 8.2 summary

A surface is a transformation of a plane. A closed surface is a transformation of a specific piece of a plane. A standard square is a good way to determine this specific piece and start building your surface.

8.3 Using nonpolynomial spatial interpolation in geometry

So far, we've learned how to interpolate data points with functions that were mostly polynomial. Rational Bézier curves and NURBS (nonuniform rational basis splines) were also derived from polynomials although enriched by new properties from rationalization. For the single-variable functions, this tool set was more than enough.

For the multivariable functions, the problem of interpolation opens to new approaches. Let's familiarize ourselves with one of them.

8.3.1 Inverse distance interpolation

Let's start with the simplest single-variable case. We have a pair of points (x_1, y_1), (x_2, y_2), and we want to interpolate between them. We already know how to do this by using linear interpolation, and we've seen that linear interpolation is a good starting point for learning about both polynomial interpolation and Bézier functions. As it turns out, it's also a starting point for a completely different approach called *inverse distance interpolation.*

Let's introduce our interpolating function as a sum of weighted values y_1 and y_2 divided by the sum of weights k_i, where the weights are functions of x themselves:

$$F(x) = \frac{y_1 k_1(x) + y_2 k_2(x)}{k_1(x) + k_2(x)}$$

Now, in the general case, the weight functions $k_i(x)$ are defined like this:

$$k_i(x) = \frac{1}{|x - x_i|^n}$$

So $k_i(x)$ is an inverse distance from x to x_i power of n. That's where the name comes from; the weight functions are literally inverse functions of distances.

Here's how it works. The $F(x)$ equation has two values, y_1 and y_2. Each value has its own weight function as a scale factor, $k_1(x)$ and $k_2(x)$, respectively. When x is getting close to x_1, the weight function $k_1(x)$ goes fast toward infinity. The smaller the denominator is, the greater the weight. This dwarfs the other weight function $k_2(x)$ by comparison. And because we divide the weighted sum by the sum of coefficients, the infinity-like k_1 from the numerator gets divided by the same scale k_1 from the denominator, and the whole function becomes close to y_1.

I know that this is hard to understand; normally, our intuition isn't trained to operate on "close to infinity" values. So let's take a look at a specific example. Suppose that we want to interpolate between points (1, 3) and (3, 1). Also, let's start by appointing 1 as our power n. Our weight functions, then, are

$$k_1(x) = \frac{1}{|x - x_1|}, \; k_2(x) = \frac{1}{|x - x_2|}$$

and the whole interpolation is

$$F(x) = \frac{y_1 \frac{1}{|x-x_1|} + y_2 \frac{1}{|x-x_2|}}{\frac{1}{|x-x_1|} + \frac{1}{|x-x_2|}}$$

or, in specific numbers,

$$F(x) = \frac{3\frac{1}{|x-1|} + 1\frac{1}{|x-3|}}{\frac{1}{|x-1|} + \frac{1}{|x-3|}}$$

In $x = 1$, the first weight function suffers from a division by zero, and in $x = 3$, the same occurs to the second one. This means that for $x = 1$ and $x = 3$, the function $F(x)$ remains undefined. It's fine, though. We only want to interpolate between the points in the range (1, 3), excluding 1 and 3 themselves. Let's compute $F(x)$ for a series of *x*s that come closer and closer to 1 to see whether the interpolant converges toward 3, as it should (table 8.2).

Table 8.2 Convergence of F(x) in x ≈ 1

x	F(x) approximately
1.1	2.89
1.01	2.98
1.001	2.99
1.0001	2.999
1.00001	2.9999
1.000001	2.99999

Yes, it does. You can do the same experiment for $x = 3$ yourself and make sure that $F(x)$ converges toward 1 there, too. What's interesting is that if we gather values between 1 and 3 and make them into a graph, we see a straight line (figure 8.16).

Was all this a fancy way to reinvent linear interpolation? Of course, not. As polynomial interpolation and Bézier functions start from linear interpolation and evolve into

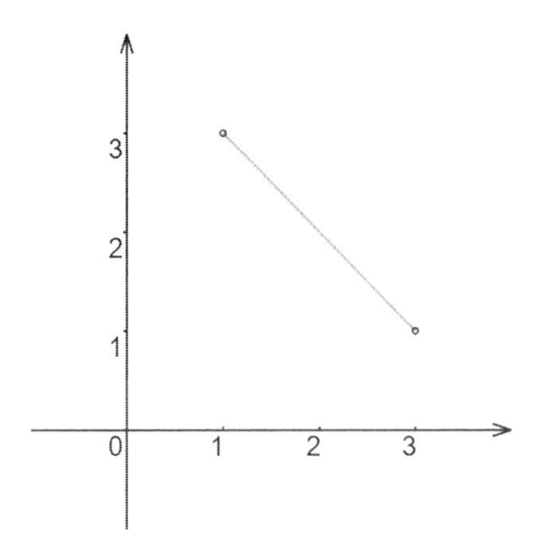

Figure 8.16 Inverse distance interpolation of two points with n = 1 is also linear.

something much more powerful, inverse distance interpolation evolves into a potent instrument, too.

First, you don't have to stick to the power of the weight functions being 1. As with polynomials, as you raise the power, you change the shape of the interpolant (figure 8.17).

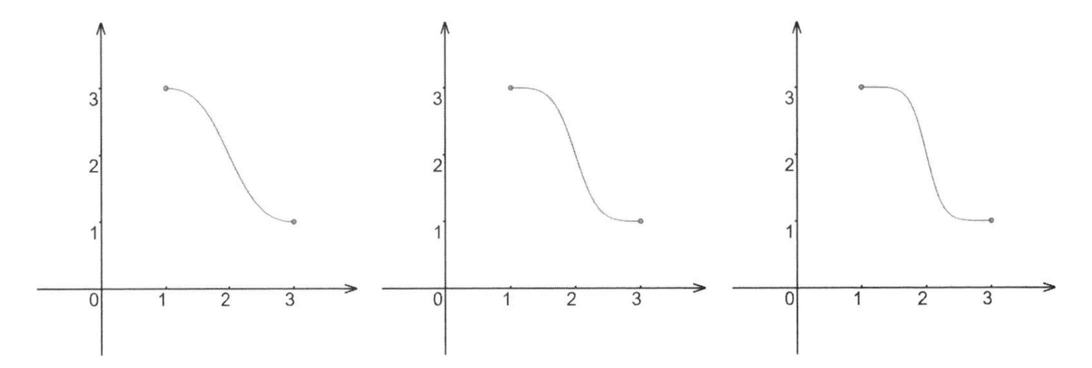

Figure 8.17 Here is the same interpolation with (from left to right) *n* = 2, *n* = 3, and *n* = 4. Raising the power of *n* changes the shape of the interpolant.

You can also add more points to the mix. For n points, the formula is a sum of all the weighted values divided by the sum of weight functions:

$$F(x) = \frac{\sum_{i=0}^{N} y_i k_i(x)}{\sum_{i=0}^{N} k_i(x)}$$

With this formula, we can interpolate as many points as we want and have ourselves an almost continuous and, with $n > 1$, even smooth function (figure 8.18).

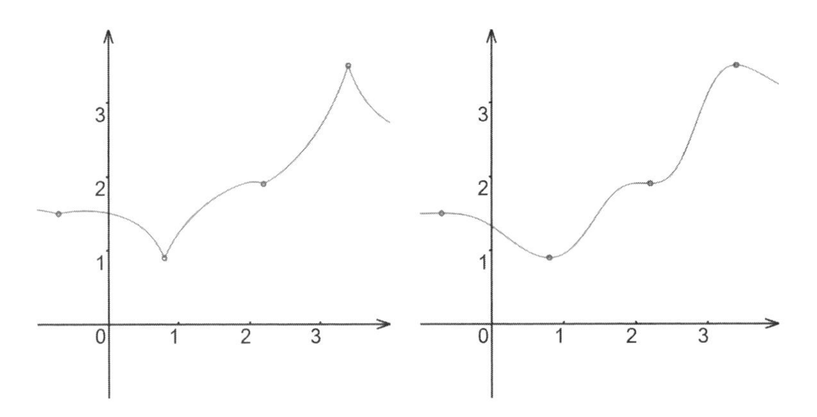

Figure 8.18 Inverse distance interpolation of four points, with the weight functions degrees 1 (on the left) and 2 (on the right)

But wait—what does *almost* in the preceding paragraph mean? Well, of course, the $F(x)$ still wouldn't be defined at $x = x_i$ because computing the weight function $k_i(x)$ in x_i still results in zero division. Not to worry—we know the function's target value in every x_i without any computation. It's y_i by the very definition of interpolation! So the full formula for the smooth and continuous interpolant is

$$\text{For all } x \text{ where } x \neq x_i : F(x) = \frac{\sum_{i=0}^{N} y_i k_i(x)}{\sum_{i=0}^{N} k_i(x)}; F(x_i) = y_i$$

Now $F(x)$ is defined for all the numbers, x_i included.

NOTE Unfortunately for us programmers, this formula opens a whole new can of worms. Sure, in math, the formula works for every x no matter how close to x_i it is. In programming, however, computing weight functions results in overflows in some nonzero proximity of x_is, so we already have to substitute specific values for the formula in this nonzero proximity. And computing the "width" of this proximity is a formidable problem on its own. In practice, this problem is usually solved by appointing some "small enough value that seems to work" voluntarily.

You may have noticed that when the weight functions' coefficient is more than 1, the function tends to make a plateau near the interpolation points. In other words, its derivative goes toward 0 in the proximity of the points. Sometimes, this effect is tolerable, but at other times, we want something more splinelike with nonzero derivatives.

The good news is that with inverse distance interpolation, we can set a derivative for every point we want. Moreover, we can set a derivative of any degree for every point we like. Instead of interpolating values, we can interpolate functions:

$$\text{For all } x \text{ where } x \neq x_i : F(x) = \frac{\sum_{i=0}^{N} y(x)_i k_i(x)}{\sum_{i=0}^{N} k_i(x)}; F(x_i) = y_i$$

Now $y_i(x)$ are functions that go through the (x_i, y_i) points, and these functions "lend" their derivatives to the interpolant. The functions could be of any origin, and of course, they could also be constants, linear functions, or polynomials. In figure 8.19, the image on the left shows four weighted linear functions; the middle image shows four quadratic ones; and the image on the right shows two quadratic and two constant functions in the same interpolant. That's right—we can mix functions of different kinds in a common interpolant!

Inverse distance interpolation is a versatile instrument. It gives you full control of your target function's look and feel, as well as its mathematical properties. The interpolant isn't constrained by the Runge phenomenon, either, so you can add as many points as you like.

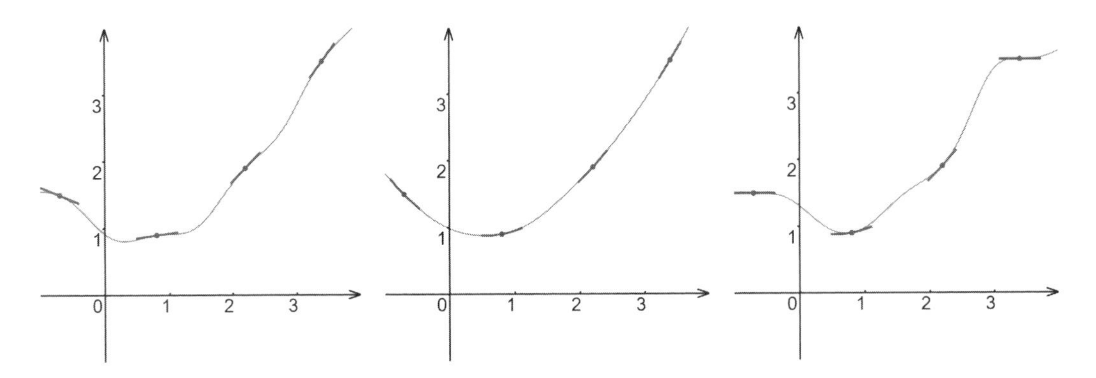

Figure 8.19 Inverse distance interpolation with different functions defined in x_i instead of just values

What's also important in the context of this chapter is that you can easily modify the weight function and make it work with multivariable functions. This generalization makes interpolation spatial, as in "working in n-dimensional space," and opens a whole new field of possibilities to us. We can use inverse distances interpolation to build geometric transformations, deformation fields, and surfaces.

8.3.2 Inverse distance interpolation in space

Adopting the inverse distance interpolation to 2D or 3D is easy. The trick is in the name. We should make the weights functions inverse powers of distances. In our usual Euclidean 2D space, the distance between two points is

$$d((x_1, y_1), (x_2, y_2)) = \sqrt{(x_1 - x_2)^2 + (y_1 - y_2)^2}$$

In 3D it's

$$d((x_1, y_1, z_1), (x_2, y_2, z_2)) = \sqrt{(x_1 - x_2)^2 + (y_1 - y_2)^2 + (z_1 - z_2)^2}$$

So the weight functions for 2D and 3D, respectively, are

$$k_i(x, y) = \frac{1}{\sqrt{(x - x_1)^2 + (y - y_1)^2}^n}$$

and

$$k_i(x, y, z) = \frac{1}{\sqrt{(x - x_1)^2 + (y - y_1)^2 + (z - z_i)^2}^n}$$

Now, the old formula for a single variable applies to the two-variable interpolation with a little change here:

$$\text{For all pairs } (x, y) \text{ where } (x, y) \neq (x_i, y_i) : F(x, y) = \frac{\sum_{i=0}^{N} v_i k_i(x, y)}{\sum_{i=0}^{N} k_i(x, y)}; F(x_i, y_i) = v_i$$

and to the three variables:

$$\text{For all triplets } (x, y, z) \text{ where } (x, y, z) \neq (x_i, y_i, z_i) :$$
$$F(x, y, z) = \frac{\sum_{i=0}^{N} v_i k_i(x, y, z)}{\sum_{i=0}^{N} k_i(x, y, z)}; F(x_i, y_i, z_i) = v_i$$

As you can see, adding a variable to the formula is easy. If you extend your distance formula for 4D, for example, you can enjoy a four-variable interpolation.

Now let's take a look at a bit of code. Suppose that we have this data:

```
xs = [2,9,8,3,5]
ys = [1,2,9,8,5]
vs = [1,2,3,4,5]
```

We want to build a two-variable function $F(x, y)$ such as $F(xs[i], ys[y]) = vs[i]$. Here it is:

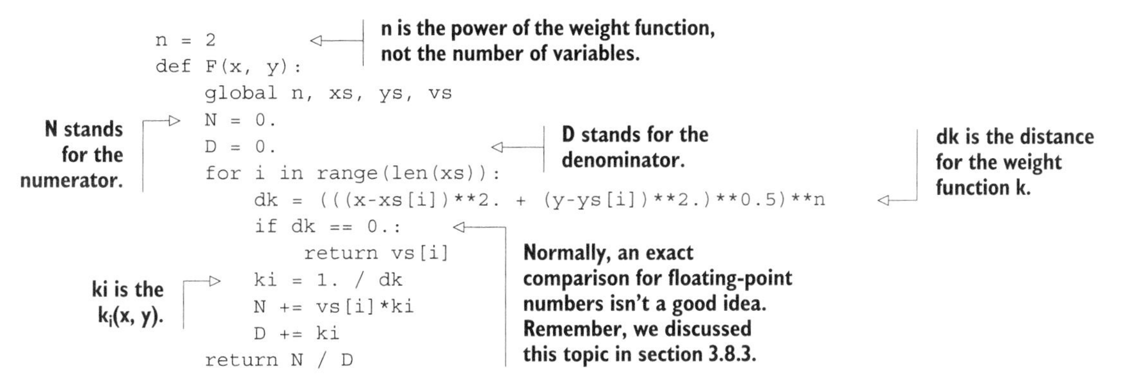

```
n = 2
def F(x, y):
    global n, xs, ys, vs
    N = 0.
    D = 0.
    for i in range(len(xs)):
        dk = (((x-xs[i])**2. + (y-ys[i])**2.)**0.5)**n
        if dk == 0.:
            return vs[i]
        ki = 1. / dk
        N += vs[i]*ki
        D += ki
    return N / D
```


The plot for this specific function on those specific data points looks like figure 8.20.

NOTE By the way, this is how the terrain was generated in the Procedure Project game.

The interpolation keeps all its properties from a single-variable case. The power of the weight functions affects the interpolant similarly, making the plateaus more pronounced as the power rises (figure 8.21).

Multivariable interpolation can work with as many points as you like. It can also be adapted to work with functions $v_i(x,y)$ instead of values v_i. It's every bit as versatile as the single-variable version.

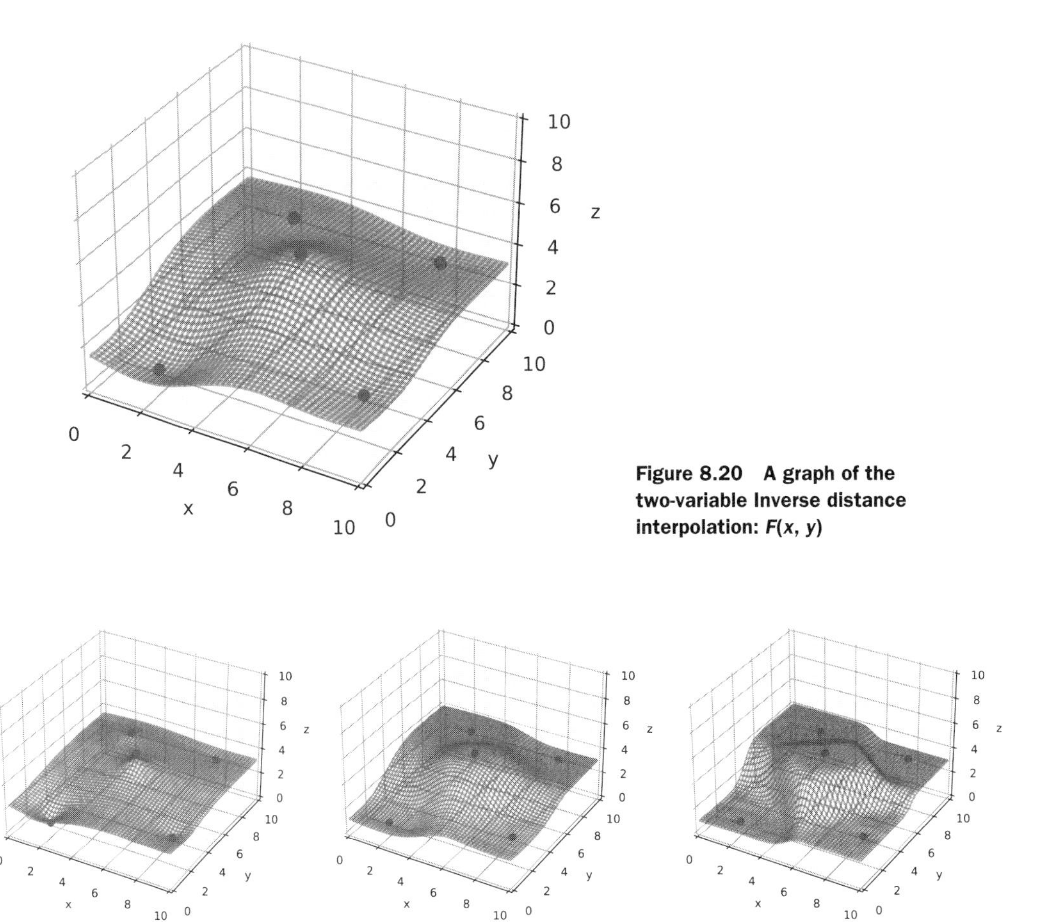

Figure 8.20 A graph of the two-variable Inverse distance interpolation: *F*(*x*, *y*)

Figure 8.21 Inverse distance interpolation with (from left to right) weight function powers 1, 3, and 7

SEE ALSO For a survey on spatial interpolation methods, see "Scattered Data Interpolation in Three or More Variables," by Peter Alfeld. It's a little bit outdated because it was published in 1989, but it's rather comprehensive and (also important) available online at https://www.math.utah.edu/~alfeld/papers.html.

8.3.3 *Practical example: Using localized inverse distance method to make a bitmap continuous and smooth*

So far in this chapter, we've talked mostly about transformations and surfaces, but multivariable interpolation is applicable on its own. Like single-variable interpolation, it defines new values between already-defined values, and we can use this property directly.

Let's say we want to enlarge a bitmap or a piece of a bitmap. Suppose that we want to look the little fella pictured in figure 8.22 in the eye.

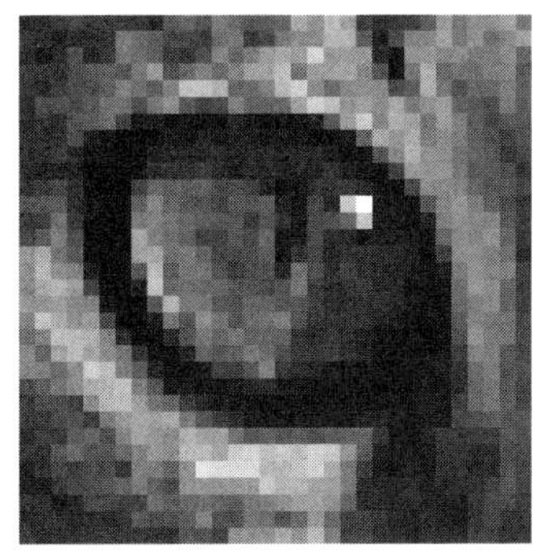

Figure 8.22 An upscale of a cat's eye. (The cat is cropped from original work by Von.grzanka, CC BY-SA 3.0, via Wikimedia Commons.)

A *bitmap* is a 2D array of colors. Each pixel has a single color, so if you upscale the bitmap exactly as it is, you get an image made of colored squares. For some reason, people don't much like looking at colored squares, so usually, the software that does the upscaling also does the interpolation, filling the image not with the exact pixel colors, but with their continuous approximation.

Let's see how we can adapt our inverse distance interpolation to help us with this task. First, we need to *localize* it. When we take all our values and put them in a single sum, each value influences the whole function, even if only a little bit. For us, this means that the black dot in the middle of the cat's eye will make the top-left corner of a picture a bit darker. This doesn't make sense in images. But if we define the sum per pixel separately, each square will be colored based on the colors of four neighboring pixels. The black dot will remain black, and the gray pixel will remain a gray pixel. This is what localization is about.

So let's define our interpolation formula for a point (x, y) that belongs to the $[i, j]$ pixel. Remember, the convention here is transposed, so index i corresponds to the y coordinate and j to the x. In mathematical notation, the "belonging" could be expressed as

$$j < x < j + 1,\ i < y < i + 1$$

Now let's introduce four per-coordinate distances, which would be distances from the point (x, y) to the pixel's "walls":

$$d_{x0} = x - j$$

$$d_{x1} = j + 1 - x$$

$$d_{y0} = y - i$$

$$d_{y1} = i + 1 - y$$

The same reasoning as with single-variable interpolation applies: the closer the point gets to a "wall," the closer the value gets to the value on the "wall." But we don't have any values on the "walls"—only values in the "corners."

We know that the blue channel in pixel (60, 125) is 0x57, and in (61, 125), it's 0x67. We don't know the value in (60.5, 125) or (60.1, 125), or anywhere on (*a*, 125) apart from when *a* is an integer number. We don't know the values in (*a*, 126), (60, *b*), or (61, *b*) for noninteger *a* and *b* either. And these values form the "walls" for our interpolation (figure 8.23).

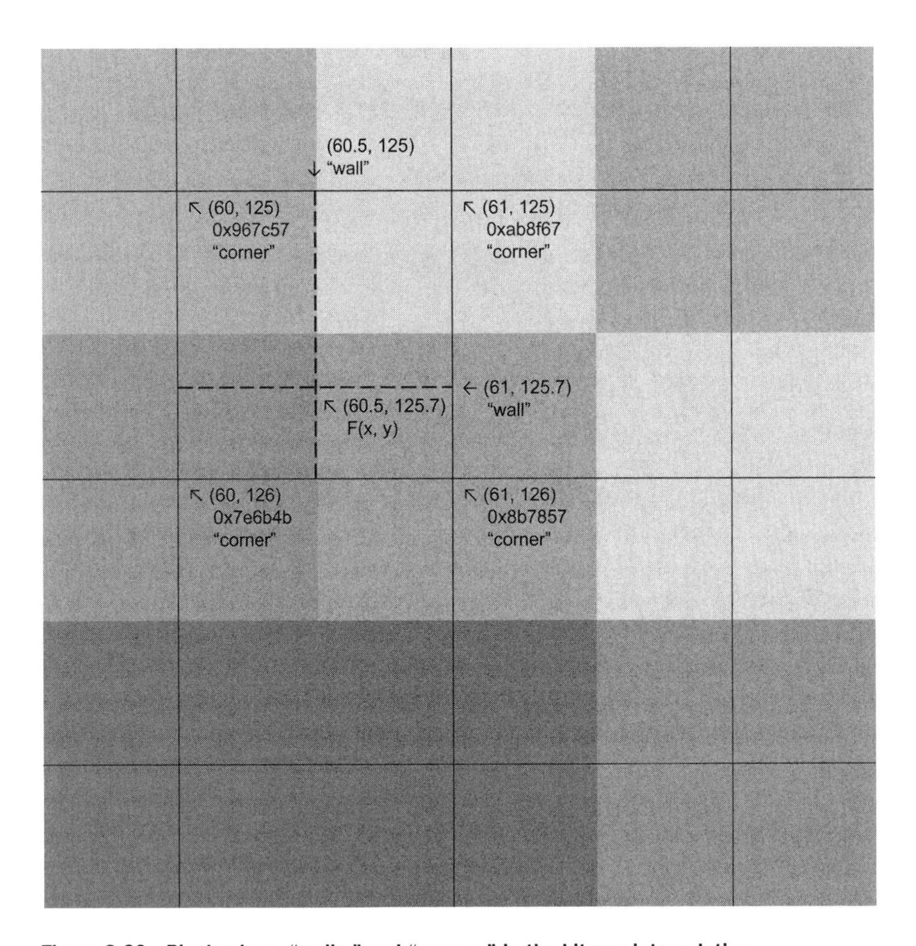

Figure 8.23 Pixel values, "walls," and "corners" in the bitmap interpolation

We can't simply compute distances to the corners and put them into the formula. Sure, this approach will work within a pixel, but our goal is to make a continuous representation of a whole bitmap, so each pixel, when turned into a square with changing colors, should align perfectly with all four neighbors.

Instead, we'll use a weight function that's a product of two per-variable distances. This way, the color on each "wall" will be computed from the values of the wall's own corners, so pixels that share that "wall" will align automatically. Then the interpolation formula will look like this:

For all points (x, y) where x lies in the range $(j, j+1)$ and y in the range $(i, i+1)$:

$$F(x, y) = \frac{b_{[i,j]}\frac{1}{(d_{x0}d_{y0})^n} + b_{[i+1,j]}\frac{1}{(d_{x0}d_{y1})^n} + b_{[i+1,j+1]}\frac{1}{(d_{x1}d_{y1})^n} + b_{[i,j+1]}\frac{1}{(d_{x1}d_{y0})^n}}{\frac{1}{(d_{x0}d_{y0})^n} + \frac{1}{(d_{x0}d_{y1})^n} + \frac{1}{(d_{x1}d_{y1})^n} + \frac{1}{(d_{x1}d_{y0})^n}}$$

Here, $b_{[i,j]}$ is the pixel color value. We assume that it's something that could be scaled and added but usually it isn't. Technically, a color value is a vector of some color space and usually represented by something more mundane, such as a 32-bit integer with 4 channel values wired in as RGBA or ARGB. *R* stands for "red," *G* for "green," and *B* for "blue." *A* stands for "alpha," and it governs transparency. With this representation, you can't scale one 32-bit number and add it to another 32-bit color number. But you can split the number into four eight-bit color channels, and apply the interpolation to all of them one by one.

In our case, this task is even easier because we have only three color channels. Our image doesn't have transparency; it is all opaque. When we interpolate the image color channel by color channel, we get the result shown in figure 8.24.

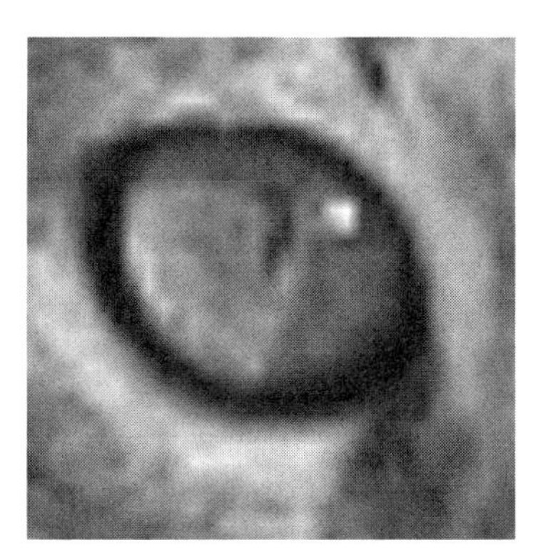

Figure 8.24 An upscale of a cat's eye, now with inverse distance interpolation

This upscale doesn't have squares anymore. Now the bitmap is a continuous function of color:

$$F(x, y) \rightarrow (\mathrm{R}, \mathrm{G}, \mathrm{B})$$

One more thing: because the interpolation behind the upscaling is a variant of inverse distance, we still get to keep our usual levers. We can define functions instead of raw pixel values, and we can vary the power of coefficient n for the weight functions.

We don't get many benefits, though. The upscale in figure 8.24 was done with $n = 1$. In figure 8.25, the same upscale was done with $n = 2$.

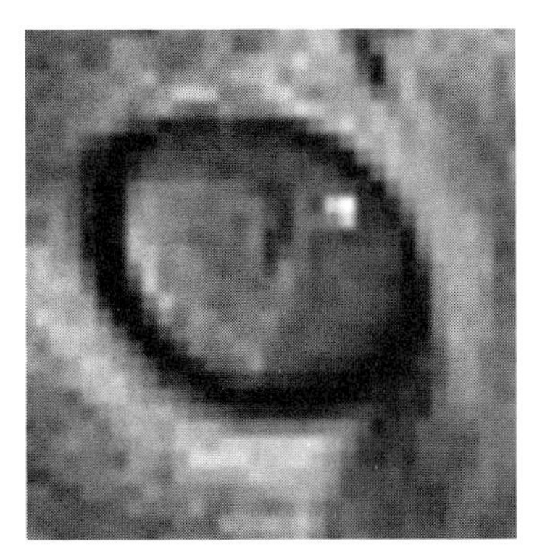

Figure 8.25 An upscale of a cat's eye, interpolated with *n* = 2

We're back to square pixels again, but this time, the squares are slightly smoothed. Adding a degree to the weight functions didn't add any value to the algorithm. So yes, sometimes it's best to keep things simple. The mere fact that we have the tools doesn't mean that we should use them all the time. With n being constrained to 1, the final formula simplifies and becomes

$$\text{For all points } (x, y) \text{ where } x \text{ lies in the range } (j, j+1) \text{ and } y \text{ in the range } (i, i+1):$$

$$F(x, y) = \frac{b_{[i,j]}\frac{1}{d_{x0}d_{y0}} + b_{[i+1,j]}\frac{1}{d_{x0}d_{y1}} + b_{[i+1,j+1]}\frac{1}{d_{x1}d_{y1}} + b_{[i,j+1]}\frac{1}{d_{x1}d_{y0}}}{\frac{1}{d_{x0}d_{y0}} + \frac{1}{d_{x0}d_{y1}} + \frac{1}{d_{x1}d_{y1}} + \frac{1}{d_{x1}d_{y0}}}$$

Well, technically, this isn't the whole formula yet. To get a truly continuous bitmap function, we still have to define it on the "walls" and in the "corners." So when $x = j$ and/or $y = i$,

$$\text{For all points } (x, y) \text{ where } x = j \text{ but } y \neq i : F(x, y) = \frac{b_{[i,j]}\frac{1}{d_{x0}d_{y0}} + b_{[i+1,j]}\frac{1}{d_{x0}d_{y1}}}{\frac{1}{d_{x0}d_{y0}} + \frac{1}{d_{x0}d_{y1}}}$$

$$\text{For all points } (x, y) \text{ where } x \neq j \text{ but } y = i : F(x, y) = \frac{b_{[i,j]}\frac{1}{d_{x0}d_{y0}} + b_{[i+1,j]}\frac{1}{d_{x1}d_{y0}}}{\frac{1}{d_{x0}d_{y0}} + \frac{1}{d_{x1}d_{y0}}}$$

$$\text{For all points } (x, y) \text{ where } x = j \text{ and } y = i : F(x, y) = b_{[i,j]}$$

This is it. This formula effectively turns a bitmap (a 2D array of colors) into a continuous and smooth function of colors. If you remember, we finished the page-unbending example from section 8.1.3 by filling the resulting bitmap with colors of the original bitmap's pixels. A simple and effective way to improve that algorithm would be to use our continuous function of colors instead of a source bitmap. That's what I did in my app, too.

8.3.4 Practical example: Building a deformation field with multivariable interpolation to generate better mushrooms

A small reminder: a *deformation field* is a function that assigns a vector to every point in space. If you put a geometric object, whatever it is, in this space, you can deform the object by applying the corresponding vectors from the deformation field to all its points.

Let's reuse our mushroom example to see how deformation works in practice. Let's even say that we already have an (x, y, z) point for every (u, v). As another reminder, in the code, the formulas for x, y, and z look like this:

```
x = r * np.cos(theta)
y = r * np.sin(theta)
z = v * (1 + dz)
```

Each of these values is a 2D array in which every pair of indices corresponds to a specific (u, v) point. This is essentially the same as having a 2D array of (x, y, z) points that kind of models an f: $(u, v) \rightarrow (x, y, z)$ function.

Now let's define a random but controllable deformation. To do that, we'll interpolate some deformation vectors between the points $(-1, -1, 0)$, $(1, -1, 0)$, $(1, 1, 0)$, $(-1, 1, 0)$, $(-1, -1, 1+dz)$, $(1, -1, 1+dz)$, $(1, 1, 1+dz)$, and $(-1, 1, 1+dz)$ where dz (another reminder) is the randomized mushroom height increment. In the code, the points of interpolation are

```
dxs = [-1, 1, 1, -1, -1, 1, 1, -1]
dys = [-1, -1, 1, 1, -1, -1, 1, 1]
dzs = [0, 0, 0, 0, 1+dz, 1+dz, 1+dz, 1+dz]
```

Now let's define deformation vectors. Because we want our mushrooms to be slightly different, this will be a good place to wire in the randomness. Let's define our variability factor as some number, such as 0.7, and then define the random deformation vector as `(random of 0.7 - 0.7/2, random of 0.7 - 0.7/2, random of 0.7 - 0.7/2)`, or, in Python code:

```
ddxs = [random.random()*0.7 - 0.7/2 for i in range(len(dxs))]
ddys = [random.random()*0.7 - 0.7/2 for i in range(len(dys))]
ddzs = [random.random()*0.7 - 0.7/2 for i in range(len(dzs))]
```

NOTE It's a good practice to keep "magic numbers" such as 0.7 out of your code. That said, actual numbers take less paper space than meaningful names. Although there may be properly named constants in the code supplied with the book, in the text itself, I prefer to keep real numbers.

Now for the interpolation function. We'll use a three-variable inverse distance function with the weight function's power coefficient set to 1:

$$\text{For all points } (x, y, z), (x, y, z) \neq (x_i, y_i, z_i):$$

$$F(x, y, z) = \frac{\sum_{i=0}^{N} v_i k_i(x, y, z)}{\sum_{i=0}^{N} k_i(x, y, z)}; F(x_i, y_i, z_i) = v_i$$

$$k_i(x, y, z) = \frac{1}{\sqrt{(x - x_i)^2 + (y - y_i)^2 + (z - z_i)^2}}$$

Here's the same function in Python:

```
def F(x, y, z, dds):
    global dxs, dys, dzs, n
    N = 0.                  # Numerator
    D = 0.                  # Denominator
    for i in range(len(dds)):
        di = ((x-dxs[i])**2 + (y-dys[i])**2 + (z-dzs[i])**2)**0.5    # Euclidean distance
        if di == 0.:
            return dds[i]
        ki = 1/di           # Weight function
        N += ki * dds[i]
        D += ki
    return N / D
```

Next, we apply this function to every coordinate of every point in *x*, *y*, and *z*. The resulting triplets form deformation vectors. Then, by adding the deformation vectors to the model's points, we conduct the deformation of the whole object. Because the interpolation function is continuous and smooth, the deformation itself will be continuous and smooth, too. Here's how it looks in the code:

```
for i in range(len(x)):
    for j in range(len(x[0])):
        xij = x[i][j]
        yij = y[i][j]
        zij = z[i][j]
        ddx = F(xij, yij, zij, ddxs)
        ddy = F(xij, yij, zij, ddys)
        ddz = F(xij, yij, zij, ddzs)
        x[i][j] += ddx
        y[i][j] += ddy
        z[i][j] += ddz
```

As a result of this interpolation-backed deformation (for the full listing, see ch_08/deformed_mushrooms.py in the book's source code), we'll get even more varied and more lifelike mushrooms than before, as shown in figure 8.26.

Deformation modeling is a respectable field of applied geometry in its own right. Conceptually, the idea behind it is simple: you define the vector function, you apply it to some kind of geometric object, and you enjoy the result.

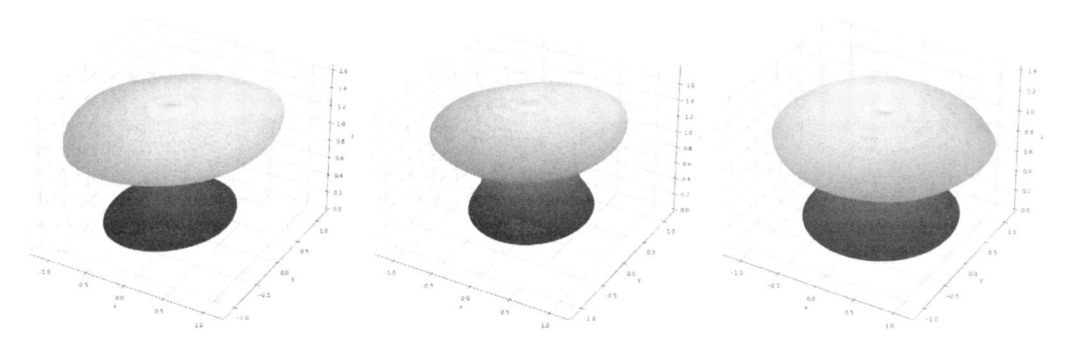

Figure 8.26 Randomly generated and randomly deformed mushrooms

SEE ALSO A survey on spatial deformation is publicly available (and a few decades younger than the one on spatial interpolation). See "A Survey of Spatial Deformation from a User-Centered Perspective," by James Gain and Dominique Bechmann, at http://mng.bz/dJjN.

8.3.5 Section 8.3 summary

Inverse distance interpolation is a powerful tool we can use on its own, as with bitmap scaling; use to build surfaces, as with procedure project terrain; or use to build deformation fields, as with randomly deformed mushrooms.

Inverse distance interpolation is only one of many spatial interpolation methods. It's easy to start with because it adapts easily to most of the tasks you can come up with, but look around, and you'll find a better-fitting method for every particular task.

8.4 Exercises

Exercise 8.1 How many different random data points do we need to confidently establish a cubic-quadratic polynomial interpolation function?

1. 20
2. 12
3. 6
4. 3.5

Exercise 8.2 How many different random data points do we need to establish a cubic-quadratic polynomial interpolation function if we allow ambiguity?

1. At most 12
2. At least 12
3. Exactly 12 points, given that the points are aligned to the axes in a 3×4 manner

Exercise 8.3 Can you theoretically model a cylinder with a polynomial transformation of a plane?

Exercise 8.4 Can you theoretically model a pair of cylinders with a single polynomial transformation of a plane?

Exercise 8.5 In multidimensional space, formulas for interpolation are getting longer, and it's getting easier to let mistakes in. Can you write a program that generalizes our pixel-enlarging exercise for every given number of dimensions by printing the N-dimensional formula code for you?

Exercise 8.6 Invent your own method of spatial interpolation. It doesn't have to be better than any of the existing ones or even practical. Have fun!

8.5 Solutions to exercises

Exercise 8.1 A single-variable cubic interpolation requires four points. Quadratic requires three. This multiplies into the 12-point requirement for an unambiguous cubic-quadratic interpolation.

Exercise 8.2 If we allow ambiguity, we can go with fewer points. The same logic applies here as with a single-variable interpolation. We can run a parabola through a pair of points; in fact, we can run an infinite amount of parabolas through a pair of points. But we can't make a straight line run through three points unless they lie on the same line. And because we expect our points to be random, we have to assume that they don't. So "At most 12" is the correct answer.

Exercise 8.3 We can make a surface that looks like a cylinder, but not a precise model of a cylinder. That's what we need NURBS for, remember? With a rational spline, we can model a cylindrical surface, but with polynomials, we can only approximate the cylinder within some tolerable error.

Exercise 8.4 No. Polynomial interpolation is continuous. To disjoint one cylinder from another, you'd need a discontinuity.

Exercise 8.5 Yes, you can!

Exercise 8.6 This is a forever-open question. No ultimate spatial interpolation method covers all possible needs. If you want something to start with, modify the inverse distance interpolation. Try new weight functions, introduce value functions, and experiment with new ways to make the interpolant local.

Summary

- Polynomial interpolation scales to all degrees and dimensions.
- If you keep your data points aligned to the grid, you can decompose your n-variable polynomial interpolation into a sequence of n single-variable ones with simpler formulas that are faster to compute.

- Polynomial interpolation is a simple and practical way to define spatial transformations.
- A surface is a transformation of a plane.
- Many methods of spatial interpolation exist. Inverse distance interpolation is one of the most versatile and adaptive.
- You can use arbitrary spatial interpolation for deformative modeling.

The geometry of vector algebra

9

This chapter covers

- Definitions and properties of vector products
- Examples of dot, cross, and triple products solving geometric problems in 3D space
- Generalizations of vector products to N-dimensional space where $N \neq 3$

Vector algebra is often seen as a staple of applied geometry. Vector products have so much to do with angles, projections, areas, and volumes that it's hard to imagine doing anything practical without them. Vector algebra is also a never-ending source of interview problems.

In practice, however, vector operations are usually concealed under utility functions' interfaces to such a degree that about half of candidates who apply for a job in my department can't tell the dot product from the cross product. These people are still doing fine professionally. But without regular practice, the math behind the interfaces fades from memory.

In the modern world, vector algebra isn't so much a must-have as it is a powerful enabler. It enables you to go beyond the utility functions your framework gives you. It enables you to write your own utility functions tailored to your own specific tasks, as well as highly efficient code that doesn't rely on utility functions at all.

9.1 Vector addition and scalar multiplication as transformations

Let's start with a small recap. A *vector* is a direction with length. The two most common notations for vectors are a tuple of coordinates of a point the vector points to and a sum of weighted basis vectors. In the standard basis, the numbers in the former and the latter notations are the same:

$$\mathbf{a} = (2, 3, 4) = 2i + 3j + 4k$$

Vectors and points are completely different entities. A vector is a direction with length, so it doesn't start anywhere in particular and doesn't end anywhere in particular either. But if you agree to start a vector from (0, 0, 0), its length and direction may indicate any point in space, which would make the vector a point vector or a position vector. This "vector is not a point, but sometimes it is" notion allows us to use vector arithmetic to do geometric work. If we model a geometric body as a set of point vectors, a vector addition (adding the same vector to all the body's point vectors) also works as a geometric transformation, namely translation (figure 9.1).

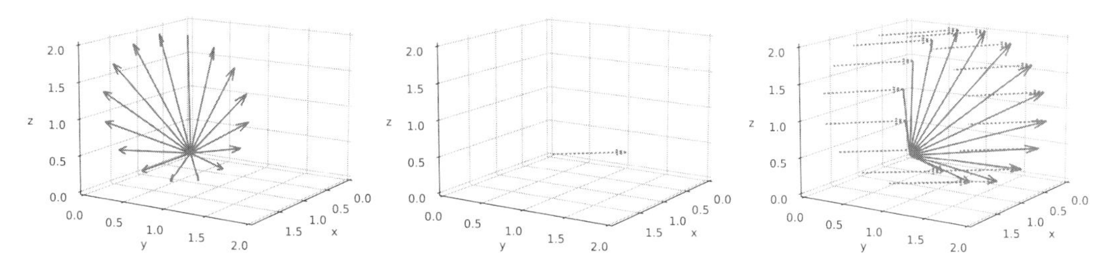

Figure 9.1 A model made of point vectors is translated by vector addition.

Similarly, scalar multiplication—multiplying a vector by a scalar—works as geometric scaling (figure 9.2).

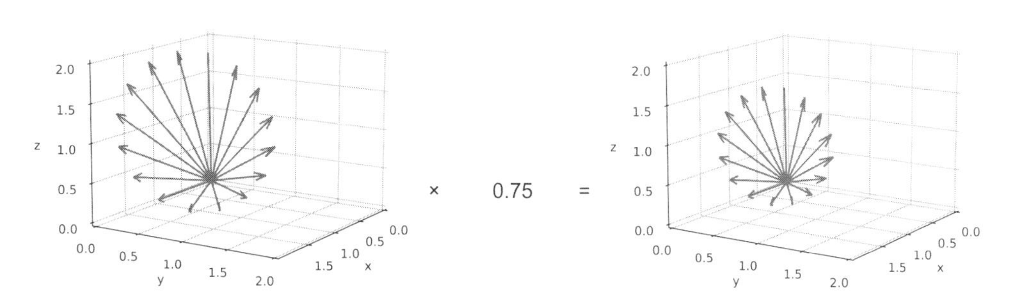

Figure 9.2 A vector point model is scaled by applying a scalar multiplication to all its vectors.

So far, these concepts play well with our intuition. We can easily visualize and understand what's happening. Geometry and algebra go hand in hand.

But apart from vector addition and scalar multiplication, you should know about three more vector operations, called vector products: dot product, cross product, and triple product. The *dot product* turns a pair of vectors into a scalar number. The *cross product* turns a pair of vectors into a vector. And the *triple product* once again produces a scalar, but it takes three vectors as its input. In this chapter, we'll examine them closely one by one.

With vector products, the geometric sense is still there, but it takes a little more effort to understand and, more important, recall whenever an opportunity arrives. When you know the geometry behind vector operations, you can use them efficiently to build your own algorithms. You can always look up the details if you forget how vector products work, but you should know when to start looking for them to begin with.

This chapter, therefore, focuses not only on the definitions, but also on the explanations, mnemonics, and practical applications of vector algebra in applied geometry.

9.2 Dot product: Projection and angle

The first vector product we'll examine is the dot product, which is commonly denoted by a dot, hence the name:

$$c = \mathbf{a} \cdot \mathbf{b}$$

9.2.1 Other names and motivations behind them

The dot product is also called the *scalar product* because the result of the dot product is a scalar, not a vector. The thin c in the preceding formula isn't a typo. We normally use a bold font to designate vectors and reserve normal weight for scalars.

Another name for the dot product is the *inner product*. Technically, *inner product* is more a classification than a name. What makes a product inner is the fact that it makes two vectors into a scalar. A dot product does this, too, but in a particular way:

$$c = \mathbf{a} \cdot \mathbf{b} = (a_1, a_2, a_3) \cdot (b_1, b_2, b_3) = a_1 b_1 + a_2 b_2 + a_3 b_3$$

When $\mathbf{a} = (2, 3, 4)$, and $\mathbf{b} = (5, 6, 7)$, for example, the product is

$$c = (2, 3, 4) \cdot (5, 6, 7) = 2*5 + 3*6 + 4*7 = 10 + 18 + 28 = 56$$

So the dot product is an inner product because it turns a vector into a scalar, but not every inner product is a dot product, although in the context of applied geometry, this formality is often neglected.

The dot product is easily generalizable for any multidimensional space. For an N-dimensional space, the dot product is

$$c = \mathbf{a} \cdot \mathbf{b} = (a_1, a_2, \ldots, a_N) \cdot (b_1, b_2, \ldots, \mathrm{b}_N) = a_1 b_1 + a_2 b_2 + \ldots + a_N b_N$$

The product also survives in 2D space, of course,

$$c = \mathbf{a} \cdot \mathbf{b} = (a_1, a_2) \cdot (b_1, b_2) = a_1 b_1 + a_2 b_2$$

which is lucky because the dot product has yet another name, which would be its fourth. That name is *projection product.* Projection is a geometrical thing, so let's dive into this name a little deeper. It's easier to inspect the properties that the name implies in a 2D space, so that's why having a dot product in 2D is bliss.

Let's say our vector **a** is (1, 0). Now, whatever vector **b** is—(2, 4), (5, 6), or (b_1, b_2)—in general the dot product **a** · **b** will always be $1b_1 + 0b_2$, so b_1. The resulting number is the length of **b**'s projection on **a** (figure 9.3).

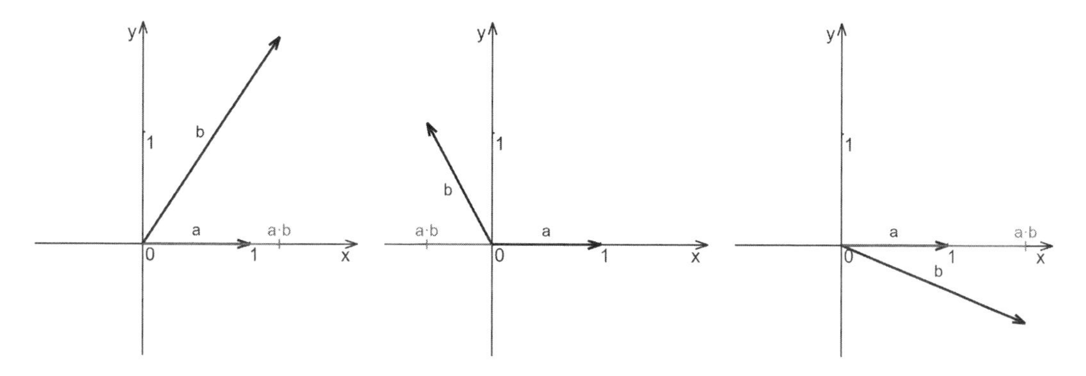

Figure 9.3 Projections of different bs on a = (1, 0)

That's trivial. The real magic starts when we move and rotate the **a** vector. Apparently, never mind the numbers, if **a** retains length 1, the dot product **a** · **b** results in the length of **b** projected on **a**—hence the "projection product" name.

From trigonometry, you may remember how to compute the length of a side given the hypotenuse and the adjacent angle (figure 9.4). The answer is the length of **b** multiplied by the cosine of the adjacent angle θ:

$$c = \|\mathbf{b}\|\cos\theta$$

In general, when **a** doesn't have to be of length 1 anymore, the formula gets scaled by the length of **a**. So the trigonometric formula for the dot vector in the general case is

$$c = \|\mathbf{a}\|\|\mathbf{b}\|\cos\theta$$

Here, $\|\mathbf{a}\|$ is the length of vector **a**, $\|\mathbf{b}\|$ is the length of vector **b**, and θ is the angle between the vectors (figure 9.5).

In practice, dot products are often used to measure angles or, most commonly, to ensure that some angle is exactly perpendicular, or exactly 90 degrees. If the angle

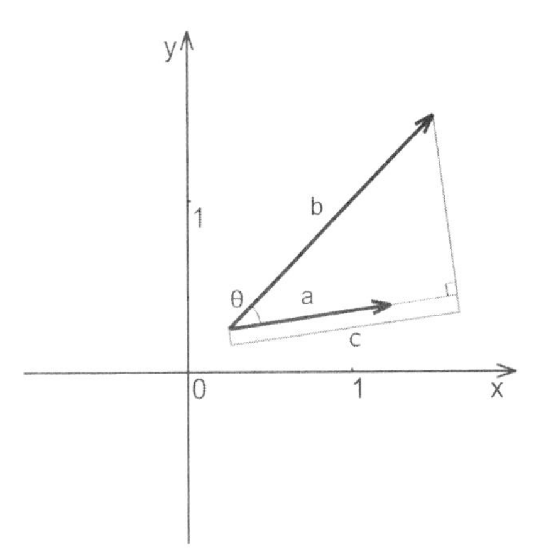

Figure 9.4 Projection is one side of the right triangle.

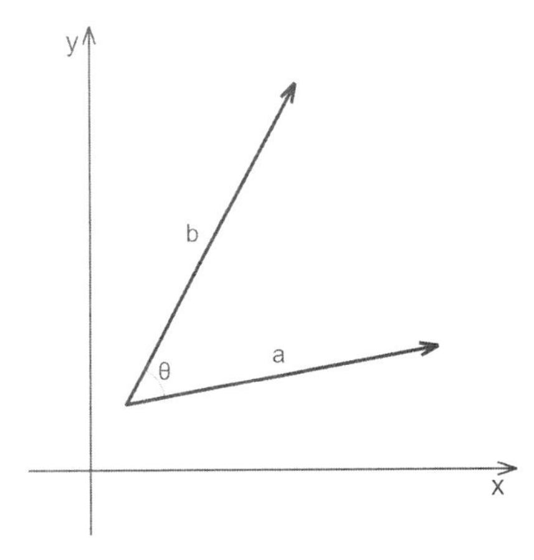

Figure 9.5 The trigonometric sense of the dot product

between vectors is right, or exactly 90 degrees, their dot product is always 0—and vice versa. If a dot product of two vectors is 0, the vectors are orthogonal. If you remember, in chapter 5 we reinvented this property for orthogonal vectors.

For all the other angles, the process is more complicated. To measure the angle between vectors **a** and **b**, you have to take their dot product, divide it by the length of **a**, divide that by the length of **b**, and then take the arccosine function (the function inverse to the cosine) of whatever remains:

$$\theta = \arccos \frac{a \cdot b}{\|a\|\|b\|}$$

If **a** and **b** are both of length 1, when they're co-directed, their dot product is also 1. When they're counter-oriented, looking opposite ways, the dot product is –1. I know that trivia like this doesn't stick in the head. Without practice, everyone who learns this in college starts mistaking cosine for sine in a few years, and mistaking 1 for –1 in a few more. So let's look at a practical application that, surprisingly, also works as a mnemonic for how the dot vector product relates to the angle between vectors.

SEE ALSO Better Explained has a detailed introduction to vector calculus, and the article on the dot product even has a video in it. See http://mng .bz/Y6aQ.

9.2.2 Vectors and their dot product in SymPy

Before we get to the mnemonics, let's see how vectors are handled in SymPy. In SymPy, a vector isn't so much a data type as it is a weighted sum of basis vectors. The basis vectors form coordinate systems, and for 3D, such a system can easily be created with a single line. For example, this is how you create a coordinate system called `N`:

```
from sympy.vector import CoordSys3D
N = CoordSys3D('N')
```

In this system, a vector (0, -1, 2) becomes

```
a = 0*N.i - 1*N.j + 2*N.k
```

Common laws apply, of course. You don't have to keep the `0*` term, you don't have to write `1*` explicitly, and you can rearrange terms if you want to. Here's the same vector from before:

```
a = 2*N.k - N.j
```

A SymPy method that does the vector dot product is unsurprisingly called `dot`:

```
a = 1*N.i + 2*N.j - 3*N.k
b = 3*N.i - 2*N.j + N.k
print(a.dot(b))
```

This denotes vector (1, 2, -3).

This denotes vector (3, -2, 1).

This results in 1*3 – 2*2 – 3*1 = –4.

A less-known abbreviation for the dot product is the `&` operator:

```
>>> a.dot(b)
-4

>>> a & b
-4
```

You can introduce your own coordinate systems, of course, and implement your own functions for vector products—and if you want to expand to an *N*-dimensional space, you have to. We'll look into that topic in section 9.5.

9.2.3 Practical example: Using a dot product to light a scene

Imagine that you want to render a 3D scene. Supposedly, every object in that scene has its own color and maybe even texture. Rendering, then, much resembles the projection problem from chapter 4. You want to find a color of a scene object that corresponds to a color of a screen pixel for every pixel on a screen.

For realistic rendering, however, this approach isn't enough. Let's say you have a poorly tesselated gray 3D sphere made of quads. When you project it onto a screen, you get . . . a gray circle (figure 9.6).

Figure 9.6 Simply bringing the object's color to the screen doesn't look 3D.

To make the sphere look three-dimensional, you have to add some shades to it. To do that, you add a lighting source with its own 3D position and compute how its lighting affects different quads of your model. When you do, you'll get a more realistic picture (figure 9.7).

Figure 9.7 Adding a lighting source and computing shading for it properly makes a sphere look like a sphere.

But how do you compute shading? It's simple! For every pixel on a screen, take the object's color as before, but also take the normal vector of an element it comes from (a vector orthogonal to the element's plane; see figure 9.8) and a 3D position of the object's point that corresponds to our particular pixel. Having a point's original position, you can calculate the lighting direction vector: a vector from the pixel's original position to the lighting source normalized. A dot product of the lighting direction vector and an object surface's normal vector will be the coefficient of lighting.

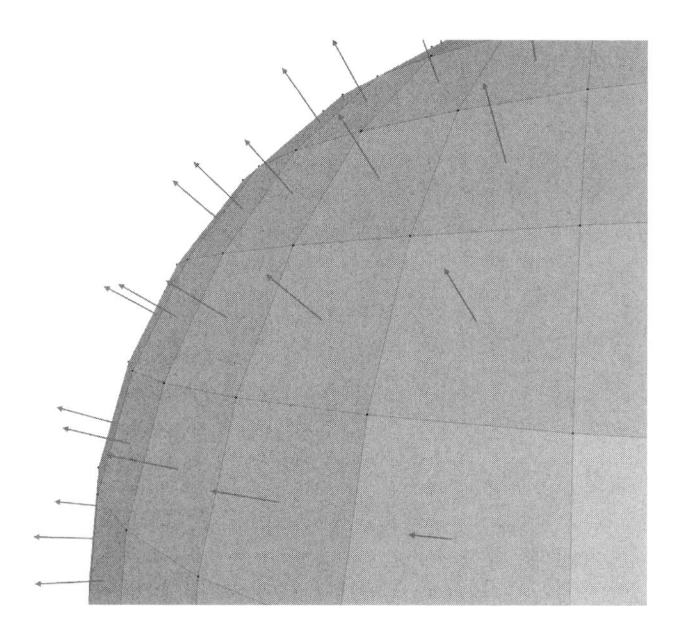

Figure 9.8 In our case, the sphere's elements are quadrilaterals (quads for short), each lying in its own plane and therefore having its own normal vector.

> **NOTE** This shading approach also works with directional lighting. When there's no particular point source of light, the light comes from some particular direction, and if you read chapter 4 carefully, you already have an idea how it's done. The directional lighting source is the point in projective space with the fourth homogeneous coordinate being 0: (x, y, z, 0). A point lighting source may even share a direction, but it would still lie in the Euclidean space with the nonzero last coordinate: (x, y, z, 1). One number makes all the difference!

When vectors are co-oriented, the lighting is maximal, so the coefficient is 1. When they're orthogonal, they have a right angle between them, so the coefficient is 0. For counter-oriented vectors, the coefficient is -1, and for lighting purposes, we can ignore all the values below 0. The light doesn't shine where the dot product is negative. These special relative positions for the lighting direction vector and the object's own normal vector are shown in figure 9.9.

Although this lighting model of a sphere is trivially simple, I find it useful to keep in mind because it also works as a great mnemonic for the dot product itself. Every time I can't recall whether there's a sine or cosine in the dot product formula, I remember this: the lightest hour on our planet is noon, and that's when the trees are almost codirected with the Sun vector. This means that the function in question is almost 1 when the input angle is almost 0. That's exactly how cosine behaves.

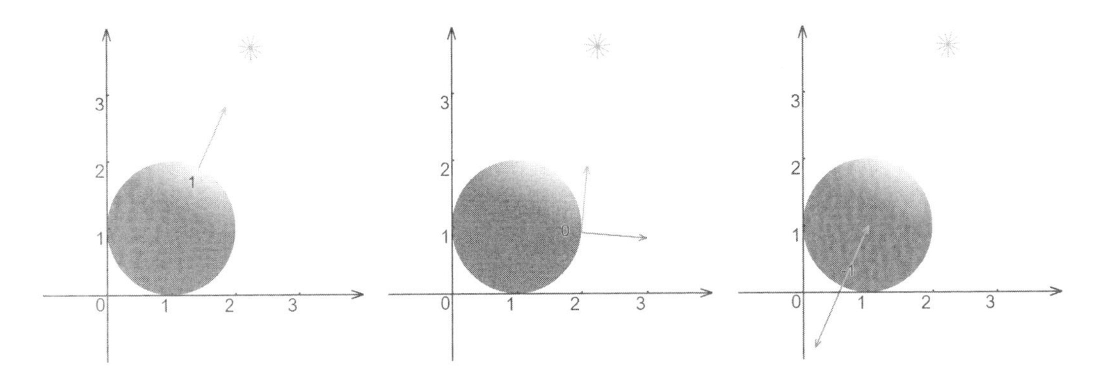

Figure 9.9 Lighting in a point is determined by the dot product of the object's normal vector and the lighting direction vector.

So the simple, effective shading technique that makes a flat circle into a spherelike thing on a screen is a dot product with a few more vector subtractions and normalizations. But wait—we're not done yet. Because the dot product is smooth and continuous, for the real sphere, the lighting changes smoothly and continuously in any point's proximity (figure 9.10).

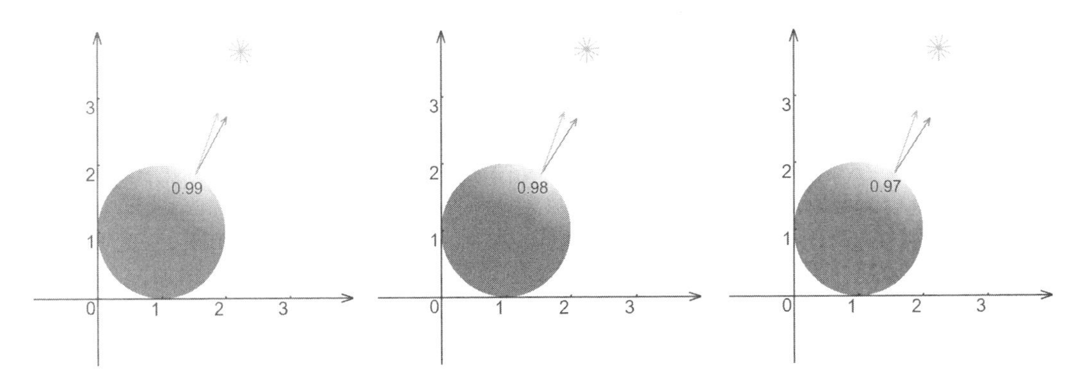

Figure 9.10 For a smooth sphere, the shading also changes smoothly and continuously.

But that's not what we saw in figure 9.6, did we? That's because our model of a sphere isn't a real sphere; the surface is made of flat quads. Never mind! We can still make it look smooth. For this task, we should find average normal vectors for every point in our model; then we can interpolate these vectors inside our quads pretty much as we did with color interpolation in chapter 8. This interpolation smooths the way the normal vector changes, which results in a smooth-looking model onscreen, too (figure 9.11).

Different methods of normal vector interpolation are available, so you don't have to get deep into details; usually, you pick a shading model that fits your taste instead of

Figure 9.11 By smoothly interpolating the normals of a sphere made of quads, you make the sphere itself look smooth.

writing it yourself. You should still understand that smooth shading doesn't change the model itself; it only interpolates the normal vectors used in the shading. The thing onscreen looks like an accurate model of a sphere but in reality is only a few hundred quads put together. That's economical.

Speaking of making things look more detailed than they are, another normal vector trick comes to mind. If we put a bitmap on an object, basically determining a UV-to-color function, we can use the bitmap's colors to encode how the normal vector of an object should be altered in the shading at any particular point. This technique, called *normal mapping*, allows us to add 3D-looking details to our 3D models without adding any new quads or triangles. In figure 9.12, a brick normal texture is applied to our sphere. Now it looks as though the sphere is made of thousands of quads, but in reality, there are only a few hundred.

Figure 9.12 Normal mapping affects the normal vectors only in the shading computation. It doesn't add any "real" geometry to the scene.

NOTE Finding a good UV parameterization, so a function that maps the model surface onto the bitmap space, is quite a problem on its own, and one that has no single one-fits-all solution. Our department even does a meta-analysis every few years, searching the papers for the next big thing: an algorithm that parametrizes more models better than the others that we have. There were no clear winners in the past ten years. But in our case (figure 9.12), the parameterization is

$UV(x, y, z) = (x, y)$, which is an x-y-plane projection—hardly the best way to parametrize a sphere but probably the simplest.

These simple shading tricks are well known and widely available. They're used everywhere rendering is used, including even boring computer-aided design (CAD) and computer-aided engineering (CAE) applications. At the same time, in the game-design industry, writing more and more intricate, powerful custom shaders is a big deal. People compete to see who can make the most realistic shader of a given effect that also runs fastest. Nobody likes to buy a new game only to look at last year's graphics, which everybody in the industry has already copied. The gamers' market is at stake!

All the 3D shading eventually boils down to the dot product. The normal vector may come solely from geometry, some interpolation scheme, a bitmap, or a custom shader code. It still gets dotted by the lighting direction vector to determine whether the particular point of a model should be lighter or darker.

SEE ALSO If you want to learn more about rendering, see this article on realistic multisource lighting, by Inigo Quilez: https://iquilezles.org/articles/outdoorslighting.

9.2.4 Section 9.2 summary

The dot product turns a pair of vectors into a single scalar. When vectors are orthogonal, their dot product is 0. Numerically, a dot product of a pair of vectors is a product of the vector's lengths and a cosine of the angle between them.

Shading is essentially about taking a dot product of the surface's normal vector and the direction toward the lighting source. You can also use the mental model of shading to remember how the dot product works.

9.3 Cross product: Normal vector and the parallelogram area

The next product we're looking into is called the *cross product*, and it's usually denoted by a cross:

$$\mathbf{c} = \mathbf{a} \times \mathbf{b}$$

9.3.1 Other names and motivations behind them

The other name for cross product is *vector product*. This name may be confusing because the dot product is also a vector product. But the name fits the cross product better because unlike the dot product, this one takes two vectors and turns them into a vector as well. This time, **c** appears in bold because it's a vector, too. The formula for this new vector is

$$\mathbf{a} = (a_1, a_2, a_3)$$

$$\mathbf{b} = (b_1, b_2, b_3)$$

$$\mathbf{c} = \mathbf{a} \times \mathbf{b} = (a_2b_3 - a_3b_2, a_3b_1 - b_3a_1, a_1b_2 - a_2b_1)$$

If you meditate on the formula for a bit, you'll see that the members in every per-coordinate subtraction are symmetrical: $a_i b_j - a_j b_i$. This means that if we switch **a** and **b** in the product, each subtraction becomes inverted: $b_i a_j - b_j a_i = -(a_i b_j - a_j b_i)$. So $\mathbf{b} \times \mathbf{a} = -(\mathbf{a} \times \mathbf{b})$.

NOTE This property is called *anticommutativity*. In chapter 2, we discussed commutativity in the context of algebraic systems. We saw that integers are commutative over addition, so $a + b = b + a$. Unlike + for integers, the cross product for vectors changes the sign when its arguments swap places, which makes it *anticommutative.*

The third name for the cross product is *directed area product*. This name reveals the geometric sense—or rather the first of the geometric senses—of the cross product. The length of the resulting vector **c** is the area of the parallelogram made by input vectors **a** and **b** (figure 9.13).

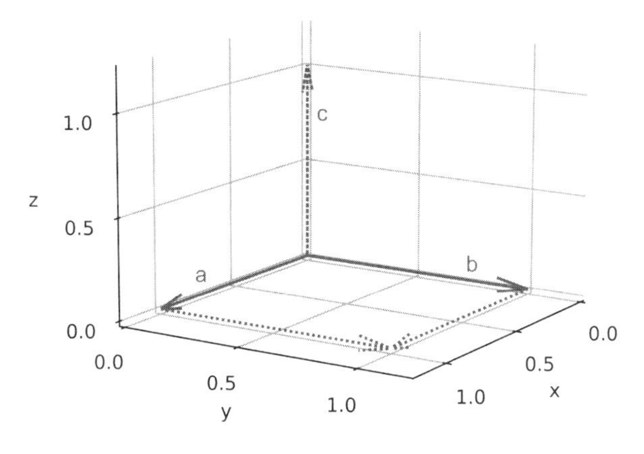

Figure 9.13 The area of the parallelogram made by (1, 0, 0) × (0, 1, 0) is 1.

If you recall some trigonometry from high school, you might come up with the formula for the parallelogram area (figure 9.14):

$$\text{area} = \|a\| \|b\| \sin \theta$$

If we use this area to scale a normalized vector **n** orthogonal to both **a** and **b**, we get an alternative formula for the cross product:

$$c = \|a\| \|b\| \sin \theta \boldsymbol{n}$$

And here comes the second geometric sense of the cross product. The cross product **c** is the direction of the normal to the plane that both **a** and **b** lie in (figure 9.15). Then the **c**, of course, is perpendicular to both **a** and **b**.

So the cross product's length is the area of the parallelogram that **a** and **b** form, and the cross product's orientation is the normal vector of the plane that **a** and **b** live

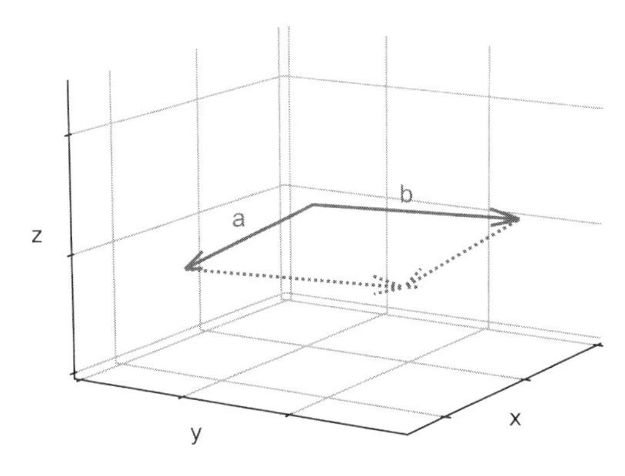

Figure 9.14 A parallelogram made by vectors a, and b

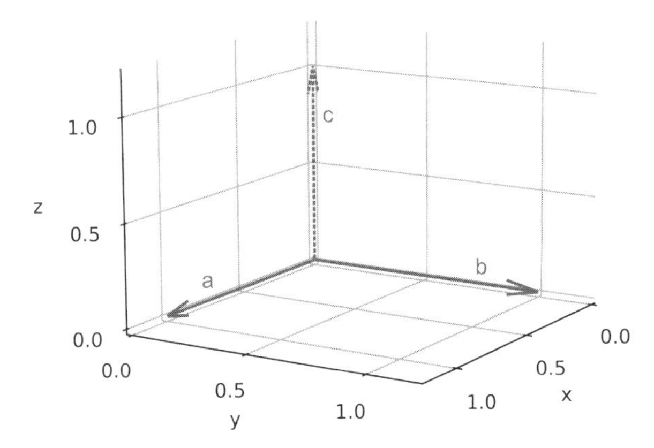

Figure 9.15 A cross product of (1, 0, 0) × (0, 1, 0) is (0, 0, 1).

in together. Both properties are often used in practice. The first one is convenient for computing triangles' areas because the triangle's area is half of the parallelogram's. The second property is useful for making equations like "**c** is orthogonal to both **a** and **b**."

A small reminder: the most important value of the dot product is 0. The fact that $\mathbf{a} \cdot \mathbf{b} = 0$ designates that the input vectors are orthogonal, which makes sense because the dot product's formula has a cosine in it. The cosine of 90 degrees is 0.

The cross product has a sine instead. The sine of 0 is 0, which means that the cross product's length is 0 when the input vectors are codirected or exactly counter-directed. In both cases, they form a degenerate parallelogram. Also, **a** and **b** share an infinite amount of planes, so their product can't have any specific direction.

The cross product doesn't really belong to dimensions other than 3. It's a 3D thing. There is, however, a generalization with the same properties that technically isn't a cross product, but it also designates a normal vector for a multidimensional hyperplane made by vectors. Further, it contains the volume of the parallelotope that the input vectors make. We'll look at this generalization briefly in section 9.5.

NOTE Another reminder: in chapter 3, we met hyperplanes in the context of linear equations. The hyperplane is the generalization of a 3D plane, like a plane but in 4D, 5D, and so on. Similarly, a parallelotop is the generalization of a parallelogram, part of an n-dimensional hyperplane bounded by parallel flat (n-1)-dimensional thingies. A parallelepiped (a part of 3D space limited by three pairs of parallel flat planes) is also a parallelotop.

Interestingly, although technically the cross product doesn't exist in 2D, it still has its application there.

9.3.2 The cross product in SymPy

Before we get to the application, let's see how cross products are computed in SymPy. Like the dot product, the cross product has a SymPy method and abbreviation, and to use them, we should introduce the coordinate system and the vectors in it:

```
from sympy.vector import CoordSys3D
N = CoordSys3D('N')

a = 1*N.i + 2*N.j - 3*N.k    <- This denotes vector (1, 2, –3).
b = 3*N.i - 2*N.j + N.k      <- This denotes vector (3, –2, 1).
```

For these vectors, taking their cross product looks like

```
>>> a.cross(b)
(-4)*N.i + (-10)*N.j + (-8)*N.k
```

or, using the abbreviation ^,

```
>>> a ^ b
(-4)*N.i + (-10)*N.j + (-8)*N.k
```

SymPy allows symbolic values for the vector components. If you want a formula for the cross product of vectors $(a_1, a_2, 0) \times (b_1, b_2, 0)$, you can produce it with this one-liner:

```
>>> (a1*N.i + a2*N.j) ^ (b1*N.i + b2*N.j)
(a1*b2 - a2*b1)*N.k
```

Why would you want such a formula, though? You'll see why in the following section.

9.3.3 Practical example: Check whether a triangle contains a point in 2D

A job interview in our department usually includes solving a small geometry problem. Sure, making people do math under stress may seem harsh, but then again, our department's whole purpose is doing math under stress, so the challenge is fair.

One particular interview problem became my favorite over the years. The problem is simple. We have three 2D points that form a triangle. We want a function that for every point in space answers whether that point lies inside or outside the given triangle.

The beauty of this problem is that no single correct answer counts as a pass; there are multiple ways to program the point-in-triangle function. Although some ways are more efficient than others, all possess merits and flaws, more often connected to programming and computational problems than to geometry. The problem motivates a candidate to show both their geometry knowledge and understanding of software engineering without writing a single line of code.

One of the best ways to solve the point-in-triangle problem—which is simultaneously geometrically correct, easy to program, and fast to compute—involves the cross product. The cross product doesn't exist in 2D, of course. No worries! We'll simply extend our space to 3D by putting our source 2D space on the 3D *x*-*y* plane. Then we will add a dimension and make the cross product viable.

Let's say we have three points—(x_1, y_1), (x_2, y_2), and (x_3, y_3)—and the point that comes as the predicate's argument is (x, y). In 3D, the same points become $(x_1, y_1, 0)$, $(x_2, y_2, 0)$, and $(x_3, y_3, 0)$, and the point to check is $(x, y, 0)$.

Now let's look into one more reason for the cross product to be called *cross*. According to the computational formula,

$$\mathbf{c} = \mathbf{a} \times \mathbf{b} = (a_2b_3 - a_3b_2,\ a_3b_1 - b_3a_1,\ a_1b_2 - a_2b_1)$$

Given that both our third coordinates a_3 and b_3 are 0, the formula simplifies to

$$\mathbf{c} = \mathbf{a} \times \mathbf{b} = (0,\ 0,\ a_1b_2 - a_2b_1)$$

Now the third coordinate of the resulting vector is the signed area of the parallelepiped that **a** and **b** make. By *signed*, I mean that if **a** lies on the right from **b**, the sign is +; if it's on the left, the sign is –. And as **a** *crosses* **b**, the resulting area changes its sign.

This "crossing" is a helpful mental model to keep in mind. Look—I even drew a piece of a road and a little crossing-sign person to make the picture more mnemonic for you (figure 9.16).

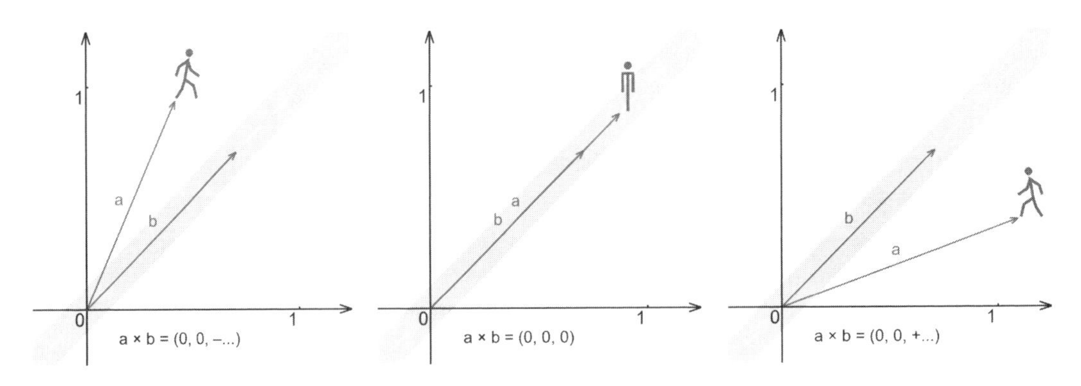

Figure 9.16 The third coordinate of the 2D cross product changes sign as a crosses b.

The right-hand rule

In 2D, if vector **a** has to go clockwise to "cross" the **b**, the third coordinate of the product is negative. If **a** has to go counterclockwise, the third coordinate is positive.

This situation has an analog in 3D called the *right-hand rule*. To see how it works, place your right wrist so that the index finger points toward the **a** vector. Now bend your middle finger so that points to the **b** vector. The direction of your thumb is the direction of the **a** × **b** product, as shown in this figure:

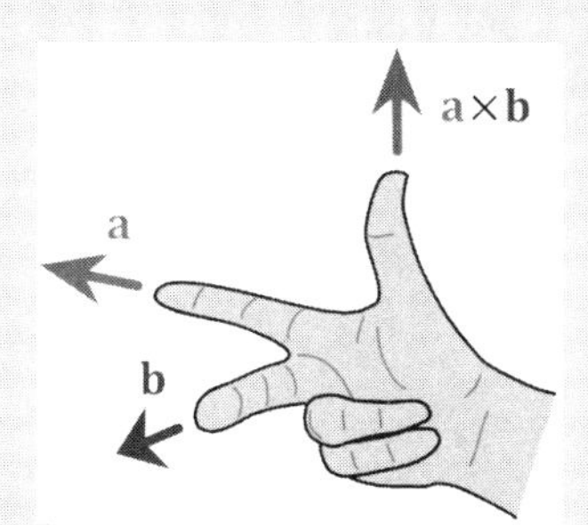

The right-hand rule (Acdx, CC BY-SA 3.0, via Wikimedia Commons)

The rule translates back to 2D. If the third coordinate is positive, you should be pointing your thumb toward yourself. If it's negative, you'd have to rotate your wrist and point your thumb away.

Now we can use the sign of the third coordinate to classify the point's relative position to all the triangle's sides. If we enumerate points as shown in figure 9.17, the point should lie on the right of all three sides to be considered inside the triangle.

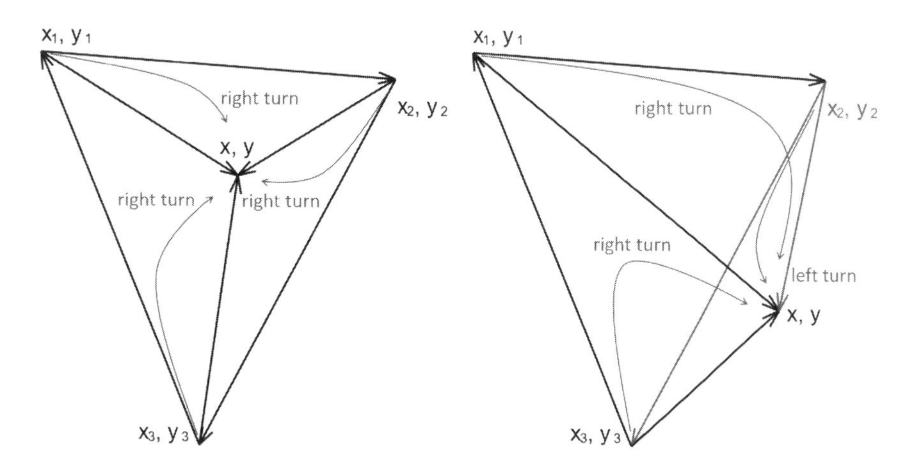

Figure 9.17 If points of the triangle are clockwise-oriented, the point that lies inside the triangle must lie on the right side of all the triangle's side vectors. If this check fails for any of the side vectors—in the figure on the right, that's for the ($x_3 - x_2$, $x_3 - y_2$) side—the point is lying outside.

We can write these criteria down as a point-in-triangle predicate:

$$(x - x_1, y - y_1, 0) \times (x_2 - x_1, y_2 - y_1, 0) > 0$$

$$(x - x_2, y - y_2, 0) \times (x_3 - x_2, y_3 - y_2, 0) > 0$$

$$(x - x_3, y - y_3, 0) \times (x_1 - x_3, y_1 - y_3, 0) > 0$$

In real code, of course, we don't have to use an actual function for the cross product. I mean, we know that two of three result coordinates are 0, so we shouldn't waste our resources computing that. The resulting predicate might look somewhat like the following:

```
def point_in_triangle(x, y):
    global x1, y1, x2, y2, x3, y2
    return ((x - x1) * (y2 - y1) - (y - y1) * (x2 - x1) > 0
        and (x - x2) * (y3 - y2) - (y - y2) * (x3 - x2) > 0
        and (x - x3) * (y1 - y3) - (y - y3) * (x1 - x3) > 0)
```

or, with three fewer operations:

```
def point_in_triangle(x, y):
    global x1, y1, x2, y2, x3, y2
    return ((x - x1) * (y2 - y1) > (y - y1) * (x2 - x1)
        and (x - x2) * (y3 - y2) > (y - y2) * (x3 - x2)
        and (x - x3) * (y1 - y3) > (y - y3) * (x1 - x3))
```

An alternative take on this problem involves computing the three areas that triangles (x_i, y_i), (x_j, y_j), and (x, y) make and comparing their sum with the input triangle's area (figure 9.18). If the sum of the "small" triangle's area is the same as the "big" triangle's area, the point is inside; if the sum of three is larger, it isn't.

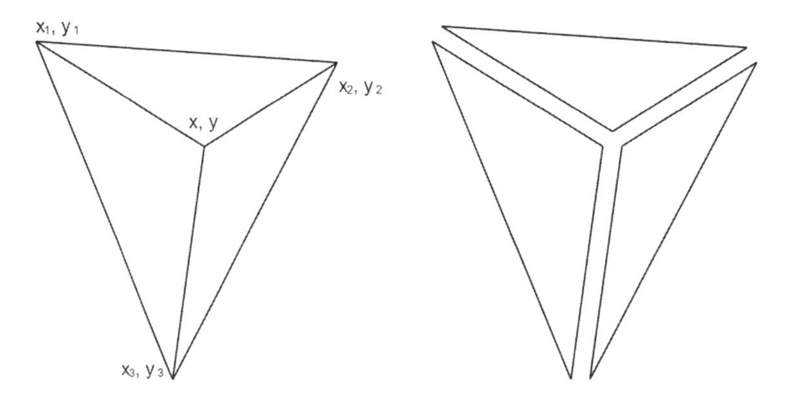

Figure 9.18 For a point inside a triangle, the area of the whole triangle should be equal to the sum of the "small" triangle's area.

If you use Heron's formula (which you may remember from school) for area computation, this approach may seem both ineffective and cumbersome:

$$A = \sqrt{s(s-a)(s-b)(s-c)}$$

$$s = \frac{a+b+c}{2}$$

$$a = \|(x_2 - x_1, y_2 - y_1)\|$$
$$b = \|(x_3 - x_2, y_3 - y_2)\|$$
$$c = \|(x_1 - x_3, y_1 - y_3\|$$

The formula itself involves more operations, and you also have to compute the triangle's side lengths. But if you start using the cross product to compute the triangle's area, the approach becomes only slightly inferior to the relative orientation criteria. In 2D, you don't have to compute all the coordinates, only the last. That's five subtractions and two multiplications per operation:

$$A = \frac{(x_3 - x_1)(y_2 - y_1) - (x_2 - x_1)(y_3 - y_1)}{2}$$

There's also a division by two, but for this problem, you can omit it for all four triangles' areas and compare the parallelepipeds' areas instead.

The only difference between Heron's formula and the 2D cross is that in the latter approach, the area is signed. Whereas for clockwise-oriented triangles, it always comes positive, for counterclockwise-oriented triangles, it's always negative. This could be a useful byproduct on its own. Now you can compute not only the triangle's area, but also the triangle's orientation. As you might see, the cross vector opens multiple ways to write efficient code in 2D where the product itself isn't even supposed to exist.

9.3.4 Section 9.3 summary

The cross product turns a pair of vectors into another vector. The resulting vector is orthogonal to both input vectors, so it serves as a normal vector for the plane that input vectors share. The length of the resulting vector is the area of the parallelogram the input vectors make. The fact that this area is 0 designates that the input vectors are co- or counter-directed.

SEE ALSO The cross-product article on Better Explained is as good as the dot-product one. You can find it at http://mng.bz/GRJ8.

9.4 Triple product: The parallelepiped volume

The last vector product we need to know is called the *triple product,* and it doesn't have a dedicated symbol. You see, it's a product of three vectors, and mathematical notation doesn't accommodate ternary operations well. So we write this product as a composition of two other products that we already know about:

$$\mathbf{d} = \mathbf{a} \cdot (\mathbf{b} \times \mathbf{c})$$

Don't worry if you forget which comes first. As long as the order of operands is kept, you can switch the dot and the cross with no effect on the result:

$$\mathbf{d} = \mathbf{a} \cdot (\mathbf{b} \times \mathbf{c}) = (\mathbf{a} \times \mathbf{b}) \cdot \mathbf{c}$$

To understand why, look at the geometric sense of the triple product.

9.4.1 The geometric sense of the triple product

Like the other vector products, the triple product has an alternative name. The triple product is also called the *box product* because the geometric sense of the triple product is the signed volume of the parallelepiped formed by the three operand vectors (figure 9.19).

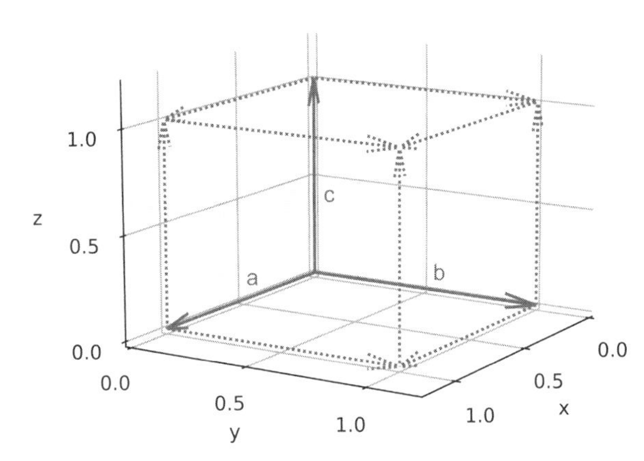

Figure 9.19 The volume of the parallelepiped formed by (1, 0, 0), (0, 1, 0), and (0, 0, 1) is 1.

The "volume" part explains why $\mathbf{a} \cdot (\mathbf{b} \times \mathbf{c}) = (\mathbf{a} \times \mathbf{b}) \cdot \mathbf{c}$. Because the dot product is commutative (meaning $\mathbf{a} \cdot \mathbf{b} = \mathbf{b} \cdot \mathbf{a}$), you can rearrange the formula in this way:

$$(\mathbf{a} \times \mathbf{b}) \cdot \mathbf{c} = \mathbf{c} \cdot (\mathbf{a} \times \mathbf{b})$$

Now while $\mathbf{a} \cdot (\mathbf{b} \times \mathbf{c})$ is the volume of a parallelepiped made by **a**, **b**, and **c**, $\mathbf{c} \cdot (\mathbf{a} \times \mathbf{b})$ is the volume of the parallelepiped made by **c**, **a**, and **b**. Wait a minute—that's the same parallelepiped!

The "signed" part makes reasoning a little bit trickier. The signed volume is the same concept as the signed area for the cross product. In fact, the sign comes to the triple product from its cross-product part. So the sign shows the relative orientation of the vectors. Although parallelepipeds made by vector triplets (**a**, **b**, **c**), (**b**, **c**, **a**), and (**c**, **a**, **b**) are exactly the same, the parallelepipeds made by three other possible triplets—(**a**, **c**, **b**), (**c**, **b**, **a**), and (**b**, **a**, **c**)—are the inverted versions of the first three. Their signed volume is exactly the opposite.

You can compute the triple product by doing its compound operations. You should start with the cross; otherwise, the result of the dot product is a scalar, and you can't cross a scalar and a vector.

But there's an alternative way: form a matrix from the three vectors and find its determinant. In chapter 3, we briefly touched on the determinant as a measure of the weight of a matrix. Now you know its geometrical sense, too, not so much a weight as a volume:

$$\mathbf{a} = (a_1, a_2, a_3)$$

$$\mathbf{b} = (b_1, b_2, b_3)$$

$$\mathbf{c} = (c_1, c_2, c_3)$$

$$a \cdot (b \times c) = det\left(\begin{bmatrix} a_1 & a_2 & a_3 \\ b_1 & b_2 & b_3 \\ c_1 & c_2 & c_3 \end{bmatrix}\right)$$

SEE ALSO The connection between determinants and volumes is explained in detail in *Interactive Linear Algebra*, by Dan Margalit and Joseph Rabinoff (http://mng.bz/zmNg).

9.4.2 The triple product in SymPy

There is no triple product in SymPy, but you can make one by using both the dot and the cross products. First, you introduce your coordinate systems and the vectors:

```
from sympy.vector import CoordSys3D
N = CoordSys3D('N')

a = 1*N.i + 2*N.j - 3*N.k
b = 3*N.i - 2*N.j + N.k
c = N.j
```

This denotes vector (1, 2, –3).

This denotes vector (3, –2, 1).

This denotes vector (0, 1, 0).

Now you can do the triple product like this:

```
>>> print(a.dot(b.cross(c)))
-10
```

The abbreviations also work. But please note that in Python, `&` takes precedence over `^`, so you have to place the brackets for the cross product explicitly:

```
>>> print(a & (b ^ c))
-10
```

9.4.3 Practical example: Distance to a plane

Let's say we have a 3D plane set by a triplet of points. We assume that the points don't belong to the same line, so one and only one plane hosts them all. Now let's say we have the fourth point, and we want to know the distance from this point to the aforementioned plane. Then we could use this distance in some practical application. In chapter 10, for example, we'll use it to determine the distance field to the whole triangle mesh.

$$\mathbf{p}_1 = (x_1, y_1, z_1)$$

$$\mathbf{p}_2 = (x_2, y_2, z_2)$$

$$\mathbf{p}_2 = (x_3, y_3, z_3)$$

$$\mathbf{p} = (x, y, z)$$

The most straightforward solution would be to Google "three-point plane equation," get the normal vector from the plane, project the fourth point on the normal vector sticking out of the plane's point of origin, and then get the distance from this projection to the plane's origin.

A more elegant solution would be to form three vectors: two that make a parallelogram on a triplet's plane and a third one that adds to them to form a parallelepiped in 3D (figure 9.20).

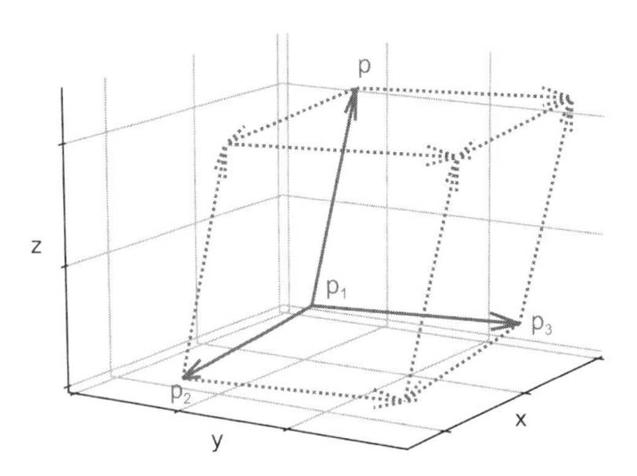

Figure 9.20 A parallelepiped formed by p_2-p_1, p_3-p_2, and p-p_1, and the parallelogram formed by p_2-p_1, p_3-p_1, too

The signed volume of the parallelepiped is the triple product $(\mathbf{p}_2 - \mathbf{p}_1) \times (\mathbf{p}_3 - \mathbf{p}_1) \cdot (\mathbf{p} - \mathbf{p}_1)$. The area of the parallelogram is the length of the cross product $(\mathbf{p}_2 - \mathbf{p}_1) \times (\mathbf{p}_3 - \mathbf{p}_1)$. The height of the parallelepiped, which is its volume divided by the area of the base, is the distance we're looking for. It's the distance from point **p** to the plane that $\mathbf{p}_1$, $\mathbf{p}_2$, and $\mathbf{p}_3$ make:

$$d = \frac{|((p_2 - p_1) \times (p_3 - p_1)) \cdot (p - p_1)|}{\|(p_2 - p_1) \times (p_3 - p_1)\|}$$

Computationally, this solution is light; we have to compute the cross product only once and reuse it in both the numerator and denominator. Geometrically, it's also the same as the straightforward solution with a Googleable plane equation. You simply have to look at it from another perspective.

Yes, if you divide the volume of the parallelepiped by the area of its base parallelogram, you get its height. At the same time, the cross product $(\mathbf{p}_2 - \mathbf{p}_1) \times (\mathbf{p}_3 - \mathbf{p}_1)$ produces the normal vector to the plane that hosts both the $(\mathbf{p}_2 - \mathbf{p}_1)$ and $(\mathbf{p}_3 - \mathbf{p}_1)$ vectors. When you divide this normal vector by its length $||(\mathbf{p}_2 - \mathbf{p}_1) \times (\mathbf{p}_3 - \mathbf{p}_1)||$, you get a normalized vector: the vector with length 1. When you get a dot product of this normalized normal vector and the vector from the point $\mathbf{p}$ to one of the points on the plane, you get the wanted distance. You may want to rewrite the formula to reflect the intention better, but it remains essentially the same computation:

$$d = \left| \frac{(p_2 - p_1) \times (p_3 - p_1)}{||(p_2 - p_1) \times (p_3 - p_1)||} \cdot (p - p_1) \right|$$

So the searchable solution still boils down to the cross and dot products, although not necessarily without extra steps. The number of extra steps depends on the specific framework or library you have to rely on.

Knowing the geometric meaning of vector products, you can opt out of third-party dependencies and implement the most efficient solution on the spot. Even if you choose to look for a solution somewhere, the formulas you'll find out there are often easier to understand when you keep their geometrical sense in mind.

I have to admit that "distance to plane" is more an interview question than a practical problem. It's a nice problem, though, because it showcases everything you want to know about all three vector products—dot, cross, and triple—and brings it into a single formula.

9.4.4 Section 9.4 summary

The triple product is a product of three vectors determined by the signed volume of the parallelepiped that the three operands make. The sign of the product depends on the cyclic order of the operands, which means that you can safely rotate the operands in the formula, but you can't swap a pair of operands without changing the sign of the product:

$$(\mathbf{a} \times \mathbf{b}) \cdot \mathbf{c} = (\mathbf{c} \times \mathbf{a}) \cdot \mathbf{b} = (\mathbf{b} \times \mathbf{c}) \cdot \mathbf{a}$$

$$(\mathbf{a} \times \mathbf{b}) \cdot \mathbf{c} = -(\mathbf{b} \times \mathbf{a}) \cdot \mathbf{c} = -(\mathbf{a} \times \mathbf{c}) \cdot \mathbf{b} = -(\mathbf{c} \times \mathbf{b}) \cdot \mathbf{a}$$

9.5 Generalization for parallelotopes

You can skip this section if you want. Technically, there's no such thing as an N-dimensional cross product. "Triple product in 4D" is an oxymoron. Yet some useful generalizations bear the geometric properties of the cross and triple product in N-dimensional space where $N \neq 3$.

9.5.1 The meaning

The N-dimensional cross product is a vector that's orthogonal to all its N–1 operands. It's like a normal vector to a hyperplane the operands make. (Chapter 3 discusses hyperplanes.)

The cross product vector's length still works like the parallelogram's area, but now it's the (N–1)-dimensional volume of an (N–1)-dimensional parallelotope.

The triple product is still a number, and the number is the signed N-dimensional volume of an N-dimensional parallelotope formed by the product's operands.

I know that all this is a little tricky to imagine. Normally, we don't exercise our intuition in high dimensions. Also, it's nearly impossible to illustrate N-dimensional things properly on a piece of paper or a screen. Besides, as practical programmers, we have to work in 2D and 3D most of the time anyway. But I still think that a brief introduction to multidimensional generalizations is good for you. If the need occurs, this chapter will give you the initial confidence boost to dive into this unpopular topic as deeply as you'll have to.

9.5.2 The math

First, we need a new notation. A cross product in 3D is a binary function, meaning that it takes only two operands. You can put these operands on both sides of the cross sign, which would work nicely in textual formulas. But in N-dimensional space, the cross product's generalization has N–1 operands. In ten-dimensional space, that's nine operands. You can't write all those operands around a cross sign and expect the formula to remain comprehensible. So for all dimensions other than three, let's agree to use the common C-like function notation:

$$\text{cross}(a_1, a_2, a_3, a_4, a_5, a_6, a_7, a_8, a_9)$$

The triple product has N arguments in N-dimensional space, so in ten-dimensional space, it should go as a function of ten variables. The name *triple* won't work anymore because in ten-dimensional space, the triple product becomes . . . *tenple*? Instead, let's adopt the nickname "the box product" for all the spaces apart from 3D. Then an example of a ten-dimensional box product would look like this:

$$\text{box}(a_1, a_2, a_3, a_4, a_5, a_6, a_7, a_8, a_9, a_{10})$$

If you remember that the triple product is the same as the determinant of the matrix its arguments make, the math of both generalizations will be simple for you.

The N-dimensional box product remains the determinant of the matrix, but now the matrix in N-dimensional space consists of N N-dimensional vectors. So a 3D box product remains as it was before:

$$box(\boldsymbol{a}, \boldsymbol{b}, \boldsymbol{c}) = det\left(\begin{bmatrix} a_1 & a_2 & a_3 \\ b_1 & b_2 & b_3 \\ c_1 & c_2 & c_3 \end{bmatrix}\right)$$

In 4D, it acquires a new argument vector:

$$box(\boldsymbol{a}_1, \boldsymbol{a}_2, \boldsymbol{a}_3, \boldsymbol{a}_4) = det\left(\begin{bmatrix} a_{11} & a_{12} & a_{13} & a_{14} \\ a_{21} & a_{22} & a_{23} & a_{24} \\ a_{31} & a_{32} & a_{33} & a_{34} \\ a_{41} & a_{42} & a_{43} & a_{44} \end{bmatrix}\right)$$

In N-dimensional space, it has N vectors, each with N elements as well:

$$box(\boldsymbol{a}_1, \boldsymbol{a}_2, \ldots, \boldsymbol{a}_N) = det\left(\begin{bmatrix} a_{11} & a_{12} & \ldots & a_{1N} \\ a_{21} & a_{22} & \ldots & a_{2N} \\ \ldots & \ldots & \ldots & \ldots \\ a_{N1} & a_{N2} & \ldots & a_{NN} \end{bmatrix}\right)$$

Computing the determinant in high dimensions isn't a trivial task. A generic formula is taught in every linear algebra course, but it's totally impractical, and practical computational schemes require a body of knowledge comparable to that of the whole business of solving linear systems.

So let's not dig into that topic at all. Let's leave linear algebra to linear algebra specialists. But let's still learn to compute the determinant for 2 × 2. The formula is simple, and it'll be enough to help us understand how the vector product generalizes because generalizations work not only in ten-dimensional spaces, but in 2D as well. So the 2 × 2 formula for the box product is

$$det\left(\begin{bmatrix} a_1 & a_2 \\ b_1 & b_2 \end{bmatrix}\right) = a_1 b_2 - a_2 b_1$$

In 2D, the meaning of the box product is still the signed volume of an N-dimensional parallelotop. A 2D parallelotop is called a *parallelogram,* and its volume becomes its area.

Now, remember how we applied a 3D cross product to 2D? We used this formula:

$$\mathbf{c} = \mathbf{a} \times \mathbf{b} = (0, 0, a_1 b_2 - a_2 b_1)$$

The third coordinate of the resulting vector is the signed area of the parallelogram formed by **a** and **b** . . . which is exactly the same as the box product of **a** and **b** in 2D!

You can choose different methods to compute the thing, but you can't escape the geometry. Whether you use a cross or box product, the formula for the parallelepiped's signed area remains the same.

Speaking of the cross product, we can use a simple trick to boil down its computation to the determinant. The product lacks input vectors to form an $N \times N$ matrix, of course, so we'll add a special pseudorow made of the basis vectors. We'll let our basis vectors take part in the computation and then use them to form a new vector. A usual 3D cross product with this trick becomes

$$cross(\boldsymbol{a}, \boldsymbol{b}) = det\left(\begin{bmatrix} i & j & k \\ a_1 & a_2 & a_3 \\ b_1 & b_2 & b_3 \end{bmatrix}\right)$$

It generalizes nicely to an N-dimensional space

$$cross(\boldsymbol{a}_1, \boldsymbol{a}_2, \ldots, \boldsymbol{a}_{N-1} = det\left(\begin{bmatrix} e_1 & e_2 & \ldots & e_N \\ a_{11} & a_{12} & \ldots & a_{1N} \\ \ldots & \ldots & \ldots & \ldots \\ a_{(N-1)1} & a_{(N-2)2} & \ldots & a_{(N-1)N} \end{bmatrix}\right)$$

and, of course, to 2D as well, although in 2D, it has only one argument:

$$cross(\boldsymbol{a}) = det\left(\begin{bmatrix} i & j \\ a_1 & a_2 \end{bmatrix}\right)$$

To understand how the trick with the basis vectors works, let's compute an example, getting a 2D cross product of the vector (2, 3)

$$\text{cross}(\mathbf{a}) = \mathbf{i}a_2 - \mathbf{j}a_1$$

in which the basis vectors are as usual:

$$\mathbf{i} = (1, 0)$$

$$\mathbf{j} = (0, 1)$$

Now let's put the vector in:

$$\mathbf{a} = (2, 3)$$

$$\text{cross}((2, 3)) = 3\mathbf{i} - 2\mathbf{j} = 3(1, 0) - 2(0, 1) = (3, -2)$$

As a result of the 2D cross product, we acquired a vector with the same length as the input one but oriented exactly orthogonally to it, which fits the geometric sense nicely.

The cross product should be orthogonal to all its operands, which is true in our case because we got only one operand. Also, the cross product should represent the volume of the (N–1)-dimensional parallelotop that the arguments make, which in 2D simply becomes a segment. The "volume" of this segment is its length.

SEE ALSO The geometric sense of the determinant, its generalization to high-dimensional spaces, and the recursive nature of its computation are all explained brilliantly on the Intuitive Math website. See http://mng.bz/0y8W.

9.5.3 The code

The main reason you don't have to keep the N-dimensional cross and box formulas in your head is that you can ask SymPy to give them to you at any time. Because we already know the secret, we already know that both products boil down to the determinant, and we can use this fact to make SymPy compose expressions for us.

For the box product, the process is as simple as declaring a matrix out of the vectors' elements and running a single `.det()` operation on it. A 2D box product looks something like this:

```
A = Matrix([[a1, a2], [b1, b2]])
print(A.det())
```

The result, as we already know, is

```
a1*b2 - a2*b1
```

To go to 3D, add another element to each vector, and don't forget to add another vector, too:

```
A = Matrix([[a1, a2, a3], [b1, b2, b3], [c1, c2, c3]])
print(A.det())
```

The result gets larger:

```
a1*b2*c3 - a1*b3*c2 - a2*b1*c3 + a2*b3*c1 + a3*b1*c2 - a3*b2*c1
```

Certainly, you can go further. You could add one more element per vector, add yet another vector, and get a 4D box product (full listing is available at ch_09/nd_box.py in this book's source code):

```
A = Matrix([
[a1, a2, a3, a4],
[b1, b2, b3, b4],
[c1, c2, c3, c4],
[d1, d2, d3, d4]])
print(A.det())
```

Its formula will be

```
a1*b2*c3*d4 - a1*b2*c4*d3 - a1*b3*c2*d4 + a1*b3*c4*d2 + a1*b4*c2*d3 -
➥ a1*b4*c3*d2 - a2*b1*c3*d4 + a2*b1*c4*d3 + a2*b3*c1*d4 - a2*b3*c4*d1 -
➥ a2*b4*c1*d3 + a2*b4*c3*d1 + a3*b1*c2*d4 - a3*b1*c4*d2 - a3*b2*c1*d4 +
➥ a3*b2*c4*d1 + a3*b4*c1*d2 - a3*b4*c2*d1 - a4*b1*c2*d3 + a4*b1*c3*d2 +
➥ a4*b2*c1*d3 - a4*b2*c3*d1 - a4*b3*c1*d2 + a4*b3*c2*d1
```

But you already see where it's heading, don't you? The 2D box product has only 2 terms, the 3D one has 6, and the 4D one has 24. We've seen this progression before. It not only rises fast, but also makes a popular interview problem. Yes, the symbolic computation of a determinant has asymptotic complexity of O(n!).

So although you can add a few more dimensions, sooner rather than later, the formulas SymPy produces for you will become impractical. For larger dimensions, you need some other approach to vector products.

NOTE The factorial complexity of the determinant goes in parallel with linear systems. If you remember, the symbolic solution for a well-defined linear system is also O(n!) in complexity, and it's more than fine for small systems. But for larger ones, you need a direct numeric solution, and for even larger ones, you need an iterative solution. Also, as with linear systems, the numeric accuracy of the straightforward determinant computation drops as the size of the matrix increases. Again, this might not be a problem for small systems, but for anything larger than 4×4, it's best to use a numeric solver. Numeric solvers for determinants are readily available. If you prefer to do your computations in Python, look up the `numpy.linalg.det` solver at http://mng.bz/KlJE.

But let's stick to symbolic solutions for a while. For the cross product, we can still use the determinant, but now we have to do the trick with basis vectors somehow. Not to worry! Because all our computation is symbolic, and SymPy doesn't segregate vector symbols from numbers symbols, we can add basis vectors the same way that we add vectors' elements:

This is the quasivector made by basis vectors. e_1 is supposed to be (1, 0, 0, 0), e_2 is supposed to be (0, 1, 0, 0), e_3 is supposed to be (0, 0, 1, 0), and e_4 is supposed to be (0, 0, 0, 1).

```
e1, e2, e3, e4 = symbols('e1 e2 e3 e4')
a1, a2, a3, a4 = symbols('a1 a2 a3 a4')
b1, b2, b3, b4 = symbols('b1 b2 b3 b4')
c1, c2, c3, c4 = symbols('c1 c2 c3 c4')
```

As before, these are three vectors: $a = (a_1, a_2, a_3, a_4)$, $b = (b_1, b_2, b_3, b_4)$, and $c = (c_1, c_2, c_3, c_4)$.

Now we can reuse the same determinant operation for the cross product. Here's the 2D cross product,

```
A = Matrix([[e1, e2], [a1, a2]])
print(A.det())
```

which results in

```
-a1*e2 + a2*e1
```

Because we imply that `e1` is (1, 0), and `e2` is (0, 1), the result reads $(a2, -a1)$. In the previous section, we saw seen that cross((2, 3)) = (3, –2), which agrees with this new SymPy-made formula perfectly. We can get the formula for the 3D cross product just as easily, so

```
A = Matrix([[e1, e2, e3], [a1, a2, a3], [b1, b2, b3]])
print(A.det())
```

will be

```
a1*b2*e3 - a1*b3*e2 - a2*b1*e3 + a2*b3*e1 + a3*b1*e2 - a3*b2*e1
```

or, in familiar notation,

```
a2*b3 - a3*b2, a3*b1 - a1*b3, a1*b2 - a2*b1
```

For the 4D (the full listing is at ch_09/nd_cross.py in the book's source code),

```
A = Matrix([
[e1, e2, e3, e4],
[a1, a2, a3, a4],
[b1, b2, b3, b4],
[c1, c2, c3, c4]])
print(A.det())
```

the formula is

```
-a1*b2*c3*e4 + a1*b2*c4*e3 + a1*b3*c2*e4 - a1*b3*c4*e2 - a1*b4*c2*e3 +
➥ a1*b4*c3*e2 + a2*b1*c3*e4 - a2*b1*c4*e3 - a2*b3*c1*e4 + a2*b3*c4*e1 +
➥ a2*b4*c1*e3 - a2*b4*c3*e1 - a3*b1*c2*e4 + a3*b1*c4*e2 + a3*b2*c1*e4 -
➥ a3*b2*c4*e1 - a3*b4*c1*e2 + a3*b4*c2*e1 + a4*b1*c2*e3 - a4*b1*c3*e2 -
➥ a4*b2*c1*e3 + a4*b2*c3*e1 + a4*b3*c1*e2 - a4*b3*c2*e1
```

Don't forget that you can use `collect` to make the result more readable:

```
>>> print(collect(A.det(), [e1, e2, e3, e4]))
e1*(a2*b3*c4 - a2*b4*c3 - a3*b2*c4 + a3*b4*c2 + a4*b2*c3 - a4*b3*c2) +
➥ e2*(-a1*b3*c4 + a1*b4*c3 + a3*b1*c4 - a3*b4*c1 - a4*b1*c3 + a4*b3*c1) +
➥ e3*(a1*b2*c4 - a1*b4*c2 - a2*b1*c4 + a2*b4*c1 + a4*b1*c2 - a4*b2*c1) +
➥ e4*(-a1*b2*c3 + a1*b3*c2 + a2*b1*c3 - a2*b3*c1 - a3*b1*c2 + a3*b2*c1)
```

9.5.4 Section 9.5 summary

We can generalize the cross and triple products from 3D to both higher and lower dimensions, retaining and generalizing their geometric properties. Technically, these products are no longer true cross or true triple products, but as programmers, we're allowed to be pragmatic and to care less about mathematical formalities.

9.6 Exercises

Exercise 9.1 What is (12.3, 0, 78.9) · (0, 45.6, 0)?

Exercise 9.2 What is (1, 0, 0) × (0, 0, 1)?

Exercise 9.3 What is box((1.2, 3.4, 0), (5.6, 0, 0), (0, 7.8, 0))?

Exercise 9.4 In section 9.3.2, we looked into a point-in-triangle problem. We used a cross in 2D trick to solve it, but the problem itself poses a trick question. A point may lie not only in the triangle, or out, but also *on* a triangle. We discussed the difference in chapter 2. The proposed solution considers a point inside a triangle only if it's "in" but not "on." Can you patch the solution so the points on a triangle will also be considered to be inside?

Exercise 9.5 In section 9.4.2, we proposed a formula for the distance between a point and a plane set by a triplet of other points. Now let's say we want to divide the whole 3D space with the plane. We want a definite criterion that tells whether any point in space lies by one side of the plane or the other. Or, of course, there is a third option: a point may lie exactly on the dividing plane. Can you make the formula from section 9.4.2 into this criterion?

Exercise 9.6 How many terms will a 5D box product have if you solve it symbolically?

9.7 Solutions to exercises

Exercise 9.1 It's 0. All the nonzero numbers in both vectors are getting multiplied by a zero from another vector before going into the sum.

Exercise 9.2 It's (0, –1, 0). Vectors (1, 0, 0) and (0, 0, 1) form a parallelepiped with area 1, so the length of the resulting vector should be 1. Also, the resulting vector should be orthogonal to their common plane, which is *x-z* plane. So it should be either (0, 1, 0) or (0, –1, 0). Using the right-hand rule, you can see that it goes against the *y*-axis, hence the minus sign.

Exercise 9.3 Well, that's exercise 9.1 with extra steps. The answer is again 0. The box product is the same as the triple product, so $\mathbf{a} \cdot (\mathbf{b} \times \mathbf{c})$. The cross product of the last two vectors is a vector (0, 0, something-something), and when you compute its dot product with the first vector, you get all zeroes.

Exercise 9.4 A point lies exactly on the triangle side when

$$((x - x_1, y - y_1, 0) \times (x_2 - x_1, y_2 - y_1, 0))_3 = 0$$

or

$$((x - x_2, y - y_2, 0) \times (x_3 - x_2, y_3 - y_2, 0))_3 = 0$$

or

$$((x - x_3, y - y_3, 0) \times (x_1 - x_3, y_1 - y_3, 0))_3 = 0$$

To extend the solution to include these points, all you have to do is to change the $>$ sign to $\geq$ everywhere.

Exercise 9.5 The formula from section 9.4.2 computes the unsigned distance. But both triple products from the numerator and the cross product from the denominator depend on vector orientation. Both the cross products, including the one in the triple product, are anticommutative, but the orientation of the cross product from the denominator gets lost when we take its length. For the triple product, we lose its sign explicitly by taking the absolute value.

But if we don't do that last step, if we don't take the absolute value of the triple product, the formula will result in a positive number when a point is on the "right-hand" side of the plane made by $(\mathbf{p}_2 - \mathbf{p}_1)$ and $(\mathbf{p}_3 - \mathbf{p}_1)$, and negative when it's on the opposite side. When the point lies exactly on the plane, the result is 0. This is our target criterion.

Exercise 9.6 It's 120. We've already seen that the complexity of determinant symbolic computation grows in factorial, so the sequence is 2, 6, 24, and 120.

Summary

- The dot product of two vectors is a scalar.
- The dot product of orthogonal vectors is always 0.
- Shading is a good mental model to use to remember how the dot product works.
- The cross product of a pair of vectors is a vector.
- The cross product is orthogonal to the input vectors.
- The length of the cross product is the area of the parallelogram that the input vectors make.
- When the input vectors are parallel, either codirected or counter-directed, their cross product is a vector of zero length.
- The orientation of the cross product adheres to the right-hand rule.
- The triple product of three vectors is a number.
- The triple product is a signed volume of the parallelepiped that the input vectors make.
- Both the cross product and triple product, although technically bound to 3D space, can be generalized to other dimensions while retaining their geometric properties by using the matrix determinant.

10 Modeling shapes with signed distance functions and surrogates

This chapter covers

- Understanding the merits and flaws of modeling shapes with signed distance functions (SDFs)
- Using typical operations on SDFs: offset, unite, intersect, and subtract
- Learning the basics of generative design with tri-periodic minimal surfaces
- Understanding metaballs as a design technique

Signed distance functions (SDFs) as instruments of geometric modeling have become important only recently. They were known for ages, but mostly in academic circles. Boundary representations and images, which we'll look into in the following chapters, were the main techniques for modeling in computer-aided design (CAD), medical imaging, and games. Now the tables are turning.

The driving force behind the prominence of SDFs is 3D printing. With a printer, you can easily produce forms so complex that chiseling them with milling tools would have been unheard of only a few decades ago. With SDFs, you can program these complex forms to follow the properties you desire. You can "program" the material to hold stress, dissipate heat, or even merge with a living tissue properly, all by programming an SDF behind the produced body (figure 10.1).

Figure 10.1 This femur model, generated from an SDF, is specifically made to be porous. Porosity is important so that implants can grow properly into living tissue.

But what is an SDF? Let's find out.

10.1 What's an SDF?

We all know from school that a distance from something to something is a positive number. That makes total sense from a mathematical perspective. From an engineering perspective, it doesn't because we let a sign bit from a floating-point number go to waste.

So let's say we have a 2D body. The body has a contour curve: a 1D object between all the space the body is made of and all the other space outside the body. Now, a *distance function* is a function that for any point in space, both inside the body and outside, tells how far the nearest point of the body's surface is. An SDF function is the same distance function, but for all the points inside the body, it's negative (figure 10.2).

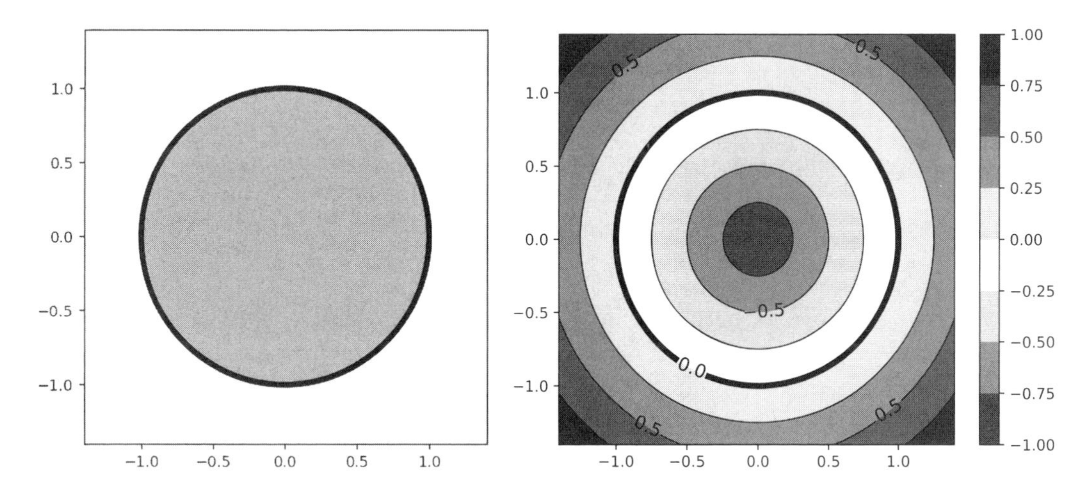

Figure 10.2 A disk and its SDF: negative inside, positive outside

NOTE The same concept translates to 3D of course. A 3D SDF is the signed distance to the 2D surface that encloses a 3D body. The principle is the same, but 1D larger.

After chapter 9, the concept of signed distance shouldn't be a shocker. We've already seen signed areas and signed volumes. These concepts may be counterintuitive because it's hard to imagine negative volume in the real world. On the other hand, ancient Greeks couldn't imagine a polynomial such as $x^2 + x$, because you can't possibly add a linear thing to a square thing in the real world. Sometimes, to understand geometry better, you should tell your intuition to depart from the real world.

10.1.1 Does it have to be an SDF?

For most possible applications, you don't need a modeling function to be a real function of distance. Usually, you only need to model a body, and for this purpose, you need the sign more than the distance. In Cartesian coordinates, for example, the formula for a circle centered at (0, 0) and with radius r is

$$x^2 + y^2 = r^2$$

Because the formula doesn't establish how one variable is a function of another explicitly, as $y(x)$ would have, it's said to be an *implicit* representation of a circle. The formula for the distance to the point (0, 0) is similar:

$$d = \sqrt{x^2 + y^2}$$

For any radius, you can turn this formula into an SDF. The distance between any point in space and the circle is the difference between the circle's radius and the distance to its center, (0, 0):

$$d_{sdf}(x, y) = \sqrt{x^2 + y^2} - r$$

But let's imagine that the circle's radius is 1:

$$d_{sdf}(x, y) = \sqrt{x^2 + y^2} - 1$$

Do we still have to spend CPU power computing the square root? The square root of 1 is 1, so the equation will work even if we omit the root. It'll be negative exactly where the original function was and be positive where the original was, too. Will the new formula

$$d_{not\_a\_real\_sdf}(x, y) = x^2 + y^2 - 1$$

also serve as an SDF? Well, yes and no. If we omit the root, the resulting function will still model the same circle. The resulting function won't be a distance function, however, because it won't reflect the true distance from any point in space to the circle we model (figure 10.3).

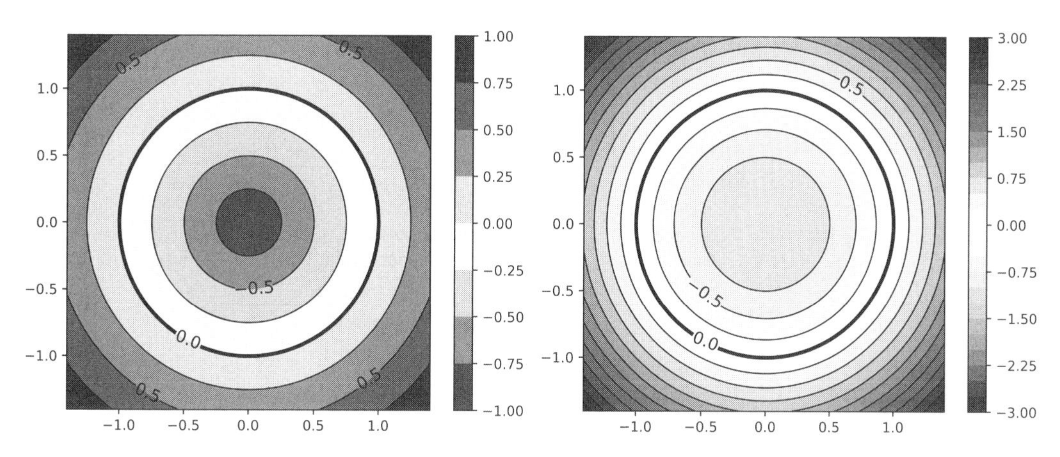

Figure 10.3 An SDF of a circle (left) and an equivalent model that isn't an SDF (right). Both functions model the same circle.

Real SDFs have valuable properties. As you'll see later in this chapter, for example, the offset operation comes cheap for an SDF. In chapter 11, we'll look into marching cubes and dual contouring algorithms; although they kind of work with non-SDFs, they're usually tailored to accommodate specifically Euclidian distances.

Also, an SDF is a common way to turn some other kind of model into an implicit mathematical representation. I mean, everybody can model a circle with its implicit formula, but what about a Stanford bunny or a Newell teapot? With SDFs, the implicit representation of a model is the signed distance to the model's surface. If you can compute the distance for any point and determine whether a point is in or out, you can turn any model into an SDF.

NOTE The Stanford bunny and the Newell teapot are the two most popular 3D models used for testing. The former is a triangle mesh, and the latter is made of Bézier surfaces. Other popular models are Suzanne (also known as the Blender monkey head), a 3DBenchy tugboat designed to test 3D printers, Spot the cow, and the Stanford dragon. You don't need to know them all by face, but they're celebrities of the 3D testing scene.

10.1.2 How to program SDFs?

In Python, an SDF is an expression. The function for a circle with a radius of 1, for example, is

```
d = (x**2 + y**2)**0.5 - 1
```

You can give it a name and make it into a function:

```
def circle(x,y):
    return (x**2 + y**2)**0.5 - 1
```

It gets a little bit more complicated if you're using SymPy to obtain an SDF in its symbolic form or NumPy for array computations. To make figures 10.2 and 10.3, I used NumPy and Matplotlib (see ch_10/sdf_circle.py in the book's source code). The SDF, along with its input data, became

```
N = 100
x = np.linspace(-1.4, 1.4, N)
y = np.linspace(-1.4, 1.4, N)

X, Y = np.meshgrid(x, y)

SDF = (X*X + Y*Y)**0.5  - 1
```

Linspace is a fancy word for an array with evenly spaced values in it . . .

. . . and a mesh grid is a 2D array. We saw both in chapter 8.

Also, if you use the vector processing capabilities of NumPy, some options in the body of the function itself may be limited for you. If you have an `if` in your function's body, for example, the branching is obvious for a scalar function but not so much for an array one. When you try to feed a branched function to NumPy, you may encounter an exception such as this:

```
ValueError: The truth value of an array with more than one element is
     ambiguous. Use a.any() or a.all()
```

This ambiguity isn't a problem, though. You can turn any function into an array-processing machine by adding `@np.vectorize` before its definition:

```
@np.vectorize
def circle(x,y):
    if x < 0:
        return (x**2 + y**2)**0.5  - 1
    else:
        return (y**2 + x**2)**0.5  - 1
```

This formula is exactly the same. You don't need any branching for a circle.

What does np.vectorize do?

Python is a dynamically typed language. In other words, an expression like `x+y` would work on numbers, arrays, or matrices as soon as the + operation for all these types is defined.

NumPy provides such operatable arrays and matrices along with the code for the operators, so normally, you don't have to bother about vectorization. Your code written for numbers works perfectly for arrays as is—until there's an `if`.

For numbers, the expression `if x < 0:` makes perfect sense. But if the `x` is an array, the `if` implies that the code execution should go both ways at the same time: an `if` clause for all the elements that are less than 0 and an `else` clause for all the elements that are not. Python is dynamically typed, not magical; it can't do that. But with `@np.vectorize`, it can.

`@np.vectorize` is a Python decorator, and decorators were specifically invented to add magic to the language. Decorators turn code into manageable data so you can modify it programmatically. In other words, they let your program modify your program. In our case, they let NumPy modify our code. Specifically, `@np.vectorize` takes our function that works on numbers and makes it into a thing that accepts arrays and matrices as input but runs our initial function on every element of this input.

So now `x+y` is not `x:array + y:array` with `+` for arrays predefined somewhere in NumPy, but an array where every element is a sum of `xi` and `yi` for `i = 0..len(x)`. Now you can add as many `if`s as you want. The vectorized function doesn't run in one go; it runs `len(x)` times, and each time, its run path is unambiguous.

`@np.vectorizes` transforms the code that has been written with numbers in mind and automatically rewrites it for you so that it works with NumPy arrays.

Another thing you may want to remember is that if you're using SymPy, you have to explicitly substitute numerics for your symbolic variables, like this:

```
def scalar_SDF(numeric_x, numeric_y):
    return symbolic_SDF.subs([(x, numeric_x), (y, numeric_y)])
```

Now SymPy, NumPy, and Matplotlib can all live together:

```
N = 100
x = np.linspace(-1.4, 1.4, N)
y = np.linspace(-1.4, 1.4, N)

X, Y = np.meshgrid(x, y)

x, y = symbols('x, y')
symbolic_SDF = (x**2 + y**2)**0.5  - 1

def scalar_SDF(numeric_x, numeric_y):
    return symbolic_SDF.subs([(x, numeric_x), (y, numeric_y)])

@np.vectorize
Def vectorized_SDF(meshgrid_x, meshgrid_y):
    Return scalar_SDF(meshgrid_x, meshgrid_y)

SDF = vectorized_SDF(X, Y)
```

NumPy arrays (x, y = np.linspace...)
NumPy 2D arrays (np.meshgrid)
SymPy symbols (symbols('x, y'))
A symbolic function (symbolic_SDF)
A numeric function: Accepts numbers as arguments (scalar_SDF)
A vectorized function: Accepts NumPy arrays as arguments (vectorized_SDF)

Now we're ready for some specific examples. We'll warm up with a theoretical one in 3D and then program a practical one in 2D.

> **SEE ALSO** Inigo Quilez keeps a collection of common 2D and 3D distance functions. The collection is supplied with common operations, too. For 2D, see https://iquilezles.org/articles/distfunctions2d. For 3D, see https://iquilezles.org/articles/distfunctions.

10.1.3 Theoretical example: Making an SDF of a triangle mesh

Let's say we want to compute the signed distance to a triangle mesh. Every triangle defines a plane with its points p_1, p_2, and p_3. To get the distance from point p to that plane, we can compute the height of a parallelepiped built of all the points p_1, p_2, p_3, and p, with the parallelogram made of p_1, p_2, and p_3 being its base (figure 10.4). By the way, we did that in chapter 9 (section 9.4.3).

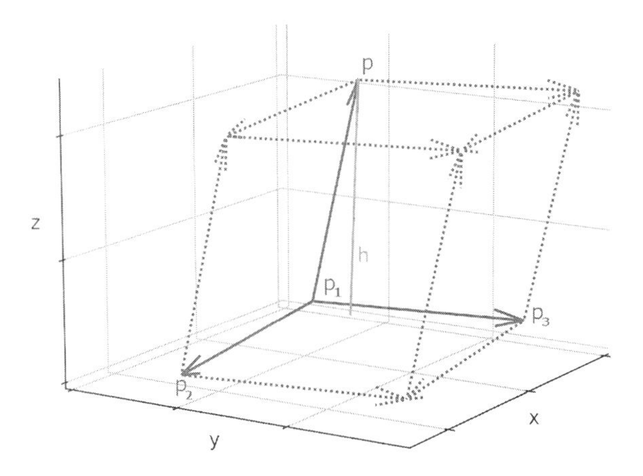

Figure 10.4 The height of the parallelepiped *h* is the distance from point *p* to the plane that p_1, p_2, and p_3 make.

The volume of the parallelogram is the *triple product* (also known as the *box product*, and if you don't remember why, please see chapter 9) of vectors $(p_3 - p_1)$, $(p_2 - p_1)$, and $(p - p_1)$. The area of the base parallelepiped is the length of the cross product of vectors $(p_3 - p_1)$, and $(p_2 - p_1)$. Now, if you remember chapter 9 well, you'll see that the volume we get from the triple product is signed. This byproduct was dubious in chapter 9, but now you see exactly what's it for! You'll keep this sign to make the distance signed, too.

The plane spans all of space, of course, but the triangle is limited by its bounding edges. So we can claim that the distance to the triangle's plane is the distance to the triangle only if the projection of point p lies on the triangle itself. If it doesn't, we should compute the distance to the triangle's nearest edge.

For the edge-distance computation, we'll use the same principle as for the volume. The distance from point p to the segment that p_1 and p_3 make is the area of the parallelogram made by p_1, p_3, and p divided by the length of the p_1, p_3 segment. The area is once again the length of a cross product of $(p_2 - p_1)$ and $(p - p_1)$ (figure 10.5).

The only thing is that the distance to the segment in 3D can't possibly be signed because unlike a plane, a segment doesn't partition the space. Segments alone, therefore, don't enclose a 3D body. You can't tell whether the point is inside the body or outside by looking at the segments that belong to the body's surface. No problem—you can still reuse the sign from the plane. Planes do partition the space; thus, they can enclose a body and establish what's in and what's out rather clearly.

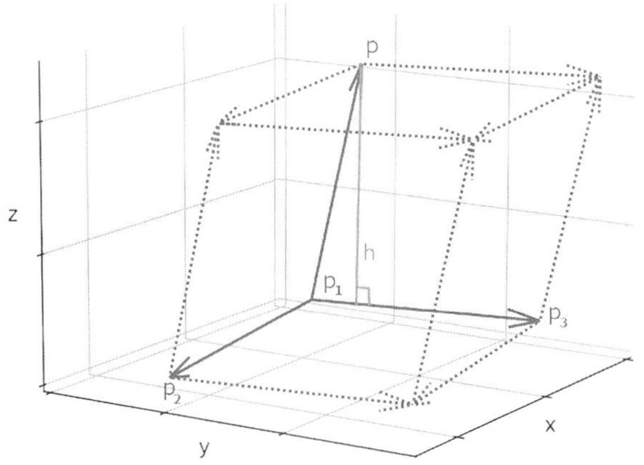

Figure 10.5 The distance from a point to a segment is the area of a parallelogram made by the three points divided by the length of the segment.

There's still one problem. If the projection of point p doesn't lie on the nearest edge per se but somewhere outside the segment, we can't claim that the distance to the nearest edge is the distance to the triangle. We'll have to compute the distance to the nearest triangle's vertex. Then the distance to the triangle is

- the signed distance to the plane if the projection point of the input point lies on the triangle;
- otherwise, the distance to the nearest edge, with the sign taken from the signed distance to the plane, but only if the projection point of the input point lies on the nearest edge;
- otherwise, the distance to the nearest triangle's vertex, with the sign taken from the signed distance to the triangle's plane.

That's it. That's the signed distance to a triangle. The signed distance from point p to a mesh, then, is the minimal absolute distance among all the distances to its triangles multiplied by the signed distance's sign.

NOTE In practice, traversing all the triangles to get the minimal distance is too expensive. We use optimized queries to find the closest triangle based on k-d trees, octrees, or bounding volume hierarchy.

SEE ALSO For a brief introduction to k-d trees, see "KD-Tree," by Yasen Hu, at https://yasenh.github.io/post/kd-tree. For a visual explanation of the bounding volume hierarchy, see the game engine development series article by Harold Serrano at http://mng.bz/51Bq.

Turning a triangle mesh into an SDF is a nice way to get rid of the mesh's inherent problems, such as degenerate, near-degenerate, or even missing triangles. But unless you've spent years dealing with all kinds of broken models, this task may not sound too exciting to you. Let's lose one dimension and make ourselves something to play with.

10.1.4 Practical example: Making an SDF of a rectangle in 2D

Making an SDF of a polygon is essentially the same as making an SDF of a triangle mesh, but in 2D. Instead of triangles, we have 2D segments. Given that the vector products generalize to 2D, as we learned in chapter 9, the mathematics remain largely the same. The code for the 2D case is much more compact, however.

So let's make an SDF for a 2D rectangle. We could use a simple formula, but for now, we'll take a longer route. Let's pretend that we want a general solution for any configuration of segments, and a rectangle is a special case to which we want to apply the generic approach.

First, we need some basic functions. We need a 2D dot and box products. We'll use the dot to project a point onto a segment and a box to compute the distance between them. Here they are:

```
def dot2d(x1, y1, x2, y2):
    return x1*x2 + y1*y2

def box2d(x1, y1, x2, y2):
    return x1*y2 - x2*y1
```

A dot product in 2D is a product of two vectors (x_1, y_1) and (x_2, y_2) resulting in a scalar.

A box product in 2D also takes two vectors and results in a scalar, but differently.

Now we need a point-to-point distance. A usual Euclidean distance goes like this:

```
def distance_p2p(x1, y1, x2, y2):
    return ((x2-x1)**2 + (y2-y1)**2)**0.5
```

The distance from (x_1, y_1) to (x_2, y_2)

You can leave the distance function as it is, but if you like code reuse, you can rewrite it as a kind of self-dot-product where the vector is being "dotted" by itself. I'm not a fan of this approach, but you should at least be aware that it's a thing:

```
def distance_p2p(x1, y1, x2, y2):
    return dot2d(x2-x1, y2-y1, x2-x1, y2-y1)**0.5
```

A distance between (x_1, y_1) and (x_2, y_2) is the dot product of vector $(x_2, y_2) - (x_1, y_1)$ and the same vector as a second argument, too.

Next, we need a signed distance to a line, not yet to a segment but only to an infinite line:

The signed distance from the point (x, y) to the segment between points (x_1, y_1), (x_2, y_2)

```
def signed_distance_p2l(x, y, x1, y1, x2, y2):
    return box2d(x-x1, y-y1, x2-x1, y2-y1) / distance_p2p(x1, y1, x2, y2)
```

As we did in the previous section, but in 2D, to get the distance from a point to a line, we take the parallelogram's signed area and divide it by its base segment's unsigned length (figure 10.6).

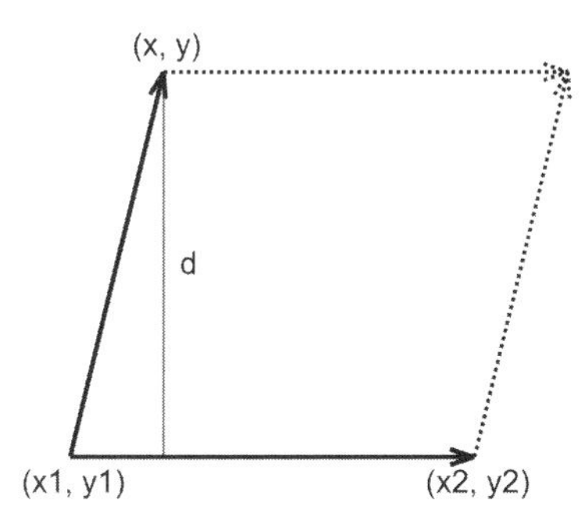

Figure 10.6 The signed distance d from the point (x, y) to the line that points (x_1, y_1) and (x_2, y_2) lie on is the parallelogram's area divided by the base's length.

We may want to get the unsigned distance and the sign of a signed one separately, so here are some utilities for that task. First, the function for the distance sign,

```
def sign_of_distance_p2l(x, y, x1, y1, x2, y2):
    return 1 if signed_distance_p2l(x, y, x1, y1, x2, y2) >= 0 else -1
```

The sign of the distance from the point (x, y) to the line set by points (x_1, y_1), (x_2, y_2)

and then the function for the unsigned distance:

```
def distance_p2l(x, y, x1, y1, x2, y2):
    return abs(signed_distance_p2l(x, y, x1, y1, x2, y2))
```

The unsigned distance from the point (x, y) to the line set by points (x_1, y_1), (x_2, y_2)

The last distance we need is the distance to a segment:

```
def distance_p2e(x, y, x1, y1, x2, y2):
    projection_normalized = (dot2d(x-x1, y-y1, x2-x1, y2-y1)
        / distance_p2p(x1, y1, x2, y2) ** 2)
    if projection_normalized < 0:
        return distance_p2p(x, y, x1, y1)
    elif projection_normalized > 1:
        return distance_p2p(x, y, x2, y2)
    else:
        return distance_p2l(x, y, x1, y1, x2, y2)
```

The unsigned distance from the point (x, y) to the segment between points (x_1, y_1), (x_2, y_2)

A segment doesn't separate space into halves—it doesn't enclose anything by itself—so the distance to a segment is always positive. The distance field to a segment looks like a stadium viewed from above (figure 10.7).

To compute the distance to the segment, first we need to understand how the input point projects on the segment's line. If the input point projects between the segment points, the distance to the segment is the unsigned distance to the line itself. If the input point projects before the first point, the distance is the distance to the first point, not the line. And if the input point projects beyond the second point, the distance to that second point is the distance to the segment as well (figure 10.8).

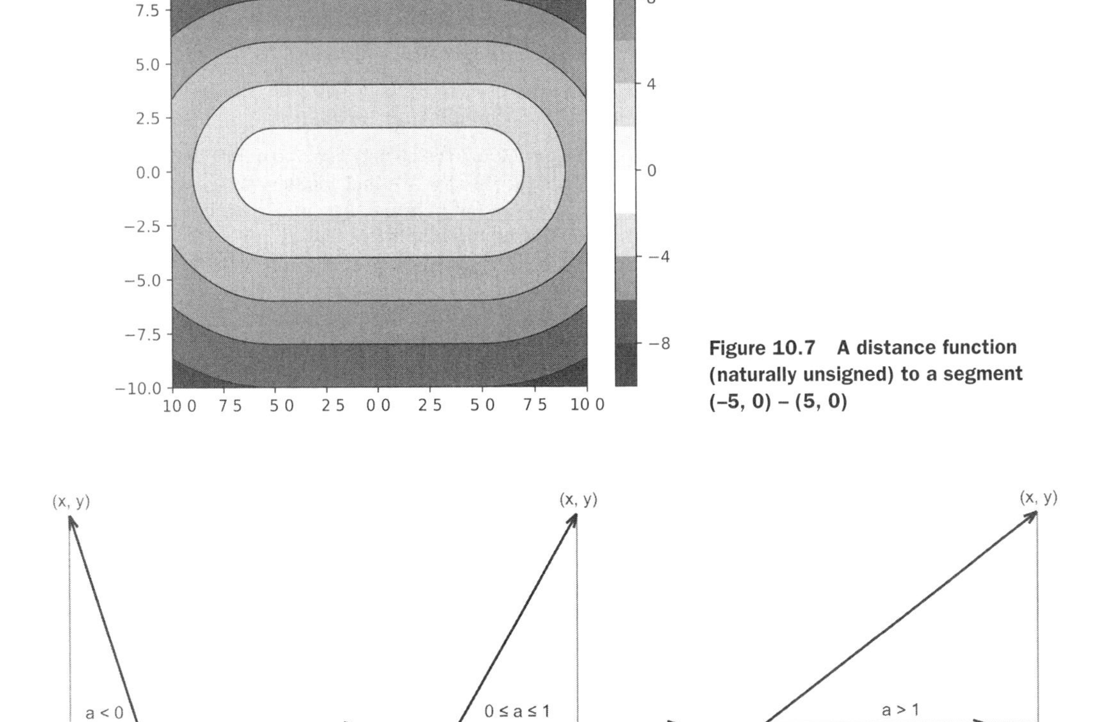

Figure 10.7 A distance function (naturally unsigned) to a segment (–5, 0) – (5, 0)

Figure 10.8 Normalized projection a can be less than 0, between 0 and 1, or larger than 1.

At this point, we have all the metaphorical building bricks we want, so let's build a signed distance function of a literal 2D brick. To do that, we'll collect the distances for the brick's (or, more formally, rectangle's) segments, find out which is closest to the input point, and then grant this distance a sign taken from the closest segment's line signed distance:

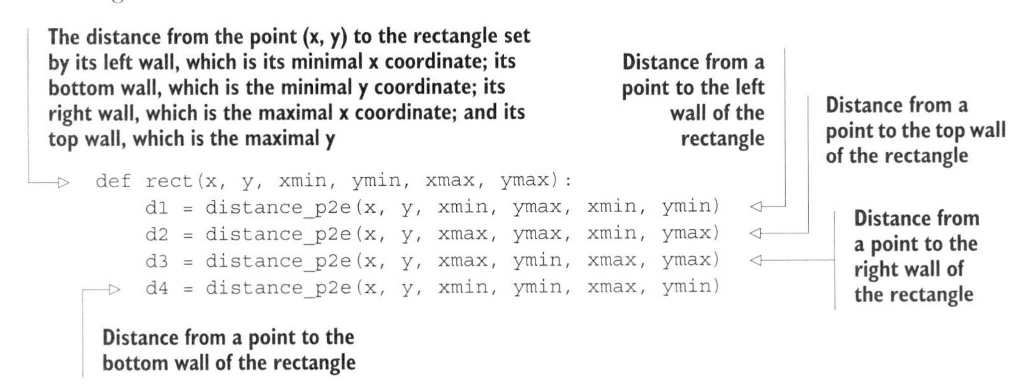


```
def rect(x, y, xmin, ymin, xmax, ymax):
    d1 = distance_p2e(x, y, xmin, ymax, xmin, ymin)
    d2 = distance_p2e(x, y, xmax, ymax, xmin, ymax)
    d3 = distance_p2e(x, y, xmax, ymin, xmax, ymax)
    d4 = distance_p2e(x, y, xmin, ymin, xmax, ymin)
```

```
    if d1 < d2 and d1 < d3 and d1 < d4:
        return d1*sign_of_distance_p2l(x, y, xmin, ymax, xmin, ymin)
    elif d2 < d3 and d2 < d4:
        return d2*sign_of_distance_p2l(x, y, xmax, ymax, xmin, ymax)
    elif d3 < d4:
        return d3*sign_of_distance_p2l(x, y, xmax, ymin, xmax, ymax)
    else:
        return d4*sign_of_distance_p2l(x, y, xmin, ymin, xmax, ymin)
```

Selects the minimal absolute distance among the four and gives it its sign

For the rectangle that starts in (–6, –4) and ends in (6, 4), the signed distance function looks like figure 10.9.

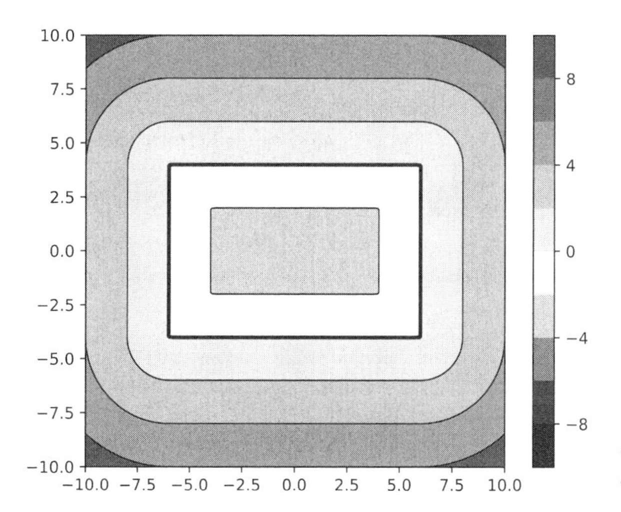

Figure 10.9 The signed distance function for the rectangle (–6, –4) – (6, 4)

If you want to double-check whether I got the formulas right, refer to chapter 9, where vector products are explained. Or run the full version of the program, plotting included, and see whether it works (ch_10/sdf_rectangle.py in the book's source code).

10.1.5 Section 10.1 summary

An SDF is a signed distance function. It's one possible way to model a shape—a body that has an interior and exterior and has a clear border between the two. In 2D, the border is a curve or a set of curves; in 3D, it's a set of surfaces.

Normally, negative values indicate the inside of a shape, and positive values indicate the outside. The border—the curve or the shape—is where an SDF equals exactly zero.

10.2 How to work with SDFs

The major advantage of SDFs over other means of modeling lies in how easy it is to move them around, make them thicker or thinner, make them into hollow shells, or compose them into more elaborate distance functions.

10.2.1 How to translate, rotate, or scale an SDF

To translate, rotate, scale, or do an arbitrary projective transformation of an SDF, we don't alter the function itself. Instead, we perform the projective transformation inverse to the given one (chapter 4) on the SDF's input coordinates. This is how the function for the translated rectangle might start:

```
def rectangle_with_translation(x, y, xmin, ymin, xmax, ymax, dx, dy):
    transformation = np.matrix([[1, 0, dx], [0, 1, dy], [0, 0, 1]])
    inverted_transformation = inv(transformation)
    xyw = inverted_transformation*np.matrix([[x], [y], [1]])
    x = xyw.item(0) / xyw.item(2)
    y = xyw.item(1) / xyw.item(2)
...
```

The transformation matrix for the direct rectangle transformation—translation by (d_x, d_y)

The distance from the point (x, y) to the rectangle set by its walls (exactly as in section 10.1.4) and translated by the vector (d_x, d_y)

This is a trick so we can easily reuse the rest of the function from section 10.1.4. We write the new transformed values while converting them from homogeneous to Euclidean coordinates on top of the input values.

Here, we're supposed to have the same code as in the rect function from section 10.1.4.

The inverse transformation being applied to (x, y) or, in homogeneous coordinates (x, y, 1)

The inverse transformation to be applied to the input point (x, y)

This could be a function for the rotated rectangle:

```
def rectangle_with_rotation(x, y, xmin, ymin, xmax, ymax, a):
    transformation = np.matrix([
        [np.cos(a), np.sin(a), 0],
        [-np.sin(a), np.cos(a), 0],
        [0, 0, 1]])
...
```

The same input as before, only instead of a translation vector, a single rotation value is supplied

The only difference in the code is in the way we form a transformation matrix.

The rest of the code, including the inverse transformation application, is the same as before.

And this could be a function for the scaled one:

```
def rectangle_with_scale(x, y, xmin, ymin, xmax, ymax, scale):
    transformation = np.matrix([[scale, 0, 0], [0, scale, 0], [0, 0, 1]])
...
```

The same input, only now we have a scaling coefficient instead of the rotation value

Again, all the difference is in the transformation matrix.

The rest of the code is the same as before.

All three examples are displayed in figure 10.10.

If you think that keeping three different functions for a rectangle under different projective transformations is a good idea, however, please refer to chapter 4. That chap-

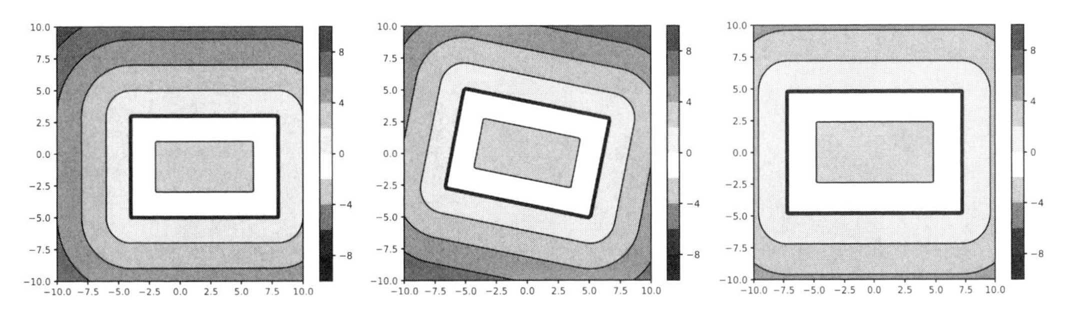

Figure 10.10 A rectangle from section 10.1.4 translated by (2, –1), rotated by 0.2, and scaled by 1.2

ter explains how projective transformations are interchangeable and composable, and how you can gain performance and reduce code size by using them properly.

> **NOTE** Also, creating and inverting a transformation matrix inside a vectorized function is a terrible, terrible idea. When you program this way, the same matrix would be recomputed for every value in the input meshgrids. What a waste!

10.2.2 How to unite, intersect, and subtract SDFs

One of the strongest points of SDF is that they're easily composable. With other kinds of surface models, getting a unification or an intersection of two objects may require a lot of computation and a lot of serious mathematics. With an SDF, both operations are one-liners.

Let's take a pair of circles to see how it works. Both circles have a radius of 6, and the first one has its center in (–2, 0) and the second – in (2, 0):

```
def circle(X, Y, r):
    return np.sqrt(X*X + Y*Y) - r

N = 100
x = np.linspace(-10., 10., N)
y = np.linspace(-10., 10., N)

X, Y = np.meshgrid(x, y)

SDF1 = circle(X+2, Y, 6)
SDF2 = circle(X-2, Y, 6)
```

This function is written as though it works with numbers, but it's intended to work on meshgrids, so 2D arrays. Don't worry; you don't have to vectorize it explicitly.

The linear spaces here are arrays of 100 elements spread evenly in a range from –10 to 10.

When we apply a circle function on these mesh grids, the result is also a mesh grid where the signed distance to a circle is written into every grid cell in [–10, 10]×[–10, 10]. The +2 and –2 are cheap ways to do translations.

The mesh grids here are 2D arrays covering [–10, 10]×[–10, 10] area each, one with x values and the other with y values.

Figure 10.11 shows the circles' plots.

A *union* of SDFs is a function that remains negative when both of the circles, or even one circle, is negative. That is, a point is inside the united shape if it's inside

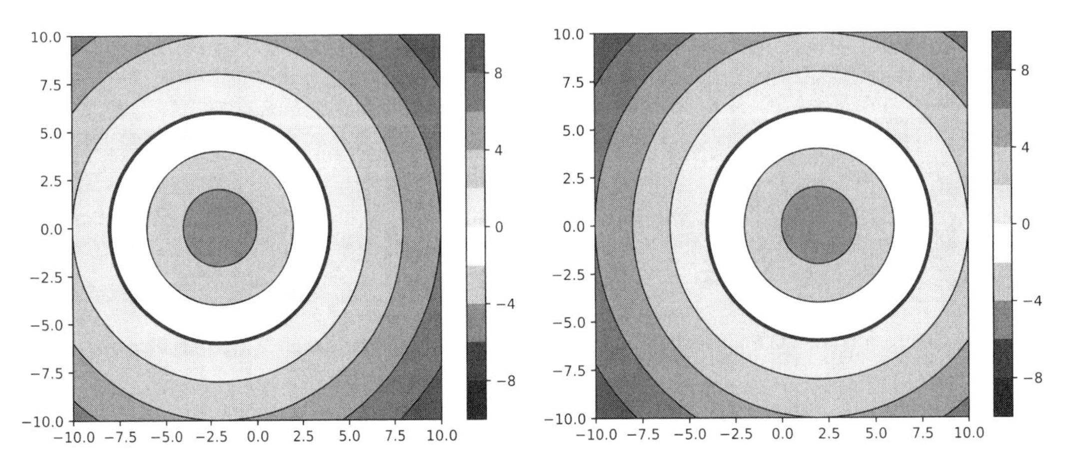

Figure 10.11 A pair of circles

either shape in the union. You can do a union of two SDFs by using a minimum function. Wherever one of the functions is negative, the result will be negative, too:

```
Z = np.minimum(Z1,Z2)
```

Figure 10.12 shows the union of our two circles (see ch_10/sdf_disks_union.py in the book's source code).

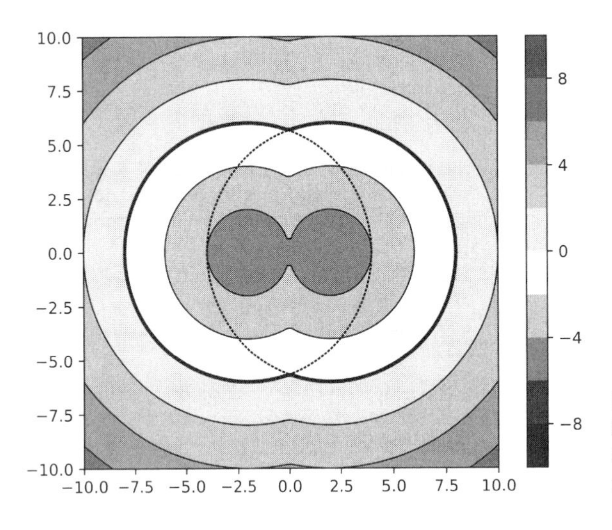

Figure 10.12 A union of two SDFs is negative when at least one of the SDFs is negative and positive otherwise.

On the contrary, the intersection requires the function to remain negative only when both input SDFs are negative. Whenever one is positive, the intersection should also be positive. A simple maximum function does exactly that:

```
Z = np.maximum(Z1,Z2)
```

That's the whole intersection, as shown in figure 10.13 (see ch_10/sdf_disks_intersection.py in the book's source code).

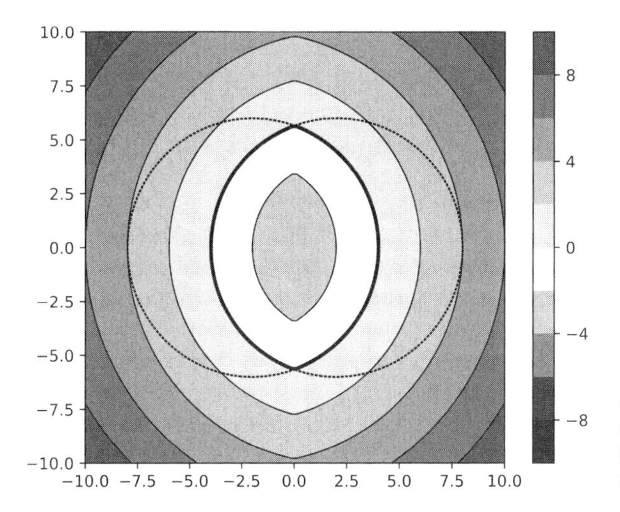

Figure 10.13 An intersection of two SDFs is negative when both SDFs are negative and positive otherwise.

You can also subtract one shape from another. The process is the same as for an intersection, but with one of the shapes inverted:

```
Z = np.maximum(Z1,-Z2)
```

Figure 10.14 shows an example of subtracting the second circle from the first one (see ch_10/sdf_disks_subtraction.py in the book's source code).

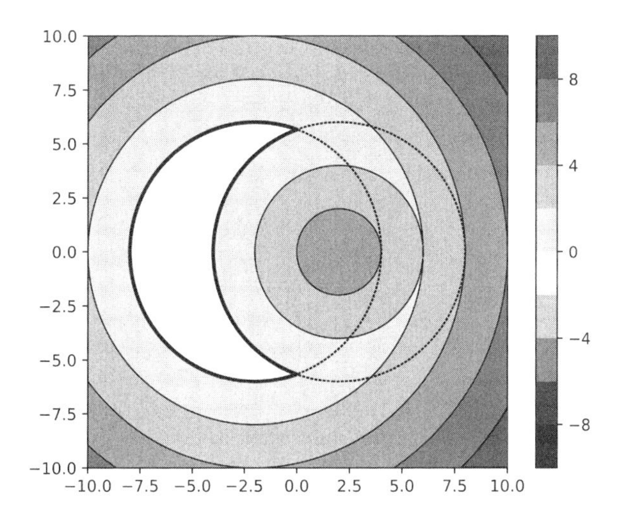

Figure 10.14 The subtraction of two SDFs is negative when the first SDF is negative and the other is positive. When you invert the second SDF, it boils down to an intersection.

At this point, you should start getting suspicious. Is this thing in figure 10.14 still an SDF? Shouldn't the isolines be less sharp? Is the union an SDF? Is the intersection an SDF? What is happening?

Ah, well, there's a catch: the combination of SDFs isn't necessarily an SDF itself. It's still an implicit function modeling the shape, and you can treat it as such. But technically, it's no longer a distance function because it doesn't reflect the true distance to the object we're supposed to model. For most practical applications, this loss of "distanceness" is tolerable. But once in a while, some algorithm relies on its input function to be a true Euclidean distance function, and this side effect might indeed become problematic. Consider this information to be a heads-up.

NOTE Operations like these are common in computational geometry. They're called *Boolean operations*, and I'm glad you're a programmer so I don't have to explain what *Boolean* means. It's like for every point in space, you do a Boolean predicate answering "whether the point is inside of a shape" based on a pair of input shapes—let's call them `a` and `b`—and the operation. For the union, the result is `a | b`. Whenever one of the shapes covers the point, the result of the predicate is `true`; for the intersection, it's `a & b`. Now the result is `true` only if it's `true` for both `a` and `b`. Then subtraction will be `a & -b`.

10.2.3 How to dilate and erode an SDF

Dilation and erosion are offset operations that are simple to perform on SDFs if your SDF is a true distance function. If you want to dilate your model, or make it bigger, you subtract a dilation distance from your SDF. If you want to erode it, or make it smaller, you add an offset (see ch_10/sdf_disks_offset.py in the book's source code) to the function (figure 10.15):

```
SDF_dilated = SDF - offset
SDF_eroded = SDF + offset
```

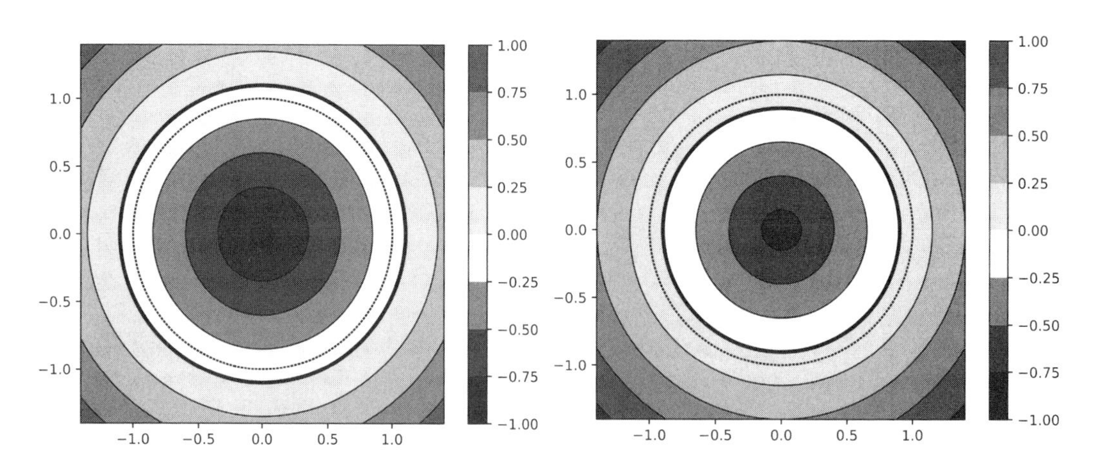

Figure 10.15 A circle with a radius of 1 dilated by 0.1 (on the left) and eroded by 0.1 (on the right)

Now let's get back to the heads-up about Booleans. Let's say you want to program a whole workflow in SDFs. You want to compose several operations in a row to solve some practical problem. At some point, you want to do a Boolean subtraction and then apply a dilation on the subtraction's result. That won't work. The problem is that as soon as you perform a Boolean operation, you lose the "distanceness" of the SDF you're operating on. It's still a function describing a shape, but it's not a true distance function anymore.

As with the Boolean subtraction example, the true offset of the crescent's "horns" should have been rounded. But the dilation of that not-really-SDF done by simple subtraction would retain the sharpness of those horns (figure 10.16).

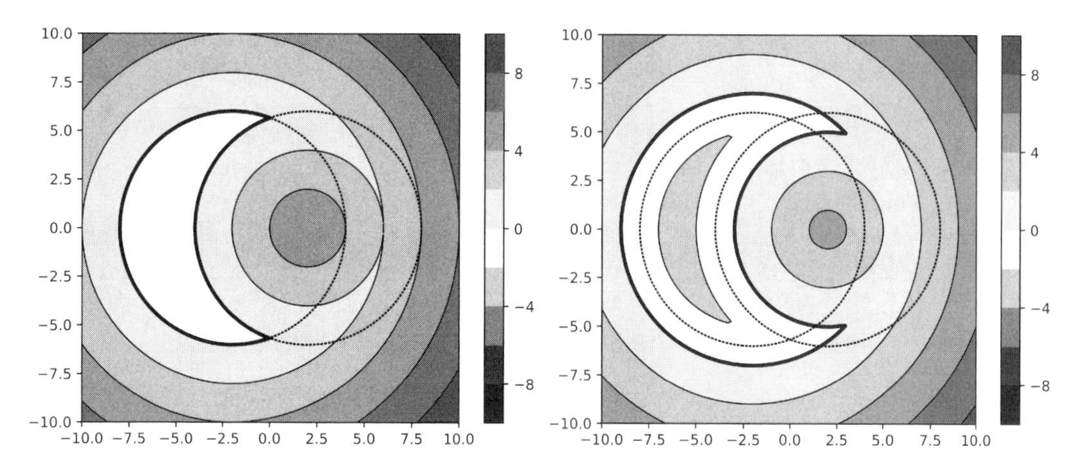

Figure 10.16 Heads up! If the model isn't a real SDF, the offset won't be the real offset either.

I have to stress this fact because losing track of when your function loses its "distanceness" is a common mistake in implicit modeling. I've done it myself on several occasions. The result of a post-Boolean offset looks almost correct, and in the simplest cases, it almost is. The incorrect algorithm may even sneak past the tests. But when the algorithm gets into production, real clients with real models start to see the difference.

10.2.4 How to hollow an SDF

Hollowing is a popular operation in 3D printing. If you're printing a normal-size Warhammer figurine, for example, you can print it as a solid object. But if you want to print a life-size orc, you'd better start saving on material costs.

The hollowing retains an outer shape of a model but turns the model into a thin shell of itself by making it hollow inside. Mathematically, it's a combination of basic arithmetic and an absolute value function.

First, we establish the shell width. Next, we erode the model by half the shell width. Then we take the SDF's sign away, making the model an empty border. Finally, we dilate the empty border by half the shell width so that the outer border of the new model matches the shape of the original one (see ch_10/sdf_hollow.py in the book's source code):

```
shell_width = 0.2
SDF_eroded = SDF + shell_width / 2
unsigned_DF = np.abs(SDF_eroded)
SDF_hollowed = unsigned_DF - shell_width / 2
```

Erodes

Turns into a 0-width border

Dilates to give the border some width

Figure 10.17 shows the three steps of a hollowing operation.

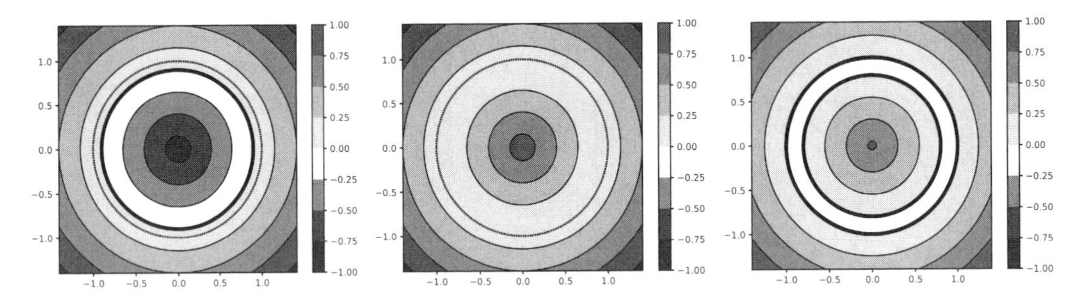

Figure 10.17 A hollow shell of a disk is a result of the erode-unsign-dilate sequence.

Once again, this sequence works properly only on true SDFs. If possible, use it before the Boolean operations, not after.

10.2.5 Section 10.2 summary

Operating signed distance functions is simple. You can translate, rotate, and scale them by using the projective transformation we've already learned. You can unify, intersect, or subtract them by using simple minimum and maximum operations. You can also dilate and erode SDFs with even simpler arithmetical operations. The only tricky thing to keep in mind is that the Boolean operations strip SDFs of their "distances," so the offset operations stop working on them properly.

10.3 Some techniques of not-really-SDF implicit modeling

As discussed at the beginning of this chapter, a true SDF is a special case of implicit representation of a geometrical object. Being a distance function, an SDF gives us privileges such as cheap offset and hollowing. But in practice, a lot of exciting automated design is being done in more generic implicit representations.

In this section, we'll take a look at a few techniques of implicit modeling that may come up in practice, such as increasingly popular tri-periodic minimal surfaces, or that may be fun things to play with, such as multifocal lemniscates. Don't worry if terms like *tri-periodic* and *multifocal* are unfamiliar; we'll get through them.

10.3.1 Tri-periodic minimal surfaces

A sine is a *periodic* function. Its graph is also a curve, and this curve is also periodic. It repeats itself over the argument axis ad infinitum. In explicit form, you can represent this curve like this

$$y = \sin(x)$$

or in implicit form:

$$\sin(x) - y = 0$$

In 3D, there are three axes. The preceding formula is still applicable, but now it describes an *x*-periodic surface (figure 10.18). A periodic surface repeats itself over an axis.

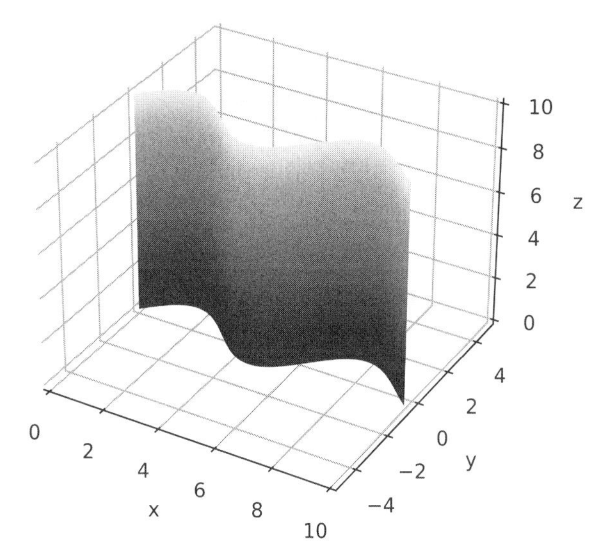

Figure 10.18 In 3D, $y = \sin(x)$ becomes an equation for a periodic surface.

A *tri-periodic* surface repeats itself over three axes at the same time. This is a tri-periodic surface:

$$\sin(x)\cos(y) + \sin(y)\cos(z) + \sin(z)\cos(x) = 0$$

This formula models a surface named *gyroid*. Gyroids are so popular nowadays that they're printed in brass and sold as art pieces (figure 10.19).

But the true value of tri-periodic minimal surfaces in 3D printing comes from their minimality. In this context, *minimal* means that the area of the surface is locally minimized, which also means that the mean curvature of such a minimal surface is 0 everywhere. Consequently, these surfaces have no 3D folds even if they look as though they do. When a minimal surface is bent one way by one axis, it's equally bent the other

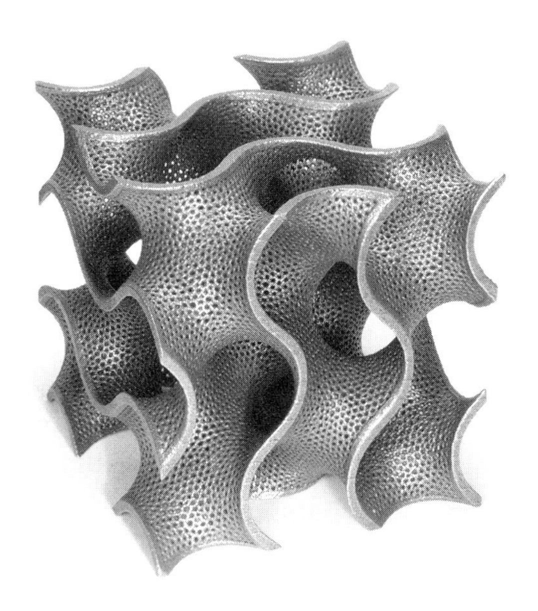

Figure 10.19 A tri-periodic surface printed as an art piece (courtesy of Bathsheba Grossman; see more mathematical art at http://bathsheba.com)

way by the other axis. In this regard, the surface remains kind of flat at any particular point while looking curvy as a whole.

We barely touched on this subject in chapter 5, so here comes a more detailed explanation. If you remember, the curvature of a curve in a point is a value inverse to the radius of a circle fitted in the point's proximity. One curve, one point, one circle. This doesn't work with surfaces. With a surface, you can fit one circle along one axis and fit another circle completely along the other axis. In fact, there could be a different circle for every possible angle. There's no single number you can take as a surface curvature. Yet.

What you want to do first is fit a tangent plane. There's only one tangent plane, given that the surface is smooth, so there's no ambiguity so far. Also, there's a single normal vector for this plane. For all the possible angles that a circle with its center somewhere on the normal vector may take, you want to fit these circles and find their curvatures. You can't merely calculate a bunch of numbers, of course, because the set of all the possible angles is continuous: $(0, 2\pi)$. You get a function instead. Then, by integrating this function and dividing the result by 2π, you get the *mean curvature* of the surface at the given point.

This process looks tedious, though. There should be another way. And there is! Thanks to Euler's theorem, which I don't quite understand myself, the mean curvature is equal to the mean value of principal curvatures, which are the minimal and the maximal signed curvatures of a given surface at a given point (figure 10.20).

SEE ALSO If you find my quick-and-dirty introduction to the curvature of surfaces too brief, see "A Quick and Dirty Introduction to the Curvature of Surfaces," by Keenan Crane, at http://mng.bz/61d5.

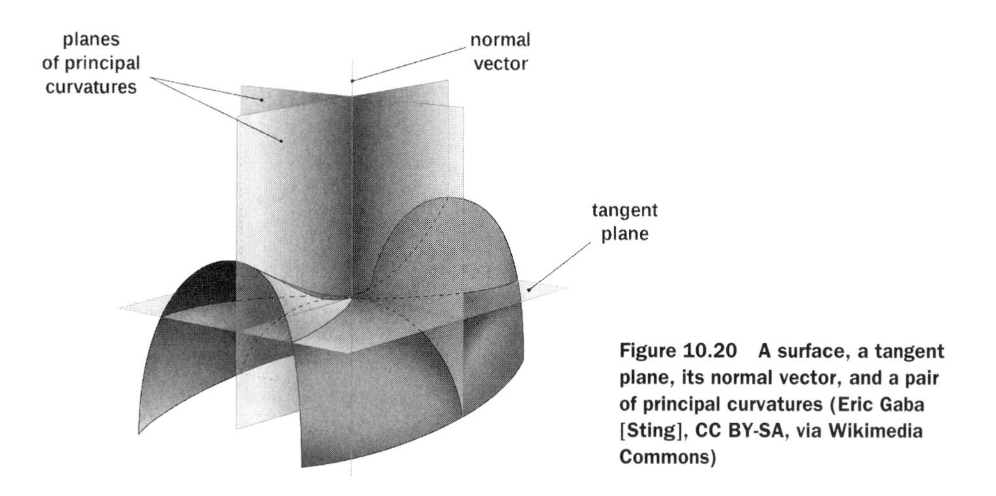

Figure 10.20 A surface, a tangent plane, its normal vector, and a pair of principal curvatures (Eric Gaba [Sting], CC BY-SA, via Wikimedia Commons)

Now back to our minimal surfaces. Because their mean curvature is 0 everywhere, when one principle curvature is positive, the other is necessarily negative. There are no bumps on the surface, neither concave (which would imply that both principal curvatures are negative) nor convex (both would be positive). As a result, when you print in powder or resin, the remains of your material don't get caught in the bumps. Your 3D-printed detail is easier to clean when the printing is done. Yes, differential geometry can be utilitarian, too.

10.3.2 Practical example: A gyroid with variable thickness

Tri-periodic minimal surfaces are often used in 3D printing as infills or support structures. Infills go inside the hollow part to make it integrally stronger, and supports go outside the model so that it comes out properly in printing. In both cases, you need some control of the mechanical properties of the body you print. Tri-periodic surfaces, being almost-SDFs, give you computationally cheap control of their thickness.

You can regulate the thickness of a body enclosed by a gyroid by adding a quasi-offset in the range of (–1, 1). This quasi-offset could be parametric. In figure 10.21, there are two cross sections of a gyroid. One cross section is a pure textbook case, in which everything is symmetrical; the other has a parametric offset that makes it thicker at the top and thinner at the bottom.

Adding a parametric quasi-offset in this case is as simple as adding a single term to the equation (for the full listing, see ch_10/sdf_gyroid_parametric.py in the book's source code):

```
X, Y = np.meshgrid(x, y)
z = 0.5
not_really_an_SDF = (np.sin(X)*np.sin(Y)
                     + np.sin(Y)*np.cos(z)
                     + np.sin(z)*np.cos(X)
                     - Y/32)
```

These meshgrids cover the part of the infinite function we want to display.

This is the z-coordinate of the x-y plane section.

This is the quasi-offset term.

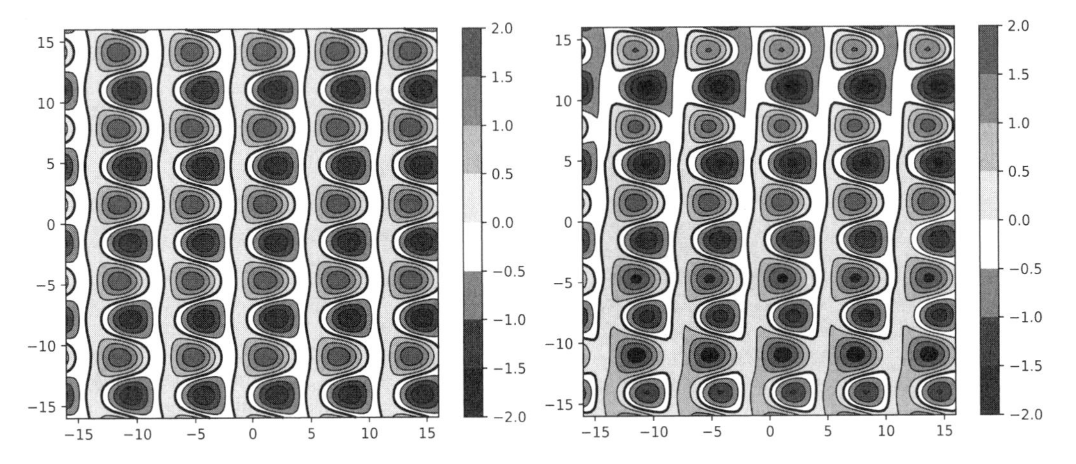

Figure 10.21 A cross section of a regular gyroid (left) and one with a parametric offset (right)

The quasi-offset isn't the true offset because the gyroid isn't an SDF—only an implicit form of a tri-periodic surface. But it does help reinforce the gyroid where it's under stress and erode it a little where it's not under stress to reduce the cost of printing.

Tri-periodic minimal surfaces (often abbreviated TPMS) aren't only a fad. They have important geometric properties that make them attractive for practical use. But they're also a fad because they look awesome and people love them.

10.3.3 Metaballs

Using metaballs is a simple yet effective way to generate a complex organic-looking shape from little input data. The shape in figure 10.22, for example, is produced by three points and a single offset number.

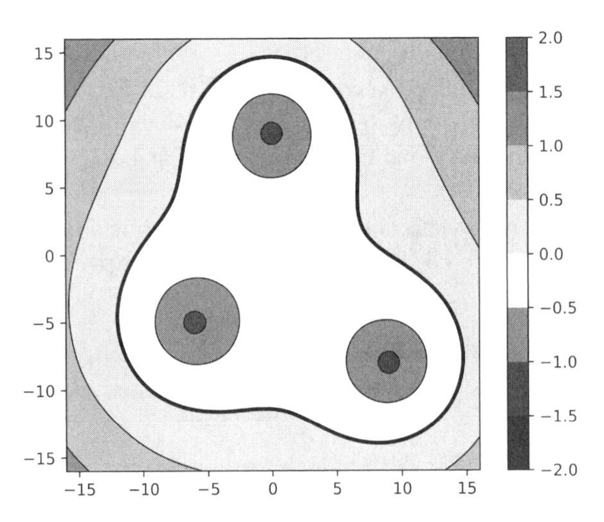

Figure 10.22 A metaball example, made of three "balls"

It's called a *metaball* because in 3D, it looks like a set of balls that merge to create an organic-looking thing with smooth borders. Here's a common formula for metaballs:

$$f(x) = \frac{\frac{1}{\sum_{i=1}^{number\_of\_balls} \frac{1}{\|ball\_center_i - x\|}}}{number\_of\_balls} - offset$$

This formula is a normalized inverse of a sum of inverse distances to the ball centers, minus the offset that governs the figure's thickness:

```
balls = [(-6, -5), (9, -8), (0, 9)]
offset = -1.25
SDF = (1.
      / sum([1 / distance_p2p(X, Y, ball[0], ball[1]) for ball in balls])
      / len(balls)
      + offset)
```

The ball's centers — balls

The predefined offset, a number making the metaballs thinner or thicker — offset

The formula — + offset)

This is fun, but who needs an organic-looking shape made of balls, anyway? Good question! A good answer would be "Probably no one." But the superpower of metaballs is that they make any SDFs into organic-looking shapes. We can make shapes out of metarectangles and metasegments, for example. We can even gather all the shapes we want—such as a ball, a segment, and a rectangle—and make them into a single smooth organic-looking shape (figure 10.23).

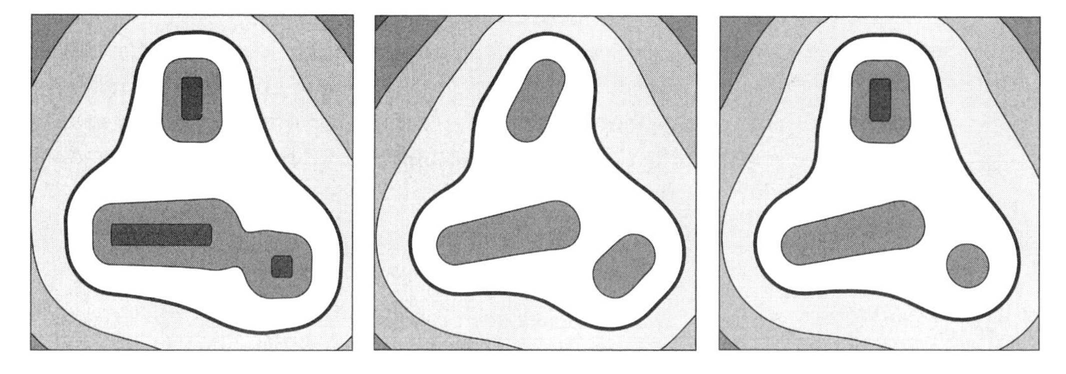

Figure 10.23 Metarectangles (on the left), metasegments (in the center), and metaeverything (on the right)

The metaballs technique is a way to glue all kinds of SDFs into a single organic-looking thing. The algorithm behind the femur model that starts this chapter is loosely based on metaballs (figure 10.24).

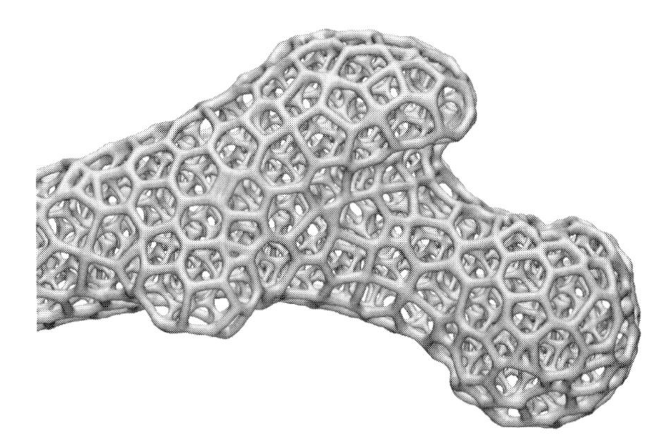

Figure 10.24 The algorithm that builds this organic-looking shape is loosely based on metaballs.

10.3.4 Practical example: Localizing the metaballs for speed and better governance

In practice, the formula with the sum of inverse distances is rarely used because of its globality. Every ball center influences the whole shape we're trying to model, which makes modeling harder because we can't adjust a few things locally and get on with it. Every change, no matter how small, changes the whole shape. For this reason, we discussed this effect in chapter 8 when we localized the inverse distance interpolation.

Also, to compute the quasi-distance function for the global thing, we have to compute all the distances to all the balls' centers every time, which makes the algorithm computationally heavy. It would be better if the balls influenced only one another in some predefined proximity, so that instead of computing all the distances all the time, we could collect a few neighboring balls that are close enough to the point in which we want to compute the function.

Instead of the inverse distance, we need something else, something that behaves a lot like an inverse function, but instead of descending infinitely into infinity, descends smoothly into some number and stays 0 from that point as the argument goes. This something will do:

$$\text{when } 0 < x \leq 1 : qi(x) = \frac{1 - 2x + x^2}{x};$$
$$\text{when } x > 1 : qi(x) = 0$$

Near zero, this function behaves like 1/0. Near 1, it becomes more like 0/1. When $x > 1$, it's explicitly said to be 0. This function emulates the inverse function (figure 10.25) but also allows us to make metaballs local, as all the neighbors that are further from a point than by 1 can be ignored in the computation.

Here's the same function in Python:

```
def quasi_inverse(x):
    if x < 1:
        return (1 - x)**2 / x
    else:
        return 0
```

This is the same as (1 – 2x + x2) / x.**

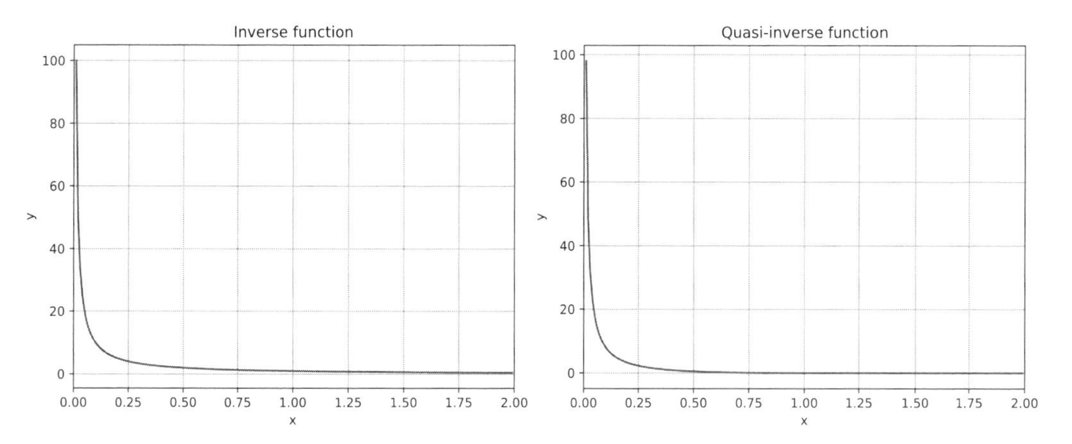

Figure 10.25 The usual 1/x inverse function and the quasi-inverse function that localizes the metaballs

We can make the locality adjustable by scaling the *x* before taking the inverse (see the full listing in ch_10/sdf_metaballs_quasi_inverse.py in the book's source code):

```
def quasi_inverse_with_locality(x, locality):
    return quasi_inverse(x / locality)
```

Figure 10.26 shows how the function looks with locality 25.

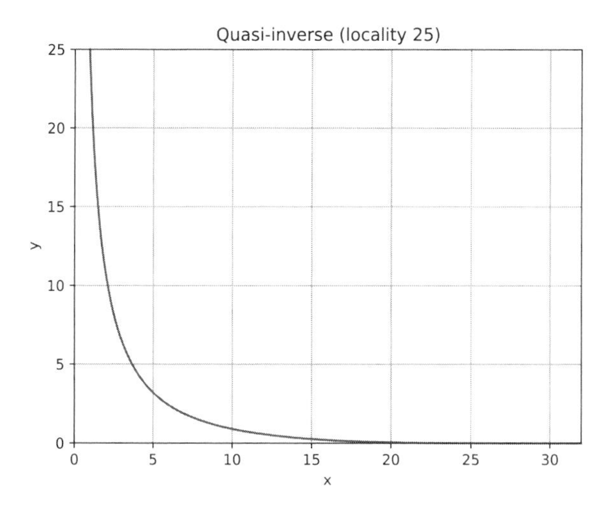

Figure 10.26 Quasi-inverse function with locality 25

Figure 10.27 shows the metaballs from before with a new localized quasi-inverse function.

Looking for the neighboring points is usually an O(log(n)) complexity problem, and gathering all the distances is O(N). Localizing the metaballs lets them into the world of practical applications because we want SDFs—functions that we have to compute a lot—to be as economical as possible.

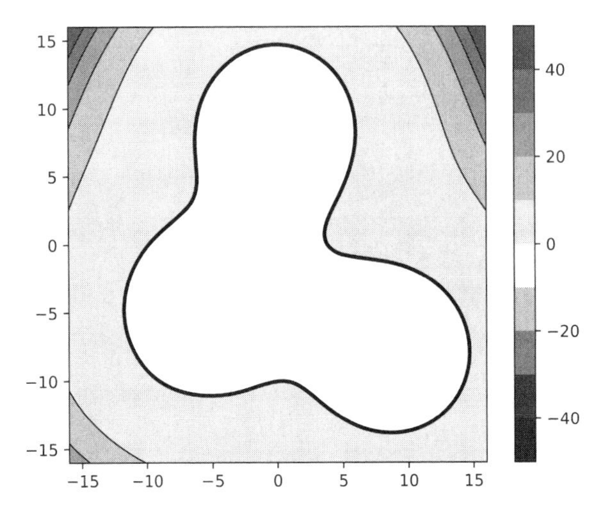

Figure 10.27 An example of metaballs with a quasi-inverse function. The function and the shape have changed, but the shape is still smooth and continuous.

10.3.5 Multifocal lemniscates

Metaballs were invented in the early 1980s by computer graphics specialist Jim Blinn, but similar techniques existed long before that, even before computer graphics itself. In Замечательные кривые (*Notable Curves*), by Alekséi Markushévich, published in 1952, a problem of approximating any curve was said to be mathematically solvable by a multifocal lemniscate.

A classic *lemniscate*, such as a lemniscate of Bernoulli, is a curve that revolves around two points named *foci* or *focuses*. This curve has an interesting property: a product of distances to its foci is constant for any point of the curve (figure 10.28).

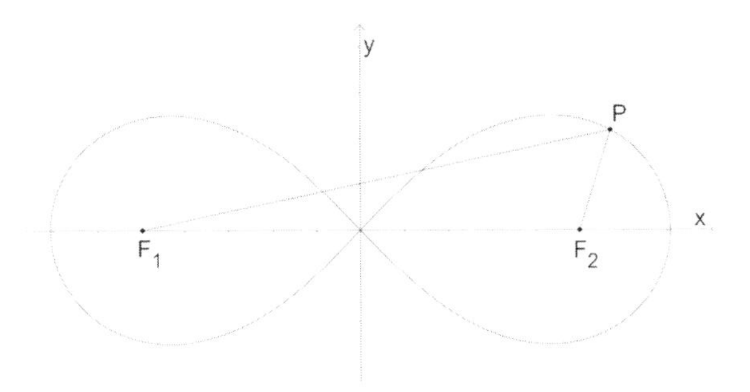

Figure 10.28 Lemniscate of Bernoulli (Kmhkmh, CC BY 4.0, via Wikimedia Commons). The product of distances from F_1 to P and from F_2 to P is constant.

We can use the curve's property to make it into an SDF-like entity (figure 10.29):

$$SDF_{Lemniscate}(x) = \|x - f_1\|\|x - f_x\| - c$$

Here, f_1 and f_2 are the points of foci, and c is a numeric constant. For the lemniscate of Bernoulli, it's quite obviously a product of distances to its foci from the zero point.

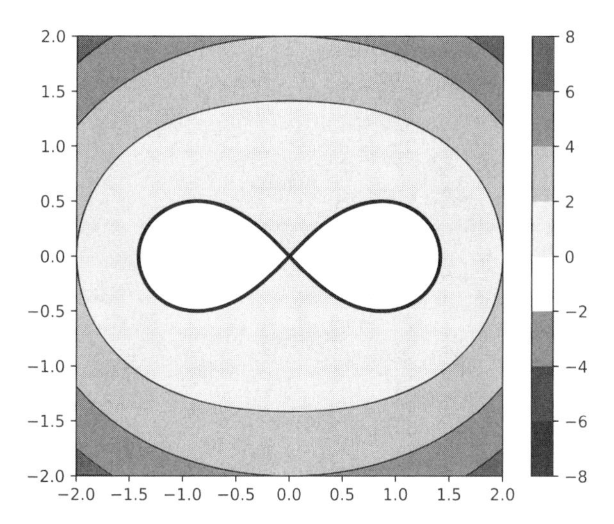

Figure 10.29 A lemniscate of Bernoulli as an SDF-like function

This "SDF" doesn't reflect the true distance to the curve, but it's negative inside the curve and positive outside, more like a smooth sign function than a distance function. We can still use it to model a lemniscate. But should we?

A simple bifocal lemniscate is indeed boring and unamusing. But what if we add more foci? As it turns out, we can add as many points of focus as we want to make the lemniscate *multifocal* and the shape it describes more complex. The formula for the multifocal lemniscate is as simple as a product of all distances to foci minus some constant:

$$SDF_{Lemniscate}(\boldsymbol{x}) = \prod_{i=1}^{N} \|\boldsymbol{x} - f_1\| - c$$

In this formula, N is the total number of focal points, and f_i is one of them. Now the constant c is free to be whatever we want. It's our lever. By making c larger we inflate the lemniscate, and by making c smaller we erode it. It works almost like an SDF offset, but not quite, because the function isn't a true distance function.

Now, why is this better than metaballs? It isn't. It's cheaper, though. Instead of computing inverse functions, you can simply multiply your distances and get a similar effect. Also, like metaballs, this process is economical with memory. We'll look at a specific example in the next section.

10.3.6 *Practical example: A play button made of a multifocal lemniscate*

To describe the Play button with a cubic Bézier spline in figure 10.30, you'd need at least 3 Bézier curves, which is 12 control points; for a multifocal lemniscate, you'd need only 6 points and a constant.

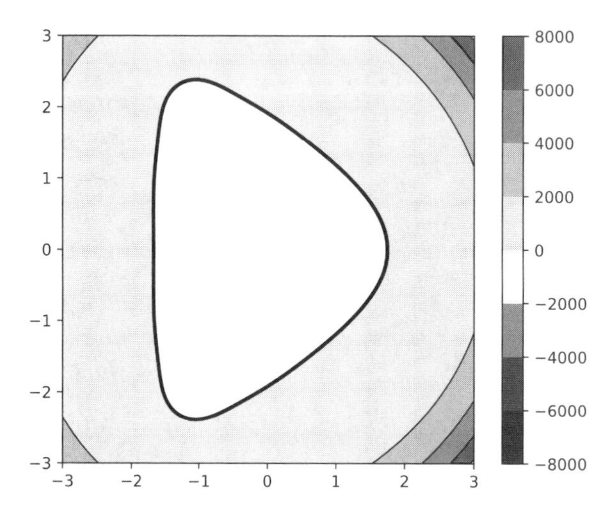

Figure 10.30 A Play button made of a multifocal lemniscate

The data and code for the button are

```
foci = [(-0.6, -1.12), (-0.6, 1.12), (-1.12, -2.06),
        (-1.12, 2.06), (0.4, 0), (1.2, 0)]
r = 2
c = r**len(foci)

def distance_p2p(x1, y1, x2, y2):
    return ((x2-x1)**2 + (y2-y1)**2)**0.5

def prod(list):
    if len(list) == 2:
        return list[0]*list[1]
    else:
        return prod([list[0]*list[1]] + list[2:])

SDF = prod([distance_p2p(X, Y, focus[0], focus[1]) for focus in foci]) - c
```

- **It's easier to govern the function with a quasi-offset by exponentiating the offset into the power proportional to the number of points.** (points to `c = r**len(foci)`)
- **All the data that describes the shape** (lines `foci = …` through `r = 2`)
- **The point-to-point distance function** (points to `def distance_p2p`)
- **The product function. Python has a similar sum function that does the summation, but for the product, we have to make our own.** (points to `def prod`)
- **X and Y form the usual coordinate mesh grid.** (points to `SDF = …`)

If you want to try out the lemniscates and maybe invent your own shapes, the full listing with plotting is available in the book's source code (ch_10/sdf_multifocal_lemniscate.py).

With multifocal lemniscates, you can do the same thing as with metaballs: create organic shapes out of primitives but now in a slightly more computationally economical way.

10.3.7 Section 10.3 summary

Implicit modeling is getting more popular with the advances in 3D printing; more and more intricate designs are becoming possible in real-life applications. Tri-periodic minimal surfaces are one example. Organic shapes made with metaballs or multifocal lemniscates are another example.

None of these functions are truly SDFs, but they borrow some techniques and approaches from the field. You can still perform cheap and simple Boolean operations on them; you can mix them with real SDFs in one shape; and as in the case of metaballs, you can use SDFs as source material for the shape you want to build.

10.4 Exercises

Exercise 10.1 In section 10.1.4, we computed distances to all four sides of a rectangle, but there was a mention of a simpler rectangle SDF. Can you propose one?

Exercise 10.2 Can you propose an SDF for a triangle?

Exercise 10.3 Can you apply any biquadratic transformation to an SDF?

Exercise 10.4 Can you propose your own quasi-inverse function? Kudos if yours appears to be computationally cheaper than mine.

Exercise 10.5 The problem with lemniscates is that they grow too fast. Even with the $c = r^N$ convenience measure, picking the right offset isn't intuitive. Can you make the lemniscate quasi-SDF more like a real SDF?

10.5 Solutions to exercises

Exercise 10.1 It's easy to make a pseudo-SDF by intersecting the four half-planes:

```
max([xmin-x, ymin-y, x-xmax, y-ymax])
```

A real SDF is a little bit trickier. You can find the distance outside the rectangle by cutting the vector from the rectangle's center to the input point vector by half width and half height, but this wouldn't work inside. Or you can use the pseudo-SDF to compute the insides, but it won't work outside then.

This situation isn't a real dilemma. The trick is to compute both the outside distance and the pseudo-SDF and cap them both to their respective signs, positive outside and negative inside, and simply add them together (see ch_10/sdf_sdf_simpler_rectangle.py in the book's source code):

```
def rect(x, y, xmin, ymin, xmax, ymax):
    center = [(xmax + xmin) / 2, (ymax + ymin) / 2]
    half_width = (xmax - xmin) / 2
    half_height = (ymax - ymin) / 2
    relative_x = x - center[0]
    relative_y = y - center[1]
```

```
    dx = max(0, abs(relative_x) - half_width)
    dy = max(0, abs(relative_y) - half_height)
    outside_d = (dx**2 + dy**2)**0.5
    inside_d = min(max(dx, dy), 0.0)
    return outside_d + inside_d
```

Exercise 10.2 Because the original approach for the rectangle in section 10.1.4 was to get the minimal distance of four edges, it looks as though we can easily modify it to work with triangles. Unfortunately, there's a hiccup: our way to get the sign works only for obtuse angles or at least, as with the rectangle, the right ones.

For the triangle, we can employ the criteria we learned in chapter 9. The inside would be where all the box products are negative (see ch_10/sdf_triangle.py in the book's source code).

Exercise 10.3 You need to find the inverse biquadratic transformation and apply it to the input point. Unlike projective transformations, biquadratic transformations and polynomial transformations in general are ambiguous. Two different points can be transformed into a single one. The inverse transformation isn't necessarily functional, so the answer is no.

Exercise 10.4 This is an open question. There are probably cheaper functions that behave conceptually the same as the quasi-inverse I proposed, but I'm not aware of any.

Exercise 10.5 This question is an open question, too—not a mathematical problem, but an engineering opportunity. If you want a hint, try getting the N-th root of the function instead of exponentiating the r.

Summary

- An SDF is one possible way to model a body or a shape.
- Normally, negative distances correspond to the model's inside and positive distances correspond to its outside. Zero indicates the model's border, which is a curve in 2D and a surface in 3D.
- There are similar ways to model a shape with a function, with similar rules, but the function isn't necessarily a distance function.
- Performing some operations—such as linear transformations, Boolean operations, and offsets—on SDFs is simple and computationally cost-effective.
- Tri-periodic minimal surfaces could be modeled with SDF-like sign functions, achieving the benefits of cheap operations.
- Metaballs and multifocal lemniscates are both useful for generating complex organic-looking shapes from primitive elements.

11 Modeling surfaces with boundary representations and triangle meshes

This chapter covers

- Understanding the pros and cons of boundary representation modeling
- Seeing the triangle mesh as a particular case of a boundary representation
- Understanding the downsides of triangle mesh modeling
- Confidently using contouring algorithms: Marching cubes and dual contouring
- Understanding the trade-off between the model precision and model size

With signed distance functions, we model a 2D shape or a 3D body by introducing a rule. The function is negative inside the shape, positive outside, and 0 on the border. Most of the time, we're not interested in anything apart from the border, which alone is enough to model a shape. So why bother modeling anything else?

With *boundary representation*, we model only the boundary of a body or shape. In 2D, we model a set of contours that represents the shape's border. In 3D, we model a set of surfaces that represents the border of a body.

In computer-aided design (CAD) applications, the smoothness of surfaces is an important property for modeling, so we usually model our things with smooth pieces of curves and surfaces. The surface pieces are in turn bound by their own boundaries in UV space.

Remember that in chapter 8, we modeled an ugly mushroom as a triangle mesh with vertices computed as a function $f(u, v) \to (x, y, z)$. Also, the UV coordinates were bound by the standard square: $(u, v) = [0..1]\times[0..1]$. In boundary representation, the $f(u, v)$ function itself is already a surface; you don't have to turn it into vertices or triangles. The UV boundaries can also be functions with their own boundaries: $f(t) \to (u, v)$, $t = [a..b]$.

Boundary representation remains the predominant way of modeling in CAD and computer-aided engineering (CAE) applications. Triangle mesh is a particular case of boundary representation that it's important on its own. Due to the way that modern hardware works, triangle mesh remains an almost-exclusive way of doing 3D modeling in applications where speed of rendering is important: games, augmented and virtual reality, and computer-generated imagery where (as we saw in chapter 9) surface smoothness could be emulated convincingly enough (figure 11.1).

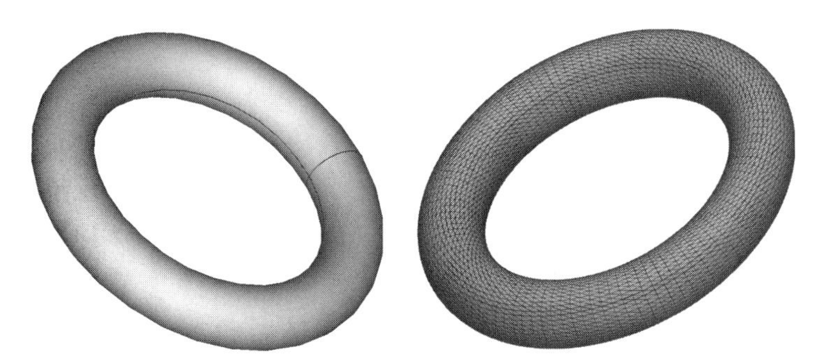

Figure 11.1 A torus described as a smooth surface (left) and as a triangle mesh of ~10,000 triangles (right)

In 3D printing, the traditional way to go was to use a triangle mesh. In fact, the name of one of the most popular triangle-based formats, STL, comes from the 3D-printing technique stereolithography. But recently, this situation has been changing. The printers are becoming more advanced and more curve-ready, and the market has shifted from design bureaus printing custom models on demand to industrial manufacturing, the domain where smooth-surface CAD data is prevalent.

The fad of the day is the hybrid model. You can have both smooth surfaces and triangle meshes working together in a single representation or interchangeably, depending on the operation you're going to perform on your model. But to understand hybrid boundary representation, we should first discuss smooth surfaces and triangles separately.

11.1 Smooth curves and surfaces

CAD and CAE, unsurprisingly, stem from unaided design and engineering—from things that people used to draw on paper. Computers added a few new tools to the box and made a few obsolete. All CAD applications, for example, support NURBS (nonuniform rational basis spline) now, because NURBS is a powerful, versatile instrument, but almost no applications support all the traditional engineering curves that we haven't even discussed in this book before: involutes, trochoids, hypocycloids, and so on. You don't have to remember any of these terms; their jobs have mostly been taken by splines.

There were splines before computers, but they were real-world splines—flexible rulers you were supposed to use as drawing tools. As discussed in chapter 7, nowadays, these splines are modeled by Bézier curves. NURBS extend Bézier curves, which are in their classical sense both uniform and not rational. There are rational Bézier curves or, rather, rational Bézier functions that help us build curves, but they also extend the splines from the real world, not simply replace them.

11.1.1 Data representation

The CAD and CAE toolbox differs from what people did on paper, but mostly in a positive way. Now we can do more things easier than before. As an example, the list of entities supported by one of the oldest data formats for 3D CAD data, IGES, includes

- Planes, lines, and points
- Circular arcs, conic arcs, spheres, toruses, and ellipsoids
- Polynomial curves and surfaces
- Rational basis splines for curves and surfaces
- Surfaces of revolution, which occur when a surface is made by rotating a curve around some axis (such as a random mushroom from chapter 8)
- Curves and surfaces obtained by offsetting other curves and surfaces
- Composite curves and surfaces obtained by running Boolean operations on other curves and surfaces

NOTE IGES is an open format; you can read more about it even on Wikipedia. Although the format, or rather its formatting, is optimized for punch cards (that's how old it is), the format is still in use today because it's one of the few nonproprietary CAD formats out there. The other popular openish format is STEP. Both formats are textual and ASCII-based, and this fact alone makes them inefficient compared with proprietary binary formats. Moreover, STEP isn't really open; it's only perceived as such because of its features are relatively easy to understand from examples and reimplement.

Before CAD, a technical drawing was a drawing that someone made on a sheet of paper. If you made a small mistake and had already laid down ink, you had to redo the drawing from scratch.

I remember struggling with drawings in college. Our professor showed us only one mistake at a time, so we had to keep coming to practices with the same exercise over and over. Once I spent several weeks trying to finish a fairly simple projection and finally made it right. The professor admitted it. He also said, "You got the part perfectly right, but your frame is 2 millimeters off, so move the frame, and I'll give you a grade."

Moving the frame, of course, meant redoing the whole drawing from scratch. You can't move stuff around on paper. But on a computer, you *can* move stuff around! And that's the neat part. Most CAD packages keep composed parts not as a final drawing or a final 3D body, but as input primitives and a series of operations that make them into the form you want. The program keeps the way the model is composed, so if you make a silly mistake at the beginning, you don't have to redo your day's work. You can simply fix the mistake, and the whole sequence of operations will reapply itself automatically.

Automatically, but not instantly. Depending on the complexity of your model, applying some operations may take quite some time. As we established in chapter 10, with signed distance functions (SDFs), doing offsets and Boolean operations is as simple as doing basic logic and arithmetic. With curves and surfaces, that isn't the case.

11.1.2 Operations are also data

Even with curves, doing a Boolean operation such as subtraction, intersection, or unification isn't easy. To perform a subtraction, for example, you need to complete the following steps:

1. Find the points of intersection.
2. Split the curves into pieces by these points.
3. Classify which chunks should be removed, and which should be inverted so that the orientation of the resulting model keeps its consistency.

None of these tasks is simple enough to be solved in a few lines of code, as is the case with SDFs, not even finding the intersections. It looks easy, and we've done intersections before. An intersection of lines is a set of equations, one per every axis. But consider this: the more entities you want to support, the more kinds of intersections you have to calculate. Even if you want to support only four entities—a line, a circle, a parabolic curve, and a cubic Bézier—you also have to support ten kinds of intersections. Also, even with cubic Béziers, a pair of two cubic curves may have up to three intersections, and to find them all, you have to solve a cubic problem—not hard, but time-consuming.

So Boolean operations on CAD curves and surfaces aren't simple and, depending on the complexity of the model, may not be fast either. But what about offsets?

An offset of a sphere is another sphere. Again, we've done offsets with SDFs, and that task was simple, so how hard could offsetting be with CAD? Well, with a single sphere, it's not hard. But even with a pair of spheres, we immediately have the same problems as with Boolean operations as soon as the spheres' offsets start to intersect (figure 11.2).

Figure 11.2 An offset on curves and surfaces becomes no simpler than a Boolean unification as soon as offsets start to intersect.

This operation is possible but doesn't come as cheap in terms of performance and complexity, as it does with SDFs. You can compose surface generation, Boolean operations, and offsets to make the shape of the body you want. Almost every operation you see in a CAD application's toolbox is a combination of these three, including chamfering and filleting (figure 11.3).

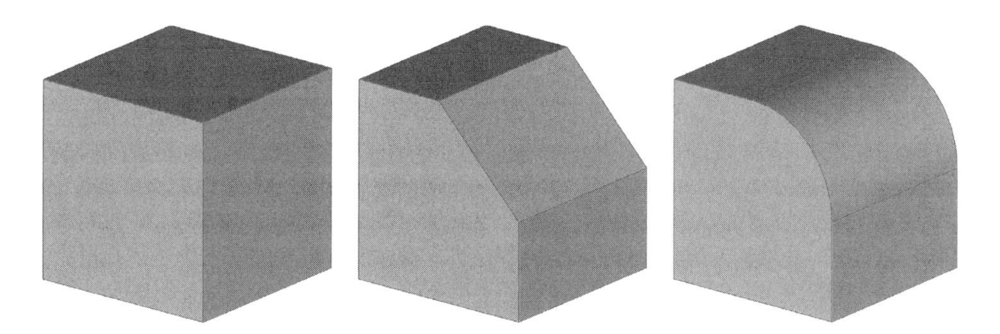

Figure 11.3 Chamfering (middle) and filleting (right) aren't independent operations on a source model (left), but a combination of offsets and surface generation via sweeping.

To apply a chamfer to a model, you pick an edge, find the edge's offset on the cube's surface, connect the offset lines with a surface, and then remove the pieces of the cube that are on the wrong side of this new surface (figure 11.4).

A fillet is almost the same, but the surface between the offset lines is guided by a Bézier function that respects the initial tangents.

When we have a robust tool set of surface generators, offsets, and Booleans, we can even invent our own custom operations, and as soon as they rely fully on the given

Figure 11.4 Chamfering is done with three operations: curve offset, surface generation, and a sort of Boolean intersection.

tool set, we can store them in a CAD file, too. CAD applications not only remember all the actions you take as you compose the model, but also save the history of these operations for later use.

As a programmer, you might find it easier to appreciate this concept if you imagine a CAD project to be a Git repository. All your committed changes are there, and if you need to introduce a change to one of the early commits, you can do that and rebase all the following changes on top of it. Also, as in Git, sometimes the merge doesn't go as expected. If you're removing a primitive you were supposed to subtract from afterward, for example, the subtraction will obviously fail.

11.1.3 Section 11.1 summary

Typical features of geometrical data representation in CAD and CAE are

- Several kinds of curves and surfaces are supported. There are universal NURBS; sweep surfaces and surfaces of revolution that are constructed from curves; and separate entities for planes, spheres, cylinders, and so on. This introduces ambiguity, because a cylinder may be a NURBS surface, a surface of revolution, or simply a cylinder. But for historical reasons, we still have to support all that.
- The Boolean operations and the offset operation on these surfaces are well known mathematically, and the algorithms for them are well developed. But in practice, their implementations are rather complex pieces of code with a nonzero chance of failure. These algorithms aren't computationally cheap, either.
- A data model in CAD incorporates operations or, rather, operation sequences. You can think of it as a text document with infinite undo capability or as a Git repository of operations and entity generations instead of commits.

11.2 Segments and triangles

Game graphics and computer-generated imagery in general don't have to be as precise as a ball-bearing ring. You can afford to approximate your smooth surfaces with pieces of planes, usually triangles, or curves with line segments (figure 11.5). This kind of representation remains a boundary representation, but a special case of it.

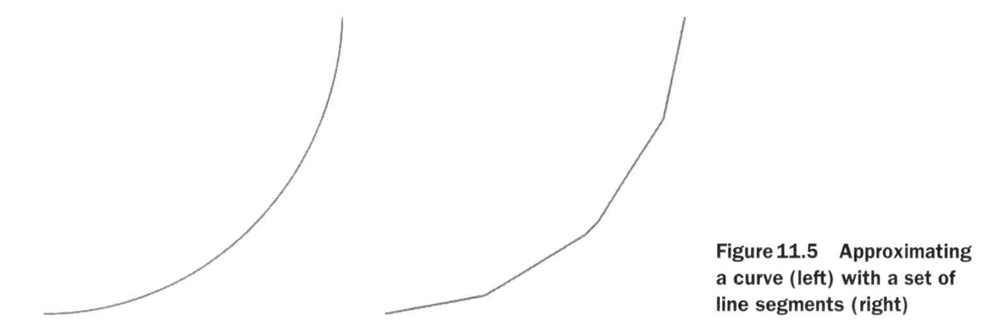

Figure 11.5 Approximating a curve (left) with a set of line segments (right)

For 2D, the data representation would be as simple as a set of polylines made of line segments. You can write the polylines as arrays of points. This representation works well both for storing and processing; it doesn't have any redundancy, and you still get to keep all the information about what segments share what points.

For 3D, things are a little more complex. You might consider having two different data representations: one for storing and another for processing.

11.2.1 Vertices and triangles vs. the half-edge representation

Because a triangle mesh is a set of triangles, you can write it down as a set of triangles: triplets of triangles' vertices' coordinates. A single triangle, then, takes $3 \times 3 = 9$ numbers to store. In a mesh, however, most triangles are supposed to be connected, so they should share most of the points. We should use this fact to our advantage.

A cube surface, for example, has 6 sides and 8 corner points. A minimal triangle mesh to represent a cube should have 8 vertices and $6 \times 2 = 12$ triangles. If you write all of them down as triplets of points, the model will take $12 \times 3 \times 3 = 108$ numbers.

But if you write the cube's vertices first and then represent triangles as triplets of vertices' indices, you need only $8 \times 3 + 12 \times 3 = 60$ numbers. See, even for a simple model, this representation is already proving to be economical.

Most file formats use this representation to store triangle meshes. This is how a standard cube is stored in Wavefront .obj format, for example:

```
v  0.0  0.0  0.0
v  0.0  0.0  1.0
v  0.0  1.0  0.0
v  0.0  1.0  1.0
v  1.0  0.0  0.0
v  1.0  0.0  1.0
v  1.0  1.0  0.0
v  1.0  1.0  1.0
```

```
f 1 7 5
f 1 3 7
f 1 4 3
f 1 2 4
f 3 8 7
f 3 4 8
f 5 7 8
f 5 8 6
f 1 5 6
f 1 6 2
f 2 6 8
f 2 8 4
```

Each `v` element represents a vertex with a triplet of coordinates, and every `f` element—a face. In our case, a face is a triangle where the numbers are the indices of its vertices. In practice, of course, you may want to extend your model with UV coordinates for texturing or normal vectors for smooth shading. Most formats support such extensions, but these extensions aren't relevant for the conceptual representation. The conceptual idea is that we store vertices and triangles separately to optimize memory intake.

This representation, however, lacks explicit topological information; it doesn't say which triangles are neighbors. You don't need this information for all the possible geometric algorithms—the topology of a mesh is irrelevant for the mesh's transformation, for example—but for a lot of other operations that you'd want to perform on a mesh, topology matters.

Mesh reduction, for instance, is the operation that reduces the number of triangles that represent a mesh while keeping the representation accuracy within reasonable limits. You don't need 65,536 triangles to represent a perfect cube; 12 will do. To reduce a mesh, you should find edges that lie on a flat surface along with all their neighboring triangles and collapse these edges to points. Each edge collapse removes a pair of triangles from a mesh, leaving their neighbors to fill the empty space.

So if you want to reduce a mesh, you need to know all the triangles that are connected to any edge to decide whether collapsing this edge to a point will affect the model too much. If you want to smooth a mesh, once again you need to know the normals of all the triangles that touch a vertex to decide whether you should move the vertex. The vertices+triangles model contains all this information implicitly, but you have to traverse all the triangles to obtain it.

In practice, we often employ another data representation for mesh processing. This new representation, called a *half-edge data model*, stores all the mesh not as triangles, but as oriented triangle edges. Normally, a single edge occurs in this model twice: in one triangle, while oriented from vertex a to vertex b and in the neighboring triangle from vertex b to vertex a. This edge is the same edge, concerning the same vertices; only the orientation changes. But because half-edges a to b and b to a belong to different triangles, you have to write them twice to form a full edge. That's why we call them *half-edges;* that's where the data model gets its name from. A half-edge (h) (figure 11.6) contains references to

- the vertex of its origin (a),
- the next half-edge in its own triangle (n), and
- its other half from a neighboring triangle (d).

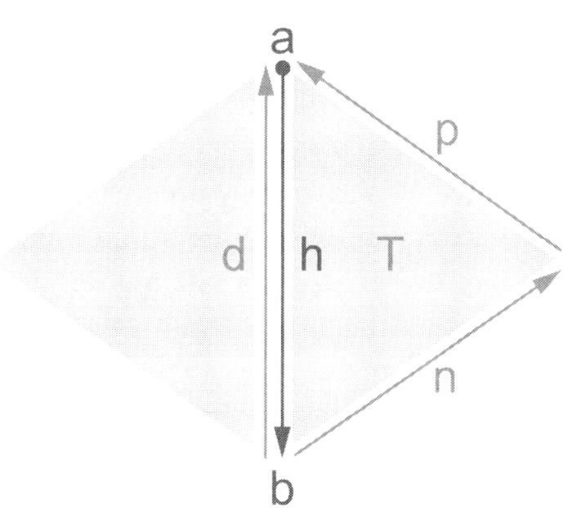

Figure 11.6 A half-edge and its references

This is already enough to represent a whole mesh as a set of half-edges—without all the possible extensions such as normal vectors and UV coordinates. Also, for the sake of convenience, a half-edge in this model can also be supplied with

- the reference to the previous half-edge in the triangle (p),
- the reference to the incident triangle's data (T), and
- the vertex of the edge's end (b).

The half-edge model contains more redundant data than vertices and triangles alone, so it's not an economical way to store meshes. It's more fitting for processing purposes, though.

> **SEE ALSO** You can find a more extensive, interactive explanation of a half-edge data structure by Jerry Yin and Jeffrey Goh at http://mng.bz/41lg.

11.2.2 Pros and cons of triangle meshes

Approximating smooth surfaces with triangle meshes drastically simplifies all the math behind Boolean operations and offsets. An intersection of a plane and a sphere is a circle, for example, but if you model both entities as triangle meshes, the intersection will be a polyline made of straight line segments. From a pure-geometry perspective, an intersection of a plane and a cone is a conic section: hyperbola, parabola, or ellipse. With triangle models, the intersection of these surfaces is also a polyline. An intersection of two polynomial surfaces is a polynomial curve. With triangles, again, it's a polyline. It's always a polyline with triangles, so you can reuse the same intersection code for all possible models.

Whatever you do, whatever shape your model is, you always operate on segments and triangles. You don't have to keep different solutions for different tasks; you need only one. But you'd better have a good one.

Although such an approximation indeed simplifies code, it has complex problems on its own. First, how good is your approximation? Technically, you can approximate any 3D body with four triangles. Make a pyramid out of them, and that's it; that's your approximation. This approximation can retain the volume and even the bounding box of the original body, but for any practical use, this approximation would be useless.

At the other extreme, you can make your approximation excessively precise. You can make sure that no triangle goes farther away from the original real-world-size model than, say, 1 nanometer. This measurement is extremely precise, but unless your model describes something as simple as a perfect cube, this approximation will also be useless. Because it requires so many triangles, any processing of a model this large will be implausible.

Pragmatically, a good approximation is an approximation that's good enough but not any better. There's no point in holding your model to 1-micrometer accuracy if you'll eventually print it on a 100-micrometer-accurate printer. At this point, modeling stops being mathematical and starts being an engineering concern. Picking a good-enough representation of a smooth surface with a triangle mesh becomes a formidable problem on its own.

Another not-quite-mathematical problem is how to preserve smoothness by modeling a surface with a mesh. Strictly speaking, you can't do that at all. You might preserve continuity, but a triangle mesh inherently isn't smooth. It doesn't have curvature, either. But a whole class of algorithms works on a triangle mesh as though it were a surface with usual differential properties. This class of algorithms constitutes a new kind of geometry: discrete differential geometry.

SEE ALSO The course notes by Keenan Crane of Carnegie Mellon University are probably the best introduction to discrete differential geometry in open access. You can find them at http://mng.bz/v1mq.

Also, there are all kinds of computational problems. We discussed the degenerate triangles in chapter 2, but the pain doesn't end there. Near-degenerate triangles, folds, intersecting triangles, and near-intersecting triangles all cause trouble in one context or another. That's why every 3D editor that works on triangles has a remeshing algorithm. Remesher is an algorithm that recovers the surface represented by a triangle mesh with another, presumably better set of triangles (figure 11.7).

The idea here is that if a processing algorithm fails to work on the mesh you got, you can remesh the model and try again. Speaking of failure, the rate of failure naturally rises as your triangle mesh gets denser. Suppose that a triangle-intersection algorithm that you use in a Boolean operation fails 1 time in 1,000,000,000 runs. That's great reliability! Any modern data center would be proud to promise you 99.999 percent uptime, and our algorithm is 10,000 times more reliable than that.

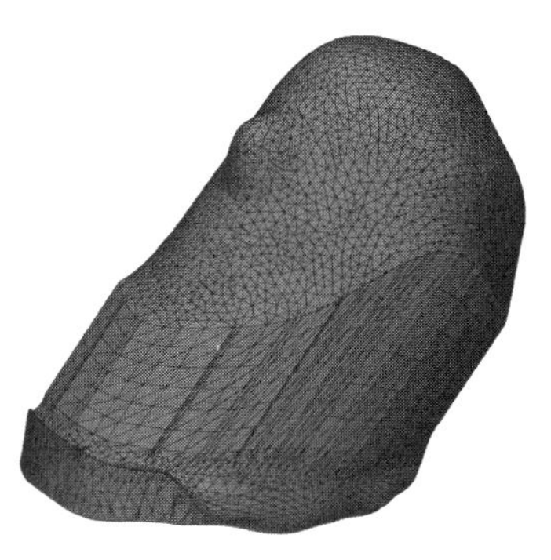
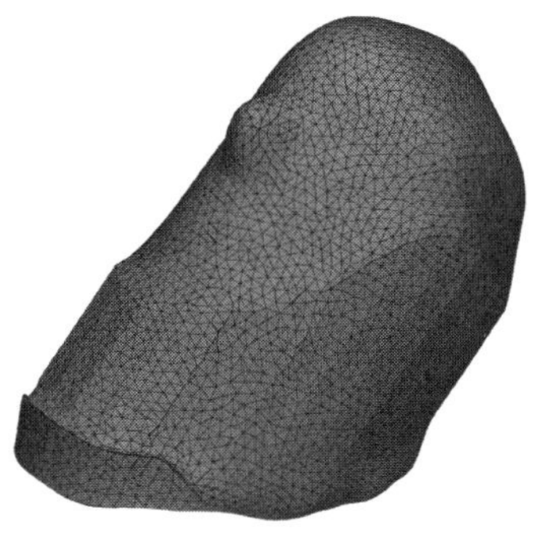

Figure 11.7 Remesher makes a source triangle mesh (left) into a similar mesh but one made with higher-quality triangles (right).

But what if our model consists of 10,000,000 triangles? Running the algorithm once on every one of them lowers the probability of success to $0.999\,999\,999^{10\,000\,000} \approx 0.99005$. So roughly speaking, now it fails once in a hundred runs. If you have to prepare a few hundred parts for printing—a normal workload for even a small prototyping bureau—you're likely to face a failure.

That's why we have a paradox in the triangle world: as hardware gets better, the software seems to get worse. With better hardware, we're ready to process more and more detailed models. But the same software that was great for the 1990s and early 2000s now fails far more often.

Speaking of reliability

This book is a geometry book, and software reliability as an expertise domain has little to do with geometry. But while we're at it, the only proven way to improve the reliability of a system is to introduce redundancy in its subsystems—proven by simple math. When you run unreliable operations in a strictly defined sequence, the probability of success drops with every added operation.

Let's say you want to hollow your model, make some holes so that the printing powder can escape from the hollow space when the part is done, and fill the hollowed space with some kind of a lattice to ensure structural integrity. That's three operations. For the sake of a mental experiment, suppose that every operation has a 0.9 chance of success. The whole workflow will have this chance of success:

$$0.9 \times 0.9 \times 0.9 = 0.729$$

Now let's presume that we have three completely different, completely independent ways to reproduce the same workflow and the probability of success for every one of them is 0.729. This makes the probability of failure 1 – 0.729 = 0.271. The probability that at least one of them will succeed is

$$1 - (0.271 \times 0.271 \times 0.271) \approx 0.9801$$

Obviously, 0.9801 > 0.9. The reliability of the system from the second example is better than the reliability of a single component from the first one. This means, that we can build reliable systems out of unreliable components! Well, independent and

(continued)

different components. The math won't work if we run the same workflow three times, hoping that it'll get better somehow.

I love this conclusion because it advocates for the invention of new algorithms and new ways to do old jobs. In programming, we're getting accustomed to the notion that there's always a single best way to do everything and that it's a bad idea to invent your own stuff. Well, even if there is a best way or a best algorithm to do something, supporting it with even slightly less efficient fallbacks makes the whole system stronger.

So please invent—if not for the fun of it, at least for the sake of reliability!

Last but not least, writing effective parallel algorithms that work with triangle models isn't always easy. There's no simple way to partition a triangle mesh properly. To remesh a huge mesh using multiple cores, for example, you need to split your model into roughly equal chunks and assign a workload to each core. If your triangle mesh consists of roughly same-size triangles, this approach is plausible. If your mesh is highly irregular, containing both extremely large and small triangles, partitioning the model evenly becomes a huge task on its own. Ironically, but not surprisingly, we usually want to remesh bad meshes to improve their quality, and not vice versa, so the remesher is the algorithm that always gets the worst input data possible.

Some highly parallel algorithms work on triangle meshes, but they either find a way to exploit a serial approach on parallel chunks of data or switch the model representation underneath completely.

11.2.3 Section 11.2 summary

Typical features of graphical data representation as a triangle mesh are

- The data model is simple—only triangles. In practice, the model is often supplemented with textures, UV parameterizations, textual labels, support structures, and so on. But the core of the model remains simple.
- The code is relatively simple due to the unified data model. You don't have to support all kinds of operations on special surfaces; you should work only with line segments and triangles.
- Your model is an approximation, and you have to balance the accuracy of your approximation and the size of your model. The size of a model affects both the performance and reliability of all the operations.

11.3 Practical example: Contouring with marching cubes and dual contouring algorithms

The operations we take for granted with SDFs—Boolean operations and offsets—become more error-prone and computationally heavy as the sizes of triangle meshes grow. At some point, converting a few meshes to SDFs, doing the operations, and converting the result back becomes a feasible alternative to doing everything on meshes alone.

With this approach, we have to compromise the precision again because the two-way mesh-SDF-mesh conversion isn't exact. But if the models we want to operate on are too large to process anyway, we'd rather have a workflow that works approximately than a precise one that works only theoretically. This conversion approach is often used in practice for fixing triangle meshes with holes, folds, and inverted triangles (figure 11.8).

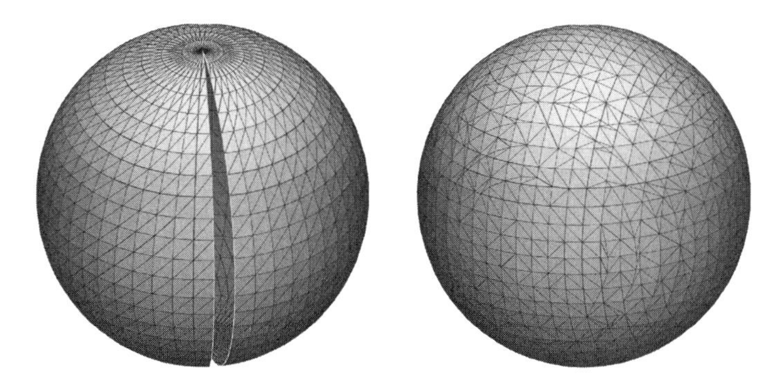

Figure 11.8 A generic fixing algorithm based on mesh-to-SDF and SDF-to-mesh conversion closes the gap in a sphere's triangle mesh.

Also, sometimes you have holes in your mesh that are too large to patch, so you have to reinvent part of the model's geometry by using interpolation. The interpolated surface comes in implicit form, so you have to turn it into triangles anyway (figure 11.9).

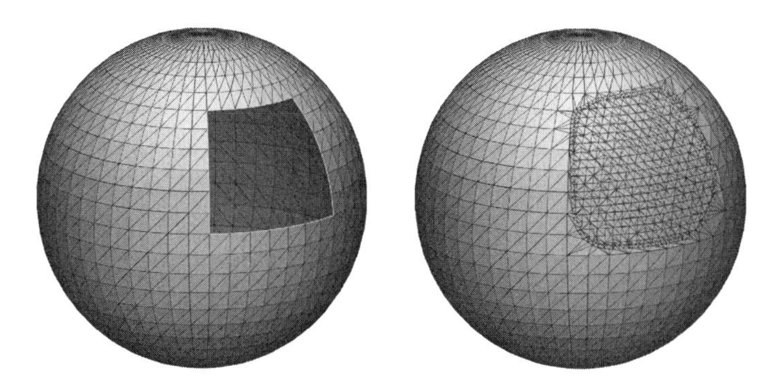

Figure 11.9 A hole-filler algorithm restores a large piece of lost geometry by using interpolation and the SDF-to-mesh conversion.

And, of course, if you need a surface offset, and if you can allow it to be somewhat imprecise, doing this offset on the model SDF and converting it back might be a good idea. The hollowing operation we discussed in chapter 10, for example, is usually done on SDFs using mesh-to-SDF and SDF-to-mesh conversions.

In chapter 10, we looked at how the mesh-to-SDF conversion is done (section 10.1.3). An SDF of a mesh is a signed distance to the nearest triangle. Now it's time to talk about SDF-to-mesh conversion or, rather, conversions because unlike the mesh-to-SDF conversion, this task isn't unambiguous.

Consider a 3D sphere once again. Let's say you want to turn the sphere's SDF into a triangle mesh. How many triangles should your resulting model have? This question isn't rhetorical; it's one that you have to answer every time you attempt a conversion because the resulting triangle mesh can have as many triangles as you want. Four triangles may already serve as a minimal model—a tetrahedron with the volume of the initial sphere. But your model might as well have 4,000 triangles—or 4,000,000.

Adding triangles makes a triangle mesh look more like a sphere, but the resulting model will never become smooth anyway. It'll never be an exact model of the sphere—only an approximation (figure 11.10).

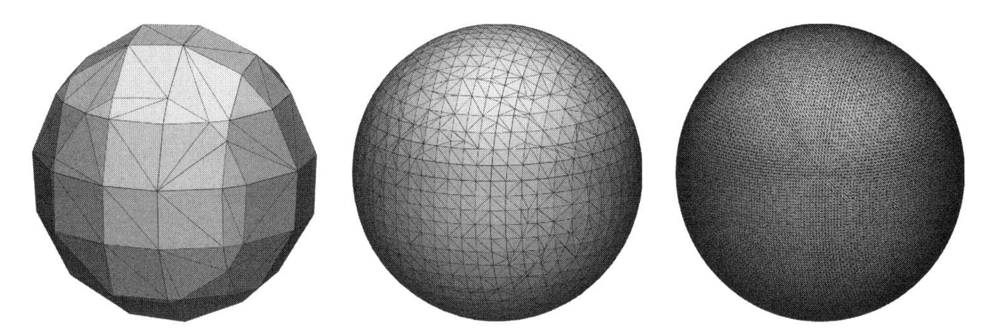

Figure 11.10 The more triangles a mesh has, the more the model resembles a sphere. A triangle mesh will never describe an exact sphere, though.

Moreover, some approximations are better than others even if they consist of the same amount of triangles, simply because the algorithms that build them work better for the task at hand. Some algorithms work better with smooth surfaces; others preserve sharp edges better. You not only have to learn the general approach, but also need to understand the merits and flaws of every algorithm.

Let's look into a pair of algorithms. Both algorithms produce a triangle mesh out of SDF or, in 2D, a contour made of line segments. For ease of understanding, we'll discuss both algorithms in 2D. Bear in mind, though, that they both generalize well to 3D because they both originally came from 3D applications.

11.3.1 Marching cubes

Marching cubes were first published in 1987 by William E. Lorensen and Harvey E. Cline as a way to visualize data obtained with computer tomography (CT). We'll discuss this data in chapter 12. For now, let's assume that we have an SDF as input, and we want a triangle mesh or, in 2D, a contour that approximates that SDF.

What Lorensen and Cline proposed is rather simple conceptually. Because we can't approximate a surface as a triangle mesh without an error anyway, let's agree on this error beforehand. We'll split our space into a grid of cells, each cell the size of the error we've agreed on, and then roughly approximate the surface in each cell with a few triangles. In each cell, the approximation will be rather poor, but because the cells themselves are small enough, the error of the whole approximation won't exceed the error we agreed upon.

Originally, the authors proposed 15 configurations of surfaces in cells (figure 11.11). Each configuration depends on which corners of a cell—in other words, vertices of a cube—are inside and which are outside the shape we're trying to approximate. As a reminder, an SDF governs this inside/outside status with its sign.

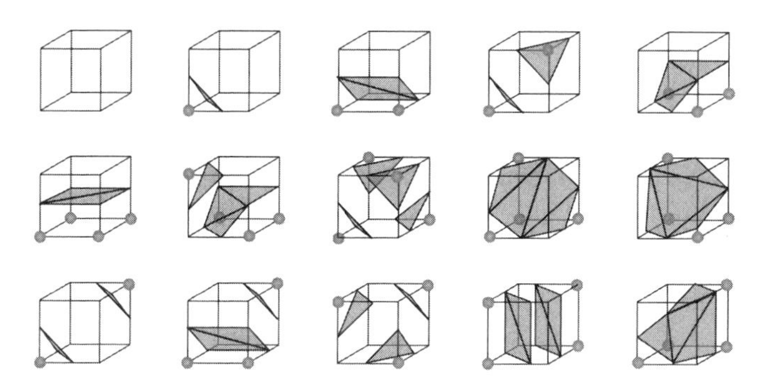

Figure 11.11 Original configurations of triangles substituting the implicit surface in cells. Note that the symmetrical configurations are omitted; otherwise, there would have been 2^8 = 256 combinations (Jmtrivial, GPL, via Wikimedia Commons).

But let's not rush into complex things right now. Let's start slow by getting back to a circle in 2D. Here's a circle:

$$x^2 + y^2 = 2^2$$

Figure 11.12 shows its graph.

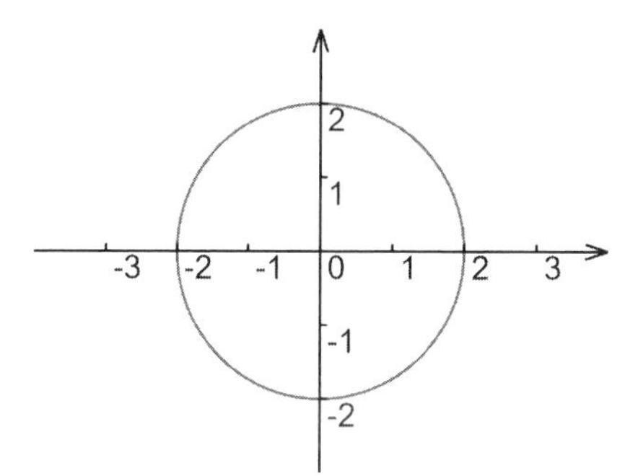

Figure 11.12 Smooth and precise graph of a circle

What we want to do is turn this circle into several line segments that describe the circle well enough—whatever this means. We want a contour of the circle like the one in figure 11.13.

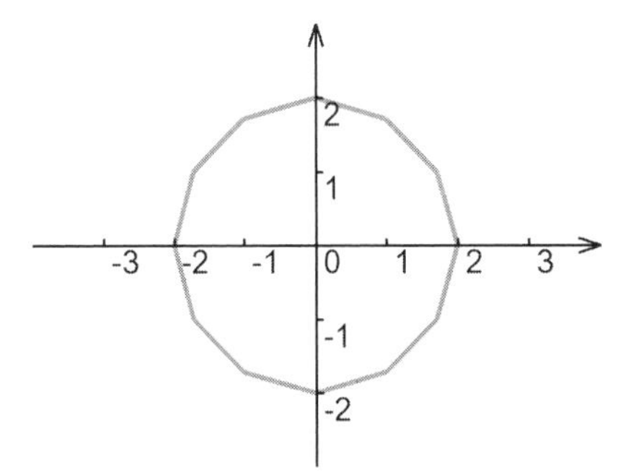

Figure 11.13 An approximate model of a circle made of line segments: a contour

Conceptually, contouring is a simple task. What we want to do is find the contour segments that segregate the positive values of SDF from the negative ones. Because we can't possibly do this without loss of precision—remember that we can't describe a smooth circle with flat line segments—we should make sure only that our error isn't too large. So let's put the circle in a fine grid and in every grid cell border put one of the following:

- A vertical line segment in which the grid value on the left has a different sign from the one on the right
- A horizontal line segment in which the grid value at the top has a different sign from the one at the bottom

These two rules already produce a contour, and the contour's error will rely on the grid-cell size, so it looks like we're nearly there. The only downside of this particular contouring result is that it has way too many corners for a circle's model (figure 11.14).

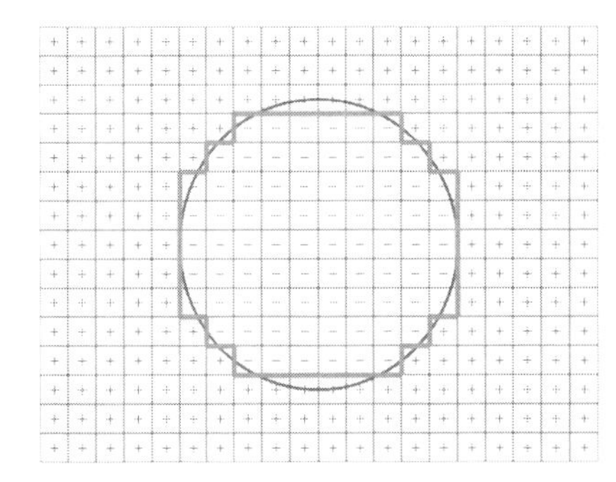

Figure 11.14 Technically, this model has an error of no more than $\sqrt{2}/2$ of a cell's width or height. Still, it's not a great model of a circle.

Sure, we can't achieve true smoothness with line segments alone, but obviously, we can do better than this. That's where the marching cubes—or marching squares, because we're in 2D—come in. Instead of putting segments on the cells' borders, let's put them inside the cells. Because a cell has four corners, and an SDF value in each corner can be positive or negative, we have to accommodate $2^4 = 16$ combinations. Each corner's sign combination gives us a possible line segment configuration in a cell (figure 11.15).

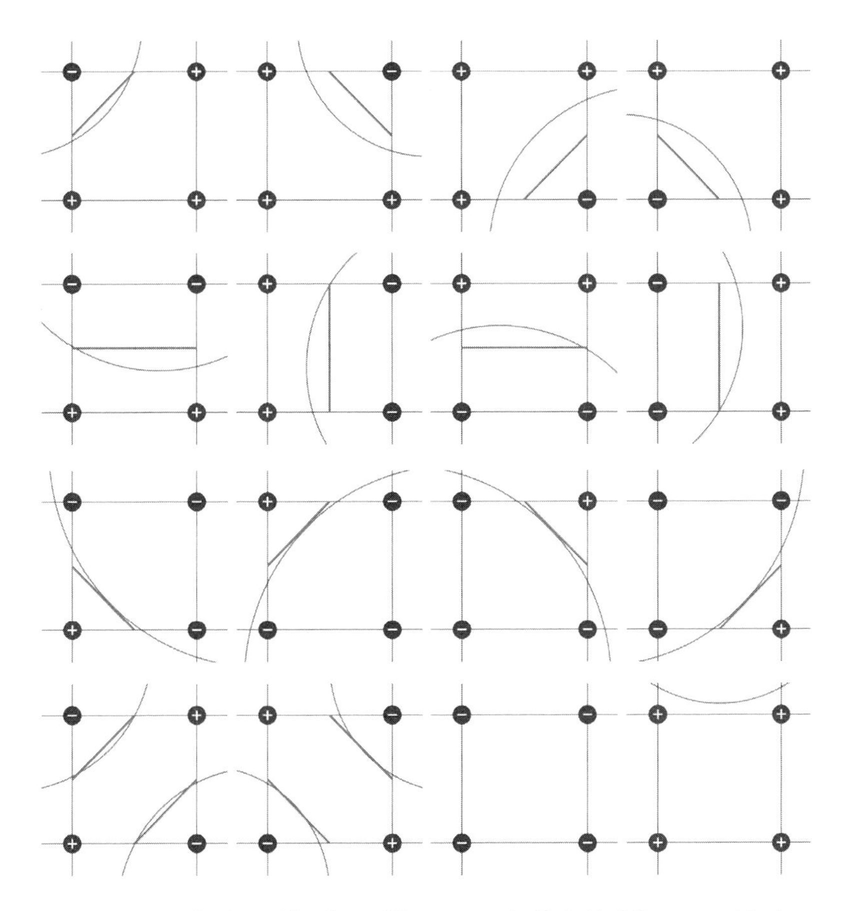

Figure 11.15 All 16 combinations of line segments. Note that the symmetrical configurations are included.

> **NOTE** SDF can be 0 in a point, too. Zero is neither positive nor negative. Pragmatically, however, we can consider it to be either. The algorithm will work either way. Besides, as you might remember from chapter 2, floating-point numbers have two signed "zeroes" instead of one true zero, so we don't even have to pick a sign for ourselves; whatever is placed on the sign bit of a number will do.

With these combinations, the result gets better than with cell contours alone. The circle becomes smoother (figure 11.16). But we're not there yet.

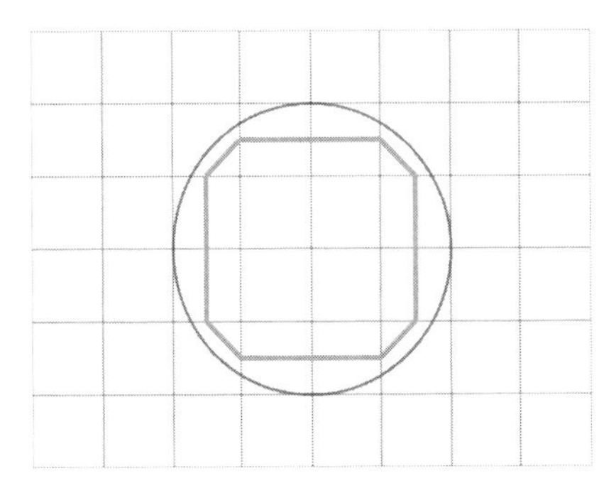

Figure 11.16 Now we place the segments in every cell according to the SDF values in the cell's corners.

We know the most fitting segment configuration for each cell, but we don't know exactly where to put each segment. We know the edge on which a segment starts and the edge on which it ends, but not at which points. This problem has several solutions, one of which is to use linear interpolation.

The SDF value reflects the distance to the closest point of a curve we want to approximate. We know that if this value is negative at one vertex of an edge and positive at another, the curve goes somewhere between the edge vertices. So using the SDF values in vertices, we can make an educated guess about exactly where the curve intersects the edge by interpolating the SDF values linearly (figure 11.17):

$$x = \frac{|SDF(x_2)|x_1 + |SDF(x_1)|x_2}{|SDF(x_1) + SDF(x_2)|}$$

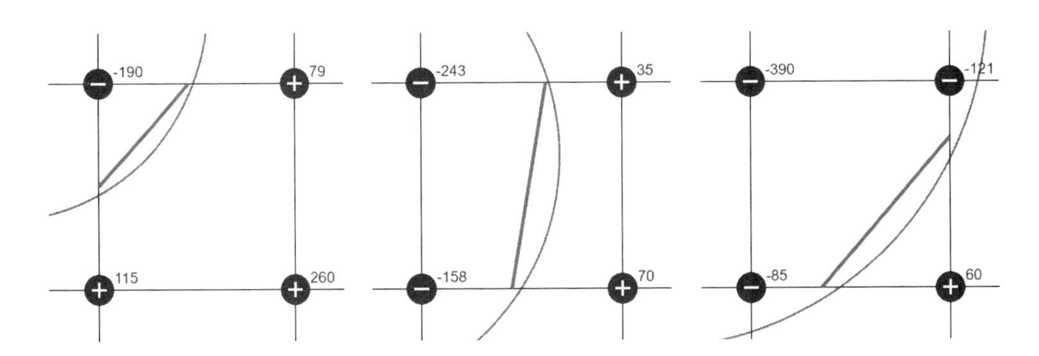

Figure 11.17 Examples of marching squares segments computed from different circle SDFs using linear interpolation. The numbers near vertices are the absolute values of an SDF; the sign marks in the vertices reflect the SDF's sign.

This estimation isn't the most exact one, of course. But linear interpolation is quick, and when you want to generate millions of triangles, every speed-up counts. Remember that this method comes from 3D. In 2D, if you want to model a 10 cm model with ~1 mm line segments, you only need 100 × 100 = 10,000 square cells, in 3D, the same precision will immediately cost you the computation of 100 × 100 × 100 = 1,000,000 cubes. With this kind of load, you have to be quick. Anyway, for most practical applications, linear interpolation works well enough (figure 11.18).

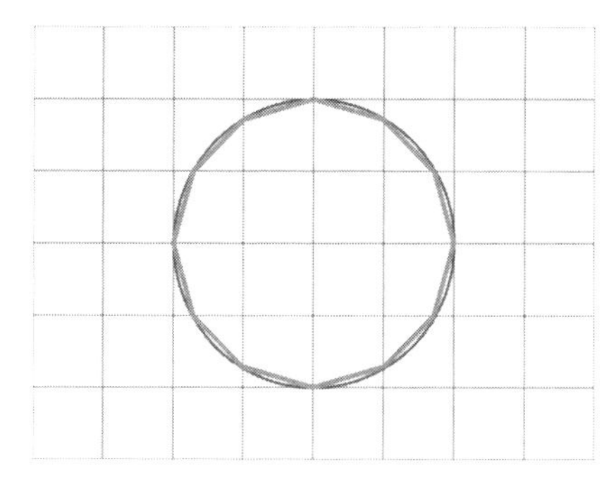

Figure 11.18 A sphere model made with marching squares and linear interpolation

Marching cubes work especially well with smooth organic shapes. The algorithm even works with implicit representations of curves and surfaces that aren't even distance functions, strictly speaking. Marching cubes play well with metaballs (chapter 10), for example, as shown in figure 11.19.

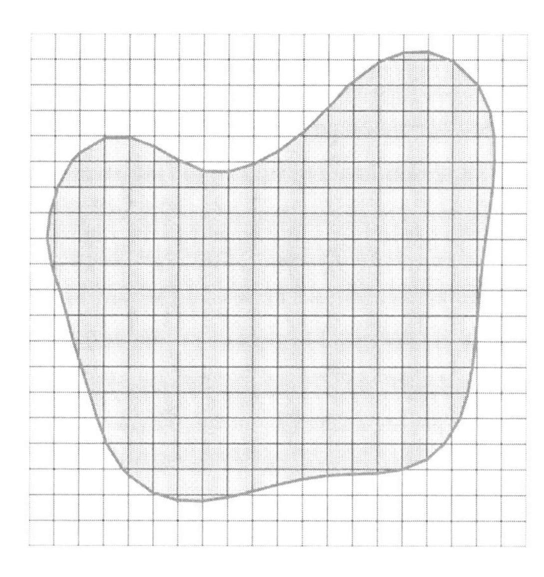

Figure 11.19 An example of a marching-squares contour obtained from metaballs' quasi-SDF

The only downside of marching cubes or squares is that ironically, the algorithm doesn't work well for cubes or squares (figure 11.20). With linear interpolation, we're cutting corners both metaphorically and literally.

Let's try something different—the same concept, but with cells and edges swapping places.

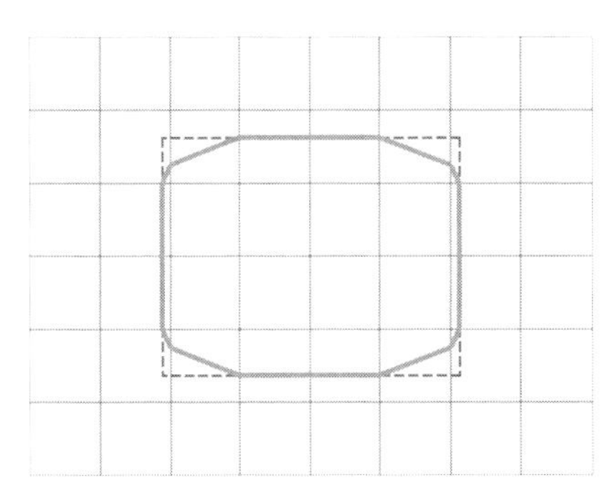

Figure 11.20 An example of a rectangle reconstructed with the marching-squares algorithm. It isn't too squarish.

11.3.2 Dual contouring

Like the marching-cubes algorithm—which while being originally a 3D technique, applies to 2D as well—contouring, as in "finding a contour that describes an SDF isoline," applies to 3D, too. The *contouring* part of *dual contouring* means modeling both a curve with line segments and a surface with triangles. But what does the word *dual* mean?

We'll get back to this question by the end of this chapter. But by the time we get there, you might have figured out the answer by yourself.

So how does dual contouring work? It starts like marching cubes; we put our SDF into a grid. The finer the grid, the more accurate our model will be. But rather unsurprisingly, the more cells we have to process, the more processing power and memory the contouring operation will take.

Then we'll put line segments where the SDF changes its sign, almost as we did before. Previously, though, we put line segments from edge to edge wherever both edges have different SDF signs on their vertices. Now when we meet an edge with different SDF signs in its vertices, we know that the curve we're going to approximate runs somewhere between these vertices, so we'll put a line segment from the middle of one of the edge's neighboring cells to the middle of the other (figure 11.21).

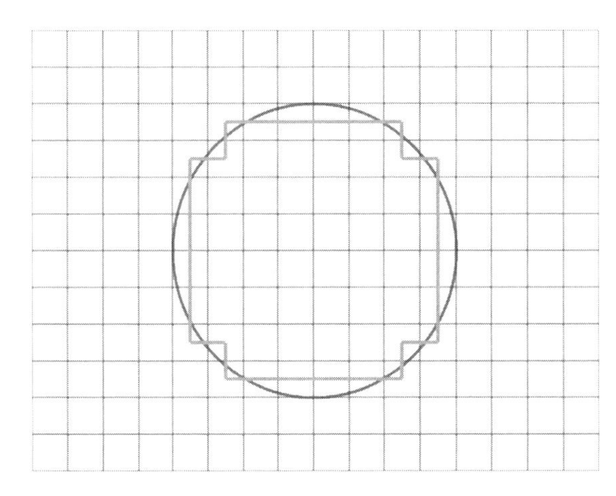

Figure 11.21 Now the contour line segments run from and to cell centers.

Certainly, running line segments from one cell center to another isn't much better than cell-border contouring (figure 11.14). But we didn't stop there before, and we won't stop now. We'll move the line segments' vertices so that they get closer to the modeled curve.

We already have a pretty good idea of how far each vertex should go. Because SDF is a distance function by definition, the SDF value in each vertex answers the question "How far is it from its optimal position?" But to answer the question "In which direction we should move each point?", we may need to learn a little bit more differential geometry.

We discussed differentiation in chapter 5. In particular, we discussed how to apply derivatives to geometrical problems involving curves, as well as how to compute derivatives symbolically. As a reminder, here are a few examples:

$$(4)' = 0$$

$$(x^2)' = 2x$$

$$(y^3 + 3y)' = 3y + 3$$

Now let's see how to apply derivatives to surfaces. A surface is a function of two parameters, so first and foremost, we have to decide how to differentiate a function like this:

$$(y^3 + x^2 + 3y + 4)' = ?$$

All the compound parts are familiar, but when they work together, we're puzzled. Should we ignore x and differentiate the function by y only? Should we differentiate by x? Or by both?

None of these problems is mathematical. We can differentiate by x, by y, or by both x and y, but the notation we use doesn't reflect which of them is a variable and which is a parameter. We can switch the notation, though. In Leibniz notation, for example, this equation says "Differentiate the function by x":

$$\frac{d}{dx}\left(y^3 + x^2 + 3y + 4\right)$$

And this equation says "Differentiate the function by y":

$$\frac{d}{dy}\left(y^3 + x^2 + 3y + 4\right)$$

Alternatively, because this book is for programmers, we can use Python's notation:

```
>>> diff(y**3 + x**2 + 3*y + 4, x)     <-- Differentiates by x
>>> diff(y**3 + x**2 + 3*y + 4, y)     <-- Differentiates by y
```

Both derivatives, by x and by y, are called *partial derivatives.* Any of them do derivate a function, but only partially. They may not be exciting by themselves, but when you combine them into a vector, you get somewhere:

$$\nabla f(x, y) = \begin{bmatrix} \frac{df(x,y)}{dx} \\ \frac{df(x,y)}{dy} \end{bmatrix}$$

This example is a *gradient* of a function. It's a vector that shows the direction in which the function grows most at a point. In our example, this is the gradient:

$$\nabla(y^3 + x^2 + 3y + 4) = \begin{bmatrix} 2x \\ 3y^2 + 3 \end{bmatrix}$$

Figure 11.22 shows how the gradient of $(y^3 + x^2 + 3y + 4)$ looks on a graph. By the way, yes, the gradient does form a vector field.

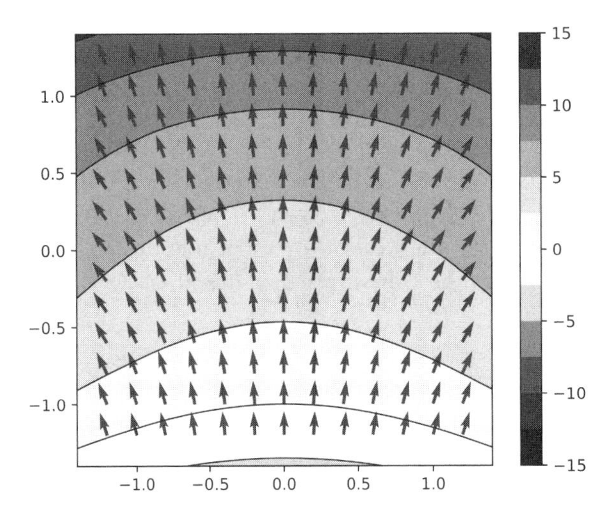

Figure 11.22 The direction of the function's gradient presented as a vector field

In our context, the direction against the gradient is where we should move a point if we want to bring it closer to the isoline.

Now that we know both the direction and the distance, we can move the vertices and make ourselves a nice smooth contour (figure 11.23).

But the true power of this particular algorithm comes not from making curves smooth, but from the ease of making corners sharp. You see, for each vertex in a contour, we can look at its neighbors' gradients to see how different they are. If they're vastly different, we can presume that the vertex in question represents a sharp feature. For sharp features, we move the vertex to the intersection of the tangent lines that its neighbors' gradients represent. This way, we can model both smooth and sharp curves and surfaces (figure 11.24).

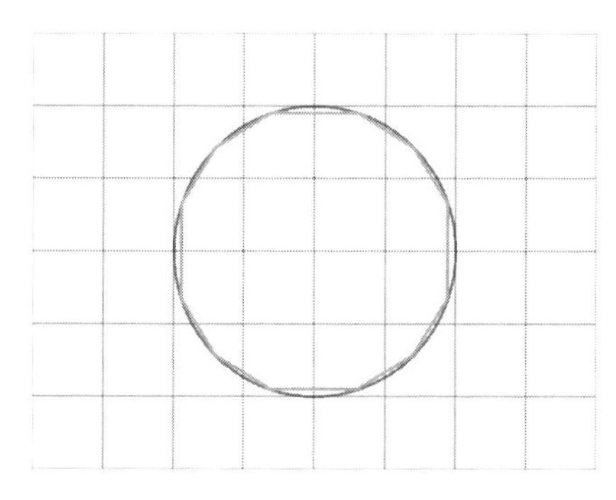

Figure 11.23 Dual contouring runs line segments not from cell centers but from a point in a cell that fits the modeled curve best to the same optimum point in a neighboring cell.

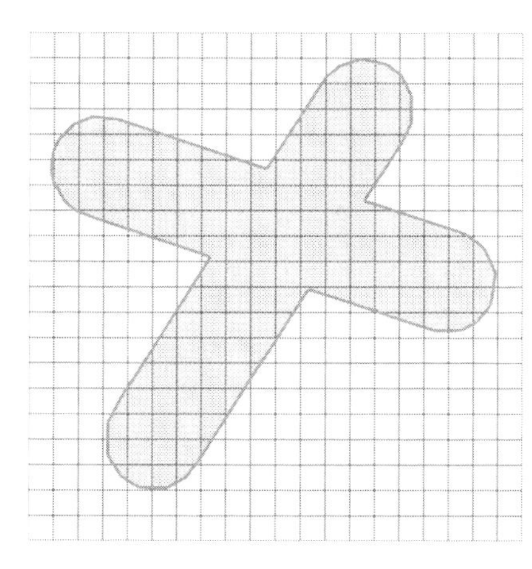

Figure 11.24 Double contouring allows us to have a generally smooth contour that has sharp features, too.

Now let's get back to the question of what *dual* means. As you may have noticed, with marching cubes, we analyzed cells to find edges that represent the modeled curve. With dual contouring, we analyze edges to find the cells. With marching cubes, we optimized vertices positions on the found edges. With dual contouring, we optimized the vertices' positions in the found cells. Conceptually, this algorithm is the same, but it's as though someone swapped cells and edges around in the description.

That's what makes dual contouring dual. It's dual to the marching cubes in the mathematical sense of complementary.

SEE ALSO Duality is a broad concept that applies to completely different fields of mathematics. You can find a simple explanation of duality with a single example from Boolean algebra on Vedantu.com at https://www.vedantu.com/maths/duality. The nLab website offers a more extended explanation with 18 examples; see https://ncatlab.org/nlab/show/duality.

11.3.3 Is it that simple in practice, too?

A few years back, a colleague gave a talk on this topic, and one of the comments from the audience was "Wow, I can't believe you use algorithms that simple in the production code." Well, these algorithms are simple conceptually, and because the talk was about the concepts, they seemed to be simple for the audience, too. But in practice, these algorithms often grow into complicated implementations, for three main reasons:

- *In practice, a model isn't merely a bucket of triangles.* A model that starts with a triangle mesh may also have textures attached, as well as normal maps, lightweight structures, labels, and generic parameterizations for whatever purpose. When we convert a model to SDFs, operate on them, and convert them back to a triangle mesh, we want all the baggage to arrive at the same airport. We want the new model to receive the original textures and parameterizations.

  This task isn't an easy one, and sometimes it's impossible. SDF operations such as inner offsets and subtractions remove parts of the model, so the original parameterizations have to be not so much translated as reinvented for the new model. Also, outer offsets and unifications make the model larger and even add new features to it. Again, the new regions didn't have original data attached to them, so we have to invent some way to apply the former data to the new surface.
- *Although this double conversion makes some operations easier computationally, the conversion itself may quickly become computationally challenging.* We've briefly discussed the fact that the number of cells rises as N^2 in 2D and N^3 in 3D. This doesn't necessarily mean that the complexity of the contouring algorithms is $O(N^3)$. Modern implementations are heavily optimized to avoid computing in empty regions, and because we're modeling a surface in 3D and not the whole body, most of the space is empty, at least from the algorithm's perspective. This makes the pragmatic complexity of the contouring algorithms closer to $O(N^2)$. Still, if you want to make your model two times more accurate, your conversion would work at least four times slower.
- *An inherent error of representation still exists.* Even if we find all the vertices' positions perfectly, and this precision comes at the cost of speed, there'll still be gaps between the triangles and the real surface because the real surface isn't necessarily flat. We can't approximate a smooth surface with flat triangles perfectly; there'll always be an error.

  This error accumulates with every new double conversion, so it's best to do as much work as possible in a single go, such as doing convert-operate-operate-operate-convert and not convert-operate-convert three times in a row.

11.3.4 Section 11.3 summary

Using naive contouring, marching cubes, or dual contouring, you can convert an SDF to a triangle mesh. This conversion supplements the mesh-to-SDF conversion, allowing you to do cheap, reliable offsets and Boolean operations on SDF representations of triangulated models.

Generally, marching cubes work better for approximating smooth surfaces, whereas dual contouring preserves sharp features better, although real-world implementations diverge far from the conceptions. There are marching-cubes implementations with sharp-feature restoration embedded, as well as dual-contouring implementations optimized to work well on smooth surfaces.

While using contouring algorithms, you have to consider the balance between computational efficiency and modeling accuracy. Normally, by making the model two times more accurate, you make the algorithm work four times slower.

11.4 Practical example: Smooth contouring

The hardest thing about learning math is waiting for all the elementary pieces to combine into something practical and exciting. It's not that elementary things aren't practical; drawing a straight line between two points is probably the most practical thing there is in geometry. But it's not exciting. It's trivial.

The algorithm I'm about to show you isn't trivial. But you've been patient enough; you've read most of the book. At this point, you may think that this algorithm is, if not trivial, at least surprisingly simple. I like this algorithm because it shows how different concepts, each of which is hard to learn by itself, combine into a practical and exciting thing. I'm talking about smooth contouring.

We'll look into the problem in 2D, where smooth contouring is similar to marching squares. We have an SDF, and we want to get the line that segregates the areas where the SDF is negative from the areas where it's positive. With marching squares or dual contouring, we get this line approximated by a set of segments. With smooth contouring, we get it as a set of Bézier curves—or, if you remember chapter 7 well, parametric cubic polynomials, which is essentially the same thing.

In 3D, this problem is extremely practical. Some companies have built their whole business on turning triangle meshes into spheres, cones, and smooth pieces of NURBS surfaces. The motivation is simple. CAD and CAE applications prefer parametric surfaces over any other representation. You don't draw a blueprint of a ball bearing in triangles; you need all the radii to match perfectly, and you need all your surfaces to be exact, not approximate. The engineering world is made of smooth surfaces. That's why the problem of making triangle meshes or SDFs into CAD-friendly pieces has the nickname *reverse engineering*.

Yes, I know—in programming, the name has already been taken by the problem of restoring the program's intention by its binary code. This fact makes licensing somewhat confusing when a reverse-engineering library comes with a clause that explicitly prohibits reverse engineering.

Anyway, let's look into a three-step algorithm that turns a 2D SDF into a set of smooth segments. This algorithm isn't trivial; in fact, explaining all its parts to someone who's completely new to geometry would be problematic. But for you, the following explanation should bring the pieces you already know into a mechanism that works. You've waited long enough.

11.4.1 The algorithm

The algorithm consists of three steps.

STEP 1: BUILD A CONTOUR ON A GRID

Let's define a grid over our SDF, find all the cells where the SDF in their central points are negative, and build a fence from all the other cells. So for every edge of every cell, if its neighboring cells have SDFs of different signs in their centers, the edge becomes a contour segment (figure 11.25).

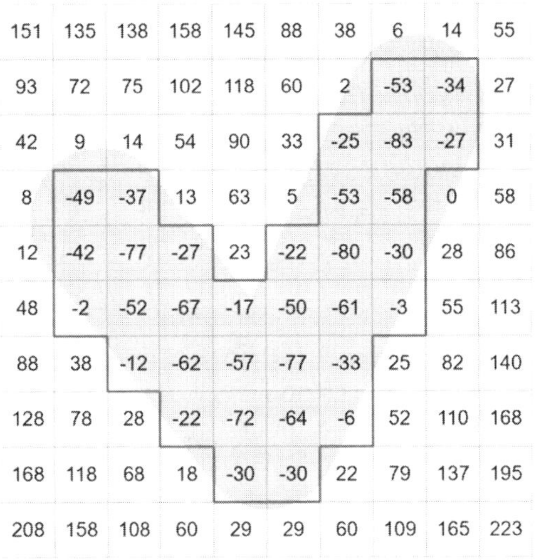

151	135	138	158	145	88	38	6	14	55
93	72	75	102	118	60	2	-53	-34	27
42	9	14	54	90	33	-25	-83	-27	31
8	-49	-37	13	63	5	-53	-58	0	58
12	-42	-77	-27	23	-22	-80	-30	28	86
48	-2	-52	-67	-17	-50	-61	-3	55	113
88	38	-12	-62	-57	-77	-33	25	82	140
128	78	28	-22	-72	-64	-6	52	110	168
168	118	68	18	-30	-30	22	79	137	195
208	158	108	60	29	29	60	109	165	223

Figure 11.25 Start from a contour that segregates the negative-valued cells from the positive-valued cells.

We saw this example earlier, where it was a simple but impractical way to get a contour made of axis-oriented segments. Now, although this type of contouring may not be practical on its own, it serves as a first step, giving us something to work with.

STEP 2: ATTRACT THE SEGMENTS TO THE EXPECTED CURVE

The contour we got from step 1 is inaccurate by design. We created a grid arbitrarily, and we measured SDF only in the cell's centers. The vertices of every line segment aren't even supposed to lie on the SDF's true border between its negative and positive areas.

But because we know that SDF is a distance function, we know exactly how far each vertex is from the expected curve. The value of SDF in each vertex is exactly the approximation error. Can we reduce this error, then? Sure! Let's shift every vertex by its own error value toward the direction inverse to the SDF's gradient. We already did this with dual contouring. The gradient explains how exactly the function changes at a point.

By doing the vertex shift, we retain the straight segments, but now they fit the expected curve much better (figure 11.26).

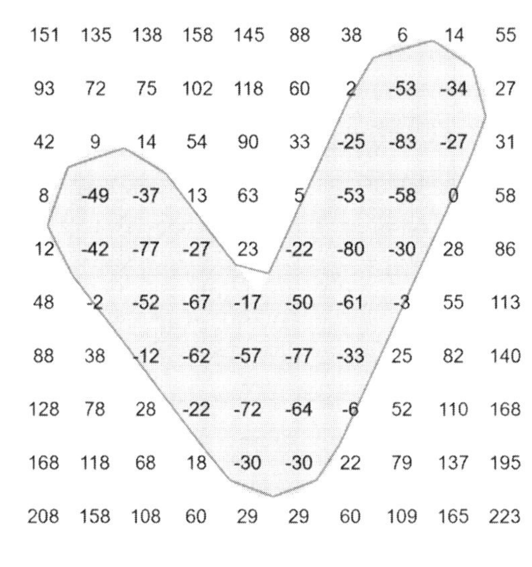

Figure 11.26 Pulling the points from the original contour by function gradients and values makes the contour fit the 0-isoline better.

Step 3: Add the curvature

The gradient says exactly how the SDF changes at a point. Its partial derivatives correspond to the function's change by x and by y. We can extract this information to build a parametric cubic curve from each line segment.

Let's say that a segment starts at a point (x_0, y_0) and ends at (x_1, y_1). Partial derivatives that we measure from the SDF in the first point are dx_0 and dy_0 for the x-axis and for the y-axis. Partial derivatives in the second point are dx_1 and dy_1.

To define a parametric cubic polynomial, we need two sets of coefficients: one for x and one for y. Each set requires four equations because a cubic polynomial has four coefficients. The segment vertices give us a pair of equations for each coordinate, and the partial derivatives provide the other pair:

$$a_x t_0^3 + b_x t_0^2 + c_x t_0 + d_x = x_0$$

$$a_x t_1^3 + b_x t_1^2 + c_x t_1 + d_x = x_1$$

$$3a_x t_0^2 + 2b_x t_0 + c_x = dx_0$$

$$3a_x t_1^2 + 2b_x t_1 + c_x = dx_1$$

$$a_y t_0^3 + b_y t_0^2 + c_y t_0 + d_y = y_0$$

$$a_y t_1^3 + b_y t_1^2 + c_y t_1 + d_y = y_1$$

$$3a_y t_0^2 + 2b_y t_0 + c_y = dy_0$$

$$3a_y t_1^2 + 2b_y t_1 + c_y = dy_1$$

If we parametrize each segment with t running from 0 to 1, the equations become much simpler:

$$d_x = x_0$$

$$a_x + b_x + c_x + d_x = x_1$$

$$c_x = dx_0$$

$$3a_x + 2b_x + c_x = dx_1$$

$$d_y = y_0$$

$$a_y + b_y + c_y + d_y = y_1$$

$$c_y = dy_0$$

$$3a_y + 2b_y + c_y = dy_1$$

Feed these equations to SymPy, and you get a nice code snippet that contains the symbolic solution for them. The final result, with cubic curve pieces instead of line segments, looks like figure 11.27.

Figure 11.27 The final result is built with the segments from before turned into cubics by adding the tangent vector equations.

11.4.2 The implementation

In Python, solving the cubic equations looks like this:

Input data: Points (x0, y0), (x1, y1) and the tangent vectors (dx0, dy0), and (dx1, dy1)

Output: The coefficients of the two polynomials Px(t) = ax*t3 + bx*t**2 + cx*t + dx, Py(t) = ay*t**3 + by*t**2 + cy*t + dy that describe the target parametric cubic.**

```
from sympy import *

x0, y0, x1, y1, dx0, dy0, dx1, dy1 = symbols('x0 y0 x1 y1 dx0 dy0 dx1 dy1')
ax, bx, cx, dx, ay, by, cy, dy = symbols('ax bx cx dx ay by cy dy')
```

```
x_coefficients = solve([
dx - x0,
ax + bx + cx + dx - x1,
cx - dx0,
3*ax + 2*bx + cx - dx1], (ax, bx, cx, dx))
```

Solving these equations results in the coefficients for the Px(t) polynomial.

```
y_coefficients = solve([
dy - y0,
ay + by + cy + dy - y1,
cy - dy0,
3*ay + 2*by + cy - dy1], (ay, by, cy, dy))
```

Solving these equations results in the coefficients for the Py(t) polynomial.

A cubic piece starts at (x_0, y_0) point and ends in (x_1, y_1). The parameter that describes the piece runs then from 0 to 1, respectively. The vector (dx_0, dy_0) sets the piece's tangent direction in (x_0, y_0), and the (dx_1, dy_1) is the curve's tangent vector in (x_1, y_1). Then the solution for each cubic piece of the contour is

```
{ax: dx0+dx1+2*x0-2*x1, bx: -2*dx0-dx1-3*x0+3*x1, cx: dx0, dx: x0}
{ay: dy0+dy1+2*y0-2*y1, by: -2*dy0-dy1-3*y0+3*y1, cy: dy0, dy: y0}
```

The code that turns SDF into a set of smooth contours looks like this:

```
border = []
for i in range(-10, 10):
    for j in range(-10, 10):
        if SDF(j - 0.5, i + 0.5) * SDF(j + 0.5, i + 0.5) < 0:
            border += [[(j, i), (j, i+1)]]
        if SDF(j + 0.5, i - 0.5) * SDF(j + 0.5, i + 0.5) < 0:
            border += [[(j, i), (j+1, i)]]

epsilon = 1e-5

def dx(x, y):
    return (SDF(x+epsilon, y) - SDF(x, y)) / epsilon

def dy(x, y):
    return (SDF(x, y+epsilon) - SDF(x, y)) / epsilon

contour = []
for segment in border:
    new_segment = []
    for i in range(2):
        x, y = segment[i]
        pdx = dx(x, y)
        pdy = dy(x, y)
        dxy_length = distance_p2p(pdx, pdy, 0, 0)
        sdf_value = SDF(x, y)
        new_p = (x - pdx / dxy_length * sdf_value,
                 y - pdy / dxy_length * sdf_value)
        new_segment += [new_p]
    contour += [new_segment]
```

Step 1: Do the cell border.

Vertical line

Horizontal line

An approximation of the partial /dx derivative

We approximate partial derivatives with differences. The epsilon variable is a different size for the approximation. Normally, the smaller the epsilon is, the better the approximation.

An approximation of the partial /dy derivative

Step 2: Fit the contour.

The x-part of the gradient vector

The y-part of the gradient vector

We'll use the gradient's length to normalize the fitting direction.

This is also the fitting vector's new length.

```
smoothened = []
for segment in contour:
    x0 = segment[0][0]
    y0 = segment[0][1]
    x1 = segment[1][0]
    y1 = segment[1][1]
    dx0 = dy(x0, y0)
    dy0 = -dx(x0, y0)
    dx1 = dy(x1, y1)
    dy1 = -dx(x1, y1)
    if dot2d(x1-x0, y1-y0, dx0, dy0) < 0:
        dx0 = -dx0
        dy0 = -dy0
    if dot2d(x1-x0, y1-y0, dx1, dy1) < 0:
        dx1 = -dx1
        dy1 = -dy1
    a_x = dx0 + dx1 + 2*x0 - 2*x1
    b_x = -2*dx0 - dx1 - 3*x0 + 3*x1
    c_x = dx0
    d_x = x0
    a_y = dy0 + dy1 + 2*y0 - 2*y1
    b_y = -2*dy0 - dy1 - 3*y0 + 3*y1
    c_y = dy0
    d_y = y0
    for i in range(10):
        t0 = i / 10.
        t1 = t0 + 1. / 10.
        x0 = a_x*t0**3 + b_x*t0**2 + c_x*t0 + d_x
        y0 = a_y*t0**3 + b_y*t0**2 + c_y*t0 + d_y
        x1 = a_x*t1**3 + b_x*t1**2 + c_x*t1 + d_x
        y1 = a_y*t1**3 + b_y*t1**2 + c_y*t1 + d_y
        smoothened += [[(x0, y0), (x1, y1)]]
```

Step 3: Smooth the contour.

The gradient is orthogonal to the tangent direction, so dx goes to the y-part and dy to the x-part of the tangent.

There are two possible directions orthogonal to the gradient. Check whether the first best choice agrees with the piece's direction; invert if it doesn't.

When we have the tangent direction, we can compute the cubic coefficients.

The last step isn't essential for the algorithm. Instead of drawing the smooth contour as a set of cubic polynomials, we refine each into a set of short flat segments. Now we can draw the border, the line segment contour, and the smoothed result in the same way.

The whole example with plotting is available at ch_11/smooth_contour.py in the book's source code. The example is in Python, but of course you're free to use any language you want. SymPy can help you by translating the solutions of the equations into the language of your choice.

The input SDF also can be anything you want. The V shape we've used throughout the chapter, for example, can be programmed like this:

```
@np.vectorize
def line(x, y, x1, y1, x2, y2, w):
    return distance_p2e(x, y, x1, y1, x2, y2) - w

def SDF(x, y):
    return np.minimum(
        line(x, y, 5, 2.5, 3, 6, 1.2),
        line(x, y, 5, 2.5, 7, 7.5, 1.2))
```

distance_p2e is the unsigned distance from a point to an edge. It was introduced in chapter 10.

Two lines form the V shape.

Figure 11.28 shows the output: an initial border, a fitted contour, and a set of cubic polynomial curves, all obtained from the same SDF.

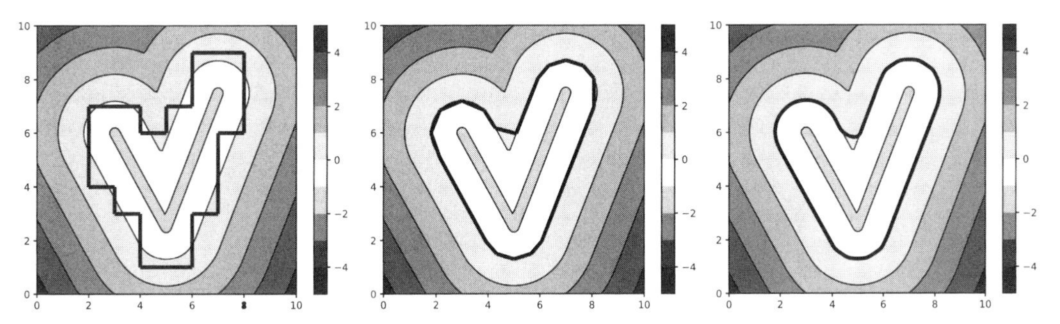

Figure 11.28 An initial border coming from the grid with cell size 1, a fitted contour, and a smoothed contour

11.4.3 Section 11.4 summary

Smooth contouring can be done with a simple three-step algorithm, which is simple only because you already know how all the compound parts work.

The first step is finding a border between positive and negative values. Any border will do. A border between grid cells is a good start.

The second step is fitting the border into the actual line segregating the negative and positive areas of an SDF by shifting each vertex along the inverted gradient by the distance that is the SDF's value in the vertex.

The third step is turning each line segment into a parametric cubic by composing and solving two systems of equations containing the segment's vertices and the SDF partial derivatives.

The algorithm shows how different concepts—grids, line segments, partial derivatives, and parametric polynomials—work together.

11.5 Exercises

Exercise 11.1 What would you presume to be the main way to represent geometric data in the following software?

- An application that helps you design chairs
- A game about a chairmaking factory
- A program that 3D-prints miniature models of chairs

Exercise 11.2 I mentioned previously that an offset is a difficult operation for boundary representations. But what's so difficult about it? If we scale a sphere, we get the offset of a sphere. Why can't we scale everything instead of doing offsets? Please answer this question with a counterexample.

Exercise 11.3 Imagine that you want to run the marching-cubes algorithm on a 2D circle. If you put the circle in a small grid—say, 2×2—you get a compact but inaccurate representation of a circle, namely a square made of four segments. Can you estimate how many segments you get when you put the same circle in a grid of $N \times N$ elements?

Exercise 11.4 Can you make a similar estimation for a 3D sphere?

Exercise 11.5 The smooth contouring algorithm we discussed preserves only C^1 continuity. The contour is smooth, but its curvature is discontinuous. Can you patch the algorithm to preserve C^2 continuity?

11.6 Solutions to exercises

Exercise 11.1 Normally, CAD applications work with boundary representation based on smooth surfaces: cylinders, cones, and of course NURBS. Production loves precise surfaces, so everything designed to be produced should be represented with as little error as possible.

A game would keep its models in triangles, which are computationally cheaper to process and simpler to render. Gamers love the frame-per-second rate, so every compromise that optimizes the engine performance is welcome.

The 3D printing application is trickier. Yes, the whole industry traditionally relies on triangles; even the name of one of the most common triangle-based file formats, STL, comes from stereolithography. But currently, 3D printing is used more often in the domains previously dominated by traditional manufacturing, usually along with milling machines. Manufacturing loves precise surfaces, but few printing machines print in smooth curves. So modern 3D printing applications usually keep data in some sort of a hybrid model that accommodates both triangles and smooth surfaces.

Exercise 11.2 One counterexample could be a pair of spheres with a small gap between them. When you do an offset larger than half of the gap distance, the gap closes. Two models merge into one. Scaling results in a pair of larger spheres with the gap between them also enlarged.

Exercise 11.3 Estimation can be approached differently. One way would be writing a program that traverses the grid and counts the number of cells in which at least one vertex has the SDF sign different from the others. This approach is a good programmer's solution. Kudos if you choose to go this way.

The other way would be to try to pin the estimate by looking at numbers. For $N = 2$, the number of segments is four, so the estimate $2N$ is adequate. For all the $N > 2$, the number of segments is higher than $2N$. It doesn't reach $4N$, though, because $4N$ is the number of segments for cell-edge bordering, and every diagonal segment in the marching-squares algorithm makes the representation more economical. So the magic number we're looking for is between 2 and 4. Last chance to guess it!

The answer is πN. As the grid cell size shrinks, the amount of line segments gets closer to the circle's circumference, expressed in grid cells instead of metric units.

Exercise 11.4 The answer is πN^2, the same as the surface area of a sphere.

Exercise 11.5 In our algorithm, C^1 continuity comes from the fact that we compute the same tangent vector for the same point, whether it's a finishing vertex of a segment or a starting vertex of its neighbor. The tangent vector comes from measuring the SDF partial derivatives. The vector—rather, the partial derivatives themselves—goes into a system of linear equations that gives us a cubic polynomial.

To patch that algorithm to preserve C^2 continuity, you need (as with usual polynomial interpolation) to extend the polynomial to the fifth degree by adding two more coefficients and adding two more equations for every set of equations. For the *x*-axis, they would be

$$20a_5t_0^3 + 12a_4t_0^2 + 6a_3t_0 + 2a_2 = dx_0^{(2)}$$

$$20a_5t_1^3 + 12a_4t_1^2 + 6a_3t_1 + 2a_2 = dx_1^{(2)}$$

where $d_{x0}^{(2)}$ and $d_{x1}^{(2)}$ are the second-degree partial derivatives, so derivatives of the derivatives in the starting and finishing vertices of the segment.

Summary

- Geometrical data in CAD and CAE is usually represented as smooth curves and surfaces that bound the modeled data—hence, the *boundary representation* name.
- This data model typically includes the history of all the operations performed.
- Data in games, computer-generated images, and most 3D-printing applications are usually represented as triangle meshes with additional structures attached: textures, labels, normal maps, and so on.
- Triangle mesh is usually an approximation of a smooth surface, so keep in mind that there's a tradeoff between the accuracy of representation and its size.
- The larger the model is, the more computational power it takes to operate on its representation. The chance of failure of any but the most trivial operations also rises as the model grows.
- Marching cubes and dual contouring are the algorithms that convert a signed distance function to a triangle mesh in 3D or to a contour made of line segments in 2D. Despite their suggestive names, these algorithms are generalized to work in both 2D and 3D. Marching cubes in 2D are also called marching squares. Double contouring in 3D remains double contouring.
- The smooth-contouring algorithm combines the grid border, gradient-based fitting, and polynomial interpolation with differential properties.

12 Modeling bodies with images and voxels

This chapter covers

- Understanding 3D images, image masks, and voxels
- Learning the pros and cons of representing 3D models as images
- Using the most common voxel operations, such as erosion and dilation
- Combining voxel operations to create more complex operations
- Making smooth contouring work on images, resulting in an image vectorization algorithm

A 3D image is an image, pretty much like any bitmap but with another dimension. As you can imagine a 2D image as an array of strings in which every string is an array of colored elements, you can imagine a 3D image as an array of 2D images (figure 12.1). 2D images are made of colored squares called *pixels,* and 3D images are made out of colored cubes called *voxels.* The term *voxel* means a pixel that has a volume: volume pixel, vo-xel, voxel.

The term *image stack* is often used interchangeably with *3D image.* The term has a slightly broader meaning, though. Images in a stack may not necessarily align

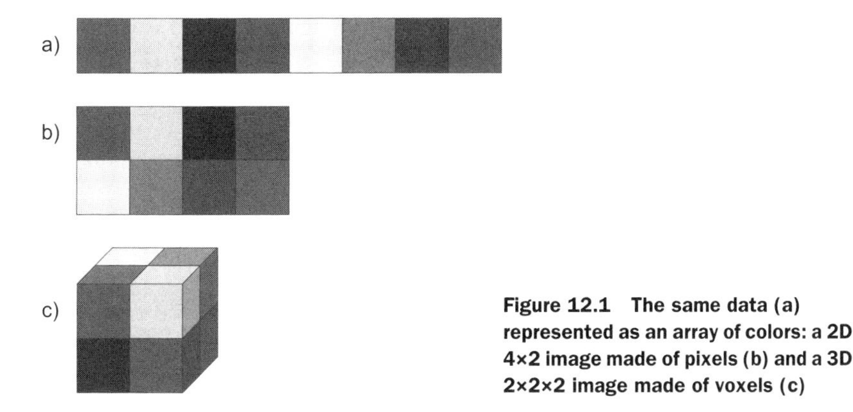

Figure 12.1 The same data (a) represented as an array of colors: a 2D 4×2 image made of pixels (b) and a 3D 2×2×2 image made of voxels (c)

with one another and may even have different sizes. A 3D image is like a perfect image stack—a special case.

In practice, 3D images rarely have colors. 2D images usually come from photography, so they capture visible light, but 3D images usually come from tomography, so they operate in other spectra, such as x-rays. A typical 3D image comes as a stack of 2D grayscale images in which the degree of grayness encodes some particular value. We'll look into how computed tomography works in the next section.

A gray-valued 3D image isn't a convenient model to operate on. How would you do a Boolean subtraction or an offset of an image? To make such operations possible, we usually extract interesting parts from an image as *image masks*: equal-size 3D images in which instead of a gray value, we keep a single bit specifying whether this particular fragment of space belongs to the interesting part. The act of detecting and masking these interesting parts is called *segmentation.* Image masks are sometimes also called *voxel models.*

Sometimes, masks are extended to keep some other piece of information for every fragment. As an example, an image mask might contain a distance value to the nearest region of interest. We'll look into this topic in section 12.3, which discusses how some of the image operations are implemented.

As with SDFs, some operations work on voxels much faster and much more reliably than they do on boundary representation. In the end, however, if we want to turn our model into something from the real world, we need smooth surfaces for traditional manufacturing or triangle meshes for 3D printing. So we'll look into a conversion technique in section 12.5.

12.1 How does computed tomography work?

Computed tomography (CT for short) is a technique that allows doctors to peek inside your body without making a single cut. Simply put, a tomograph scans your body and captures a lot of 2D x-ray images; then a computer turns all those images into a single 3D image, in which every voxel contains a value in special units estimating how much of the x-ray radiation is being absorbed by tissue in that voxel.

NOTE In CT, how much radiation a voxel absorbs is measured in *Hounsfield units* (HU), which are units of radiodensity. The air has a radiodensity of –1000 HU, and water has exactly 0 HU. Soft tissues have radiodensity of about 100 HU; bones lie in the range of ~300 to ~2000 HU; and metals can reach radiodensity as high as 30,000 HU.

When you lie in a CT scanner, an x-ray emitter moves around you and emits radiation. The detectors in the machine detect the radiation that the emitter emits, but when it comes through your body, some of the radiation gets absorbed, so the detectors see less energy than was emitted. The difference is your body's total radiodensity on the x-ray path. The radiodensity is different in bones, muscles, and airways, so each detector sees slightly different energy.

With one scan—one x-ray "photograph"—it's impossible to say exactly where the bones and soft tissues are. But if you have several scans, or a few thousand, you can use them to compute the full 3D picture by essentially solving a series of linear systems.

Let's see how this process works on a simple example. We'll see how a single minuscule 2D slice consisting of four pixels can be computed from a pair of two-pixel 1D x-rays.

Each pixel in a 2D slice reflects a certain radiodensity value in a specific position of a scanned body (figure 12.2). Let's name these values a_{11}, a_{12}, a_{21}, and a_{22}. A scanner wouldn't give us these values; we'd have to compute them ourselves. But a scanner gives us a pair of 1D x-ray images taken from different angles, reflecting the emitter's energy minus energy lost in the scanned body. Let's name the values from the first 1D scan b_{11}, b_{12} and those from the second one b_{21}, b_{22}.

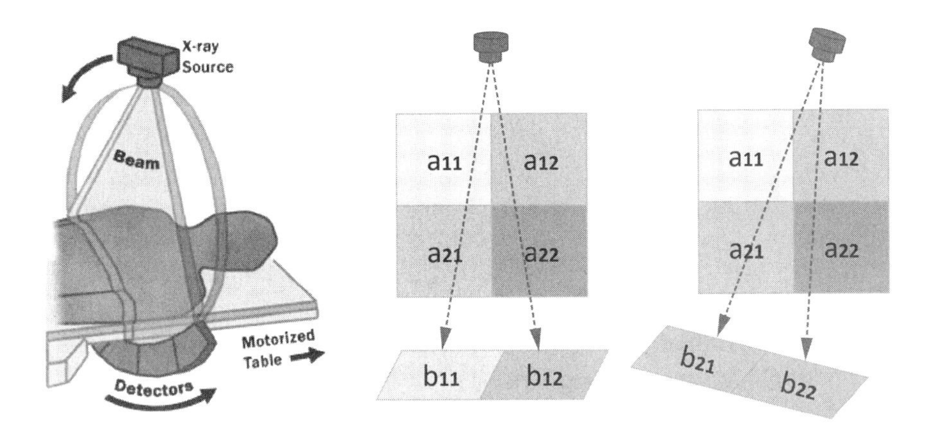

Figure 12.2 A CT scan scheme (FDA, public domain, via Wikimedia Commons), and our simplified example. We take two 1D x-ray pictures at a different angle (first b_{11}, b_{12} and then b_{21}, b_{22}) to compute a single 2D scan ([[a_{11}, a_{12}], [a_{21}, a_{22}]]).

Computing how far each ray goes through each pixel is a simple ray-rectangle intersection problem. Each number is the distance between the point where a ray enters a corresponding pixel and the point where it leaves the pixel. Let's say that we already have these numbers. The corresponding distance for the first ray going through pixel a_{11} will be 1.04.

Now let's say the initial x-ray energy is x. As the ray travels through the pixels, it loses this energy proportionally to the length of travel multiplied by the pixel's radiodensity. The rest of the energy shows itself as the degree of whiteness in the two-pixel images.

The value b_{11}, for example, is the energy x minus the loss from pixel a_{11} multiplied by the distance the ray travels through pixel a_{11}, and minus the loss from pixel a_{21} multiplied by the distance the ray travels through pixel a_{21}:

$$b_{11} = x - a_{11} \times 1.04 - a_{12} \times 1.04$$

Given that we already have the b-values and want to compute the a-values, all we have to do now is form the system that reflects how the rays travel and solve it:

$$x - 1.04a_{11} - 1.04\ a_{12} = b_{11}$$

$$x - 1.04a_{21} - 1.04a_{22} = b_{12}$$

$$x - 0.27a_{12} - 0.78a_{11} - 1.06a_{21} = b_{21}$$

$$x - 1.01a_{12} - 1.01a_{22} = b_{22}$$

As a reminder, we know x and the b-values; we want only a-values. If you remember chapter 3 well, you recall that the system we have is simply a system of linear equations, and a well-defined one, too! You can solve a small system like this with any method you like.

In practice, of course, a 2×2 image isn't too helpful; usually, we have to deal with images more like 512×512 or 1024×1024. The latter comes from a system made of more than 1 million equations. To be well-defined, it also requires not 2 but 1024 x-rays, given that each x-ray has 1024 values of data in it. And because we're dealing with real-world measurements, which usually come with some error, we normally take even more data and then use some approximate solver that finds not an exact solution, but one that's good enough. Given that these input x-rays should come from different angles, and that it takes time to reposition the emitters and detectors, this fact explains why you have to keep still for a while when being scanned.

There are other methods for looking inside things, of course. Ultrasound or magnetic resonance imagery (MRI) scanners are used along with x-ray-based tomographs. But mathematically, all these methods work alike. MRI doesn't involve radioactive *x*-rays, so the procedure may take longer without any affect on the patient, resulting in more data and, consequently, better precision. That's why MRI shows more details than traditional CT. It's not as much about the physics of the process as it is about the mathematics of data processing.

All in all, in the end, a 3D image made of gray values signifies some physical property. You can show these images to surgeons to help them plan operations, for example, so there's already a great deal of use for the technology. Surgeons, however, prefer to learn things with their hands and eyes, not their eyes alone. Wouldn't it be great to extract an organ from an image so we could 3D-print it and give it to a surgeon for examination? Let's see how.

NOTE My company prints anatomic models for surgeons. The first one I saw was a white hip joint covered in brightly colored marks and small intangible inscriptions. I was impressed; I didn't think we could print in multicolor and apply textual labels with this precision. So I asked our analyst. She replied, "Well, technically we can print in multicolor. But these marks were simply made with Sharpies." It wasn't immediately apparent to me, but to a surgeon, the mere ability to put marks on a model is already a huge help. Investing time in planning surgery on a realistic model makes real surgery go much smoother.

12.2 Segmentation by a threshold

Suppose that we have a 3D image of something from the real world. We want to model this object as a surface because, well, most 3D-printing software relies on boundary representation. At the very least, we need to find the object's boundary.

That task isn't easy, as it turns out. A human eye, or rather a human brain, does it instantly. You can easily spot bones in figure 12.3, for example. But for a computer, this task is more difficult. Sophisticated computer vision algorithms can tell bone from other tissue, and neural networks can be trained to see skulls everywhere, but in practice, none of these systems has 100% accuracy. Often, specially trained engineers polish the surface model after a computer makes an initial attempt to build one from imagery.

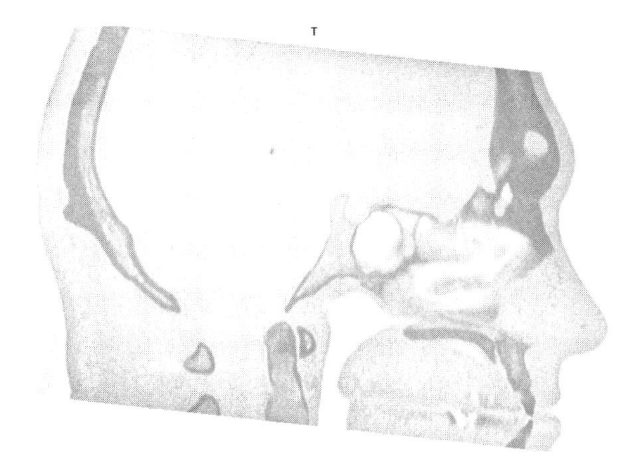

Figure 12.3 A CT scan of a head. The skull bones are instantly recognizable to the human eye, but not necessarily to a computer.

The problem of finding an object on an image, or splitting the image into several objects, is called *segmentation.* We won't look into sophisticated segmentation methods, which deserve books of their own, but we'll try the simplest one, which is still surprisingly effective: segmentation by threshold.

You see, CT measures how good a material in a voxel is at absorbing x-rays. CT isn't like photography, which captures reflected light; the same object may be dark or light depending on the lighting. Normally, high-density tissues absorb x-rays better than low-density ones do, which immediately shows up in the image. So you could say that everything between 300 and 3000 HU is supposed to be bone tissue and everything outside this range is probably fat, muscles, and air. Applying the simple rule "the voxel should have a radiodensity of 300 to 3000 HU" filters out a lot of irrelevant tissues, leaving us mostly with bones (figure 12.4).

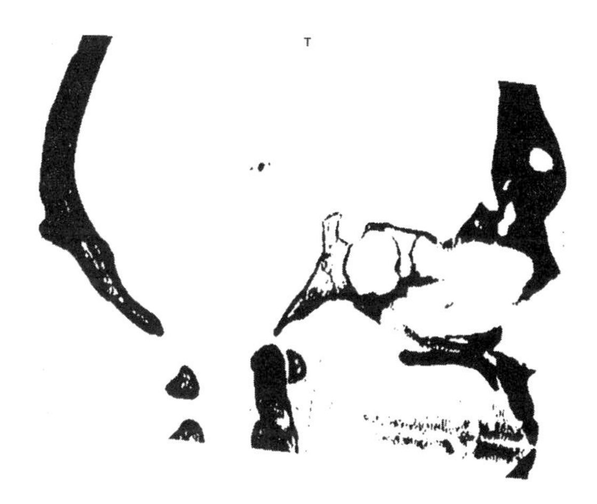

Figure 12.4 Segmentation of a skull. Now only bones remain in the picture; the rest is mostly filtered out.

This approach is called *segmentation by threshold.* The range of HU in which we're expecting to see bone tissue is our threshold.

This approach has a few problems. First, bone tissue isn't uniformly dense everywhere. Bone marrow is way less dense than cortical bone. Also, the part of the bone that comes close to a joint is usually porous, and CT isn't accurate enough to reflect porosity other than with lower density. Further, people of different ages and occupations have slightly different bone densities as well.

All that means that if you keep your threshold range narrow, you'll leave a lot of bone tissue unsegmented. And if you get generous with the threshold, you'll start segmenting not soft tissues but noise (figure 12.5).

Luckily, there are techniques that allow us to get rid of the noise in an image mask. We'll reinvent one of them in the following section.

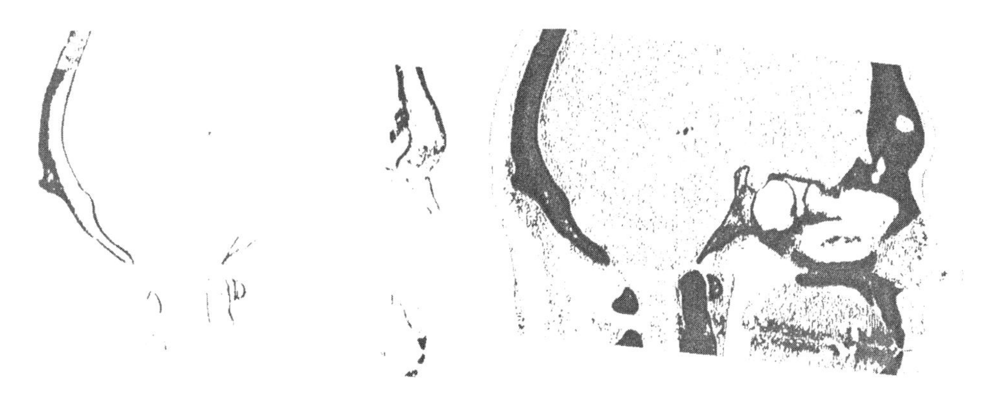

Figure 12.5 Bad segmentations: a thin threshold (left) and a thick threshold (right)

12.3 Typical operations on 3D images: Dilation, erosion, cavity fill, and Boolean

The staple operations on image masks or voxels are

- *Dilation*—Growing your image by adding voxels to its border (figure 12.6)

Figure 12.6 Dilation adds "material" to your voxel model.

- *Erosion*—Doing the opposite, taking voxels away from your voxel model's border (figure 12.7)
- *Cavity fill*—Filling the fully enclosed empty areas within your model (figure 12.8)
- *Boolean operations*—Operations like intersection, unification, and subtraction

Using these operations, you can successfully enhance your model: remove the noise, smooth the border, and fill the unwanted pores.

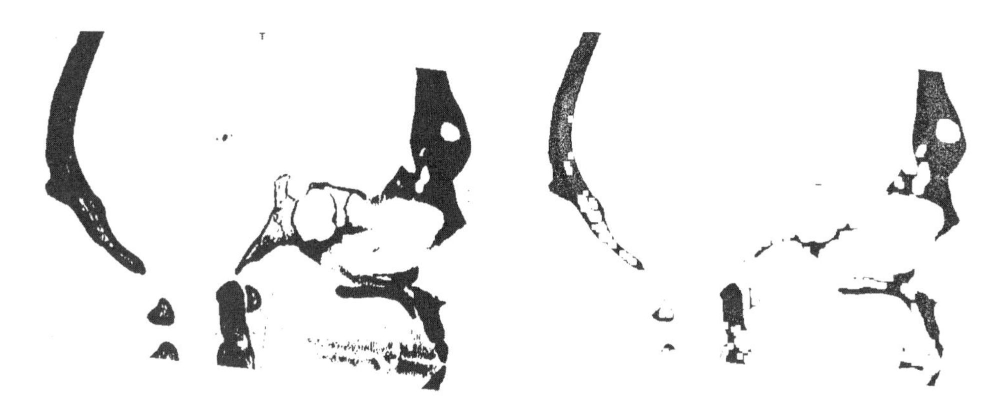

Figure 12.7 Erosion removes "material" from your voxel model.

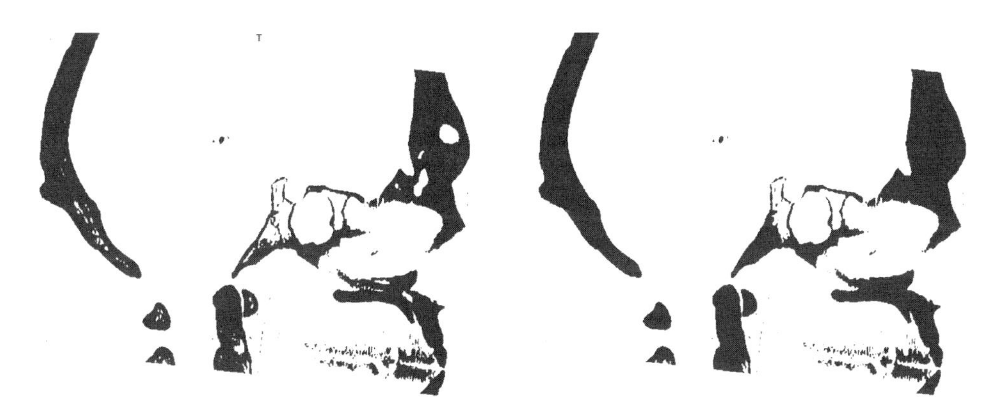

Figure 12.8 Cavity fill targets fully enclosed areas. Note that the picture you see is a 2D slice of a 3D image. A cavity may appear closed in 2D but still leak into the third dimension.

12.3.1 Dilation

Conceptually, a dilation by one voxel is a simple operation. Go through all the voxels in an image, and check their neighbors. If a voxel is empty, and one of its neighbors isn't, mark the voxel as ready to be filled, and when you're done checking the whole image, fill the marked voxels. That's it (figure 12.9).

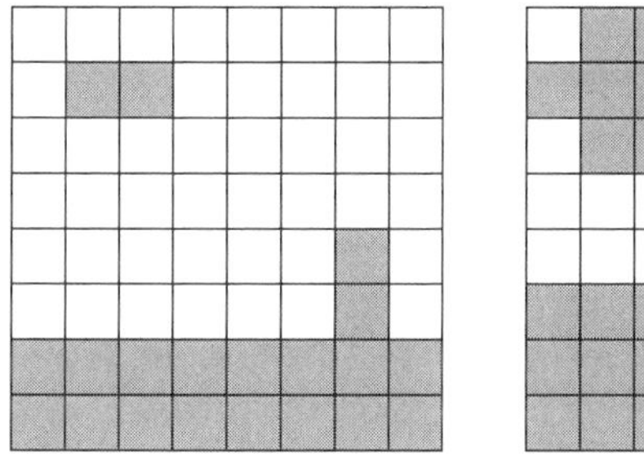

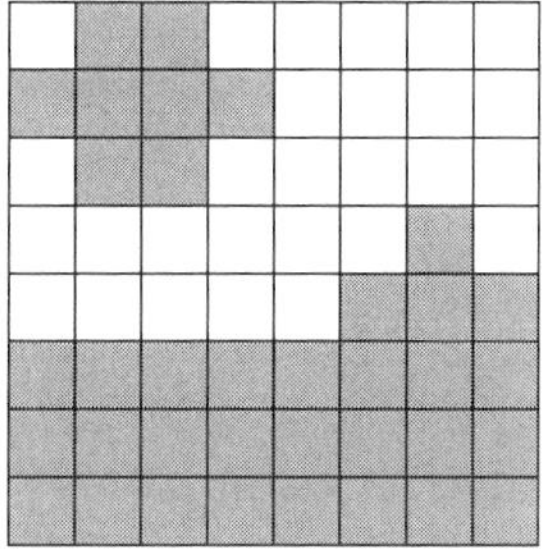

Figure 12.9 Dilation by one voxel

In practice, however, we rarely want to dilate an image by a single voxel. Well, surely, if you want to dilate a voxel model by 5 voxels, you can repeat the 1-voxel dilation 5 times. But performancewise, this approach is wasteful.

What we can do instead is compute the distance to the closest filled voxel for every empty voxel. Then we fill all the empty voxels that have a filled voxel within the dilation depth. To do that, we travel the image twice, computing the approximate distance values by incrementing the distance stored in the neighboring voxels:

- First from top to bottom, from left to right, and (in the 3D case) from front to back
- Then in reverse order: from bottom to top, from right to left, and (again, we omit this on 2D figures) from back to front

In the first run, we use a simple rule: if there is a filled voxel on the left, at the top, at the front, or at the bottom of an empty voxel, we consider the distance to the nearest filled voxel to be 1.

NOTE We can make our algorithm a bit more accurate if we also consider diagonal voxels with distance √2. This improvement will affect simplicity, though, so for now, let's omit it.

If not, we gather all the previously defined distances from the neighbors around the voxel, find the lowest value, increment it by the distance to the neighbor itself, and write it into the voxel (figure 12.10).

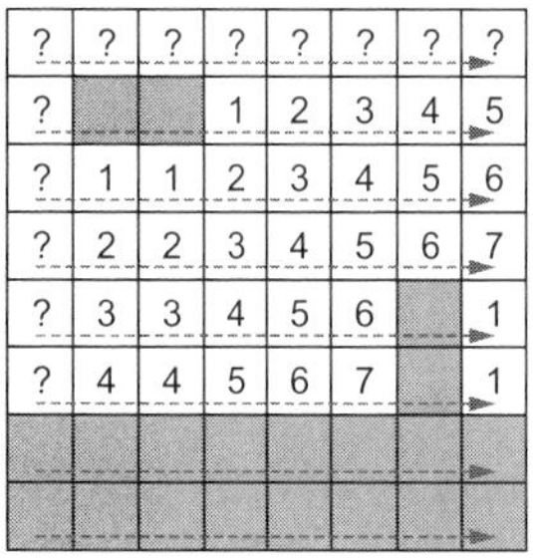

?	?	?	?	?	?	?	?
?			1	2	3	4	5
?	1	1	2	3	4	5	6
?	2	2	3	4	5	6	7
?	3	3	4	5	6		1
?	4	4	5	6	7		1

Figure 12.10 The first step goes like this: travel from left to right, from top to bottom, put 1 if there is a filled neighbor at the top or on the left, and if not, put 1 plus the minimum of the top and left neighbors' stored distances.

While traveling back, we essentially do the same thing, but now we're probing for all the neighboring voxels. Otherwise, this second step remains largely the same as the first one. We still get the minimal value and increment it by the distance to its source, but now we have to look all ways because we already have voxels filled from the first run (figure 12.11).

Now, when we have a distance map, all we have to do to dilate an image is fill the voxels that have distance values lower than the dilation depth (figure 12.12).

SEE ALSO Hypermedia Image Processing Reference is a great free online resource if you want to learn more about image processing. You can find its page on dilation at http://mng.bz/Y6JQ.

2	1	1	2	3	4	4	5
1			1	2	3	3	4
2	1	1	2	3	3	2	3
3	2	2	3	3	2	1	2
2	2	2	2	2	1		1
1	1	1	1	1	1		1

Figure 12.11 The second step: travel from right to left, from bottom to top, put 1 when there is a filled neighbor, and put 1 plus the minimum of all the stored neighboring voxel distances when there are no filled neighbors.

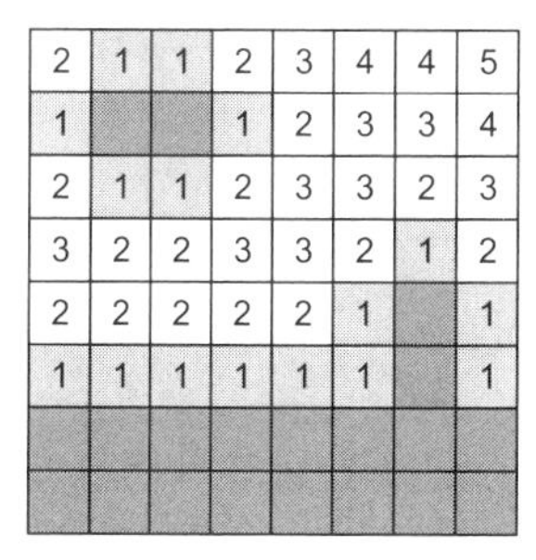

2	1	1	2	3	4	4	5
1			1	2	3	3	4
2	1	1	2	3	3	2	3
3	2	2	3	3	2	1	2
2	2	2	2	2	1		1
1	1	1	1	1	1		1

Figure 12.12 Dilating the model by 1 pixel by adding all the pixels with a distance of 1 (left); dilating the same model by 2 pixels at once by adding all the pixels with a distance ≤2 (right)

12.3.2 Erosion

Erosion is an operation dual to dilation. You can implement a similar algorithm with inverted rules like this: if a voxel is filled, write a minimal distance to an empty voxel in it. Here's a simpler approach: invert a voxel image, do the dilation, and invert the image back (figure 12.13).

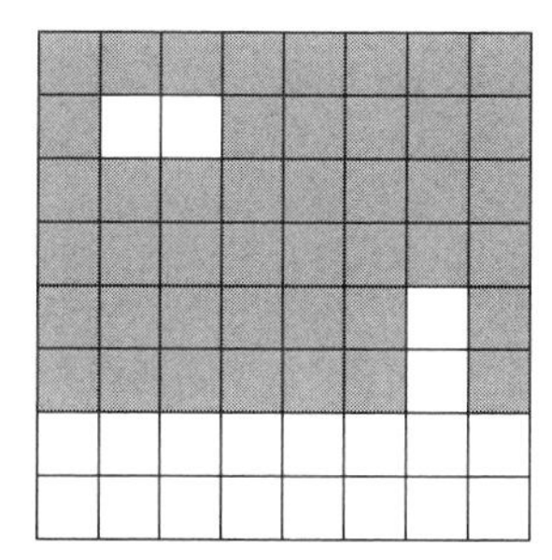

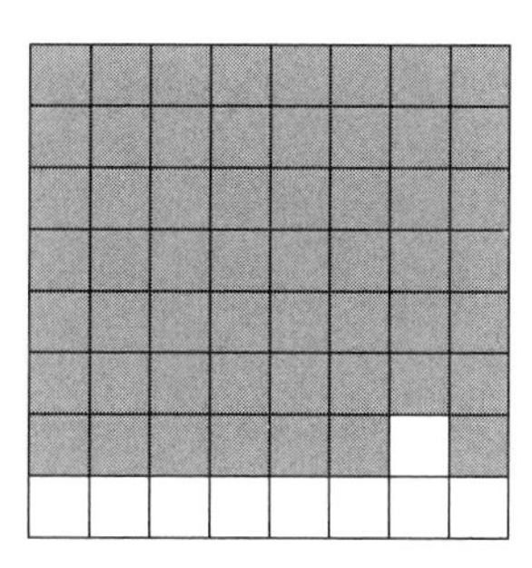

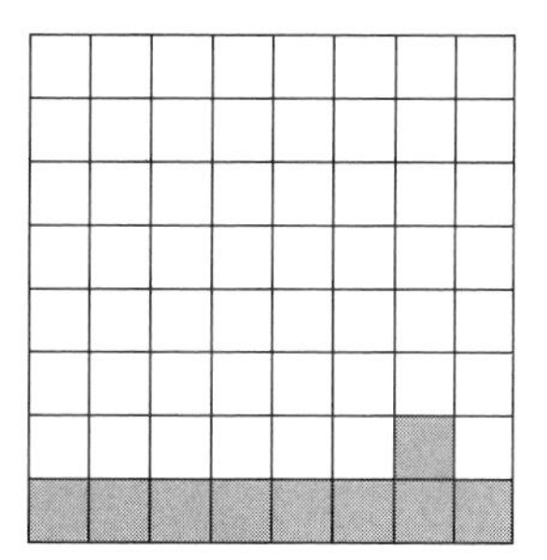

Figure 12.13 Erosion is an inverted dilation of an inverted image. This double-inversion trick helps us reuse a bunch of code.

SEE ALSO Hypermedia Image Processing Reference has an erosion page, too, at http://mng.bz/eJvZ.

12.3.3 Practical example: Denoising

Let's get back to our practical problem. We've already seen that on a real image, there may be small pieces of nondense tissue that still somehow fall within the threshold by their HU value. They may come from scanning inaccuracies or even from errors in computing the image. We call these small artifacts *noise*.

CT isn't the most accurate technology; a little noise in the images is common. We want to get rid of this noise, and now we have all the tools to do so. The trick is to erode the voxel model first and then dilate it back (figure 12.14).

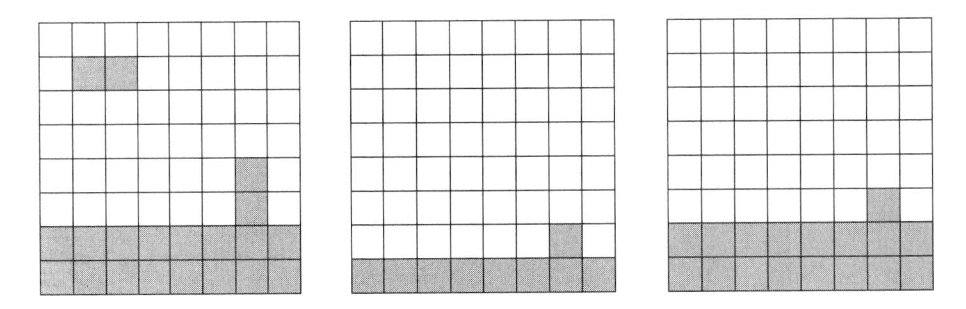

Figure 12.14 Denoising is a small erosion followed by proportionate dilation.

Denoising "eats" all the small bits and smooths the surface, reducing the noise on the border, too. It retains the features that are already smooth enough.

Let's see how denoising looks in the code. Once again, we'll use Python, but you're free to reproduce this algorithm in the language of your choice. We'll also look at the 2D example because it's not conceptually different from the 3D one, only shorter.

First, let's agree on the data. I propose marking filled voxels as 1 and empty voxels as 0. This way, we represent our model as a list of lists of ones and zeros. Here's a 2D model that has one piece of noise (top left), one thin feature (bottom right), and one thick feature that the noise filter should ignore (bottom left):

```
image = [[0, 0, 0, 0, 0, 0, 0, 0],
         [0, 1, 1, 0, 0, 0, 0, 0],
         [0, 0, 0, 0, 0, 0, 0, 0],
         [0, 0, 0, 0, 0, 0, 0, 0],
         [0, 0, 0, 0, 0, 0, 1, 0],
         [1, 1, 0, 0, 0, 0, 1, 0],
         [1, 1, 1, 1, 1, 1, 1, 1],
         [1, 1, 1, 1, 1, 1, 1, 1]]
```

Next, we need a dilation function. We'll implement one as we discussed in section 12.3.1, but we'll stick to a simpler, less accurate implementation, the one without diagonal elements. The algorithm consists of four parts:

1. Initialize the image with all the distances to compute.
2. Walk the source image from top to bottom and from left to right, computing the nearest distance in every pixel.
3. Do the reverse walk from bottom to top, and right to left, finishing the computation.
4. Mark all the voxels that lie within some given depth of dilation, according to our computed distances, as filled:

```
def dilate(img, depth):
    distances = [[inf for _ in row] for row in img]    <-- Initializes the distance map with infinite values to denote still-undefined distances

    for i in range(len(img)):    <-- Performs the first travel: left to right, top to bottom
        for j in range(len(img[0])):
            if img[i][j] == 1:
                distances[i][j] = 0
                continue
            if j > 0 and img[i][j-1] == 1:
                distances[i][j] = 1
            if i > 0 and img[i-1][j] == 1:
                distances[i][j] = 1
            candidates = [distances[i][j]]
            if j > 0:
                candidates += [distances[i][j-1] + 1]
            if i > 0:
                candidates += [distances[i-1][j] + 1]
            distances[i][j] = min(candidates)

    for i in reversed(range(len(img))):    <-- Performs the reverse travel: right to left, bottom to top
        for j in reversed(range(len(img[0]))):
            if img[i][j] == 1:
                distances[i][j] = 0
                continue
            if j < len(img[0]) - 1 and img[i][j+1] == 1:
                distances[i][j] = 1
            if i < len(img) - 1 and img[i+1][j] == 1:
                distances[i][j] = 1
            candidates = [distances[i][j]]
            if j > 0:
                candidates += [distances[i][j-1] + 1]
            if i > 0:
                candidates += [distances[i-1][j] + 1]
            if j < len(img[0]) - 1:
                candidates += [distances[i][j+1] + 1]
            if i < len(img) - 1:
                candidates += [distances[i+1][j] + 1]
            distances[i][j] = min(candidates)

    for i in range(len(img)):    <-- Makes all the voxels with a distance lower or equal to the depth value part of the model
        for j in range(len(img[0])):
            if distances[i][j] <= depth:
                img[i][j] = 1
    return img
```

We also need erosion. As we discussed, erosion is a dual operation to dilation. We don't need a separate algorithm for it; we can reuse what we got with a simple "inversion" trick:

```
def invert(img):                                  <-- A simple rule: every one
    for i in range(len(img)):                         becomes a zero, and every
        for j in range(len(img[0])):                  zero becomes a one.
            img[i][j] = 1 - img[i][j]
    return img

def erode(img, depth):                            <-- Erosion is an inverted dilation
    return invert(dilate(invert(img), depth))         of an inverted image.
```

Now denoising becomes essentially a one-liner:

```
def denoise(img, noise_size):
    return dilate(erode(img, noise_size), noise_size)
```

This is how it works on our data, given that the noise size is set to 1:

```
[0, 0, 0, 0, 0, 0, 0, 0]
[0, 0, 0, 0, 0, 0, 0, 0]
[0, 0, 0, 0, 0, 0, 0, 0]
[0, 0, 0, 0, 0, 0, 0, 0]
[0, 0, 0, 0, 0, 0, 0, 0]
[1, 1, 0, 0, 0, 0, 1, 0]
[1, 1, 1, 1, 1, 1, 1, 1]
[1, 1, 1, 1, 1, 1, 1, 1]
```

Denoising by 1 pixel removes the noise shell in the top-left corner and shortens the thin feature in the bottom-right corner. The thick feature at bottom left remains intact. The full listing is available at ch_12/denoise.py in the book's source code.

The algorithm proposed isn't terribly precise. Without the diagonal element, it considers the diagonal distance between voxels to be 2, not $\sqrt{2}$. It's not heavily optimized, either. But it still shows how useful simple algorithms can be when you have the right data model. And for voxels, even simpler useful algorithms are available.

12.3.4 Boolean operations on voxel models

Before we move further, we should discuss Boolean operations quickly. On voxel models or image masks, which are conceptually the same, these operations are as simple as those on individual true/false Boolean values. The main Boolean operations are

- Intersection (figure 12.15)
- Unification (figure 12.16)
- Subtraction (figure 12.17)

Boolean operations on voxels are simple indeed—so simple that unlike with Boolean operations on boundary representations, their implementation in the code is usually faultproof. The same goes for dilation and erosion. This simplicity and consequent

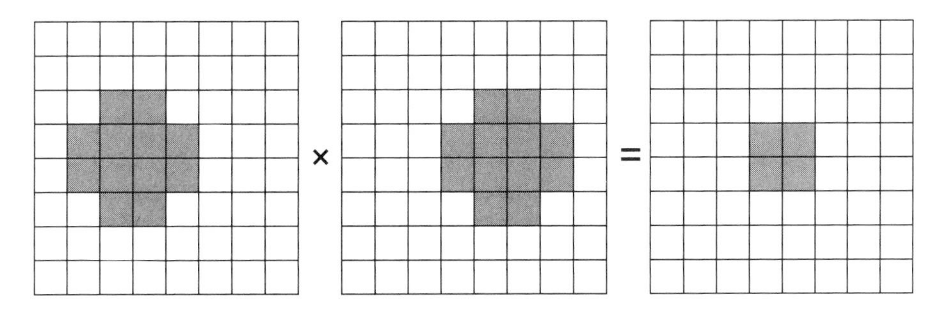

Figure 12.15 A Boolean intersection. Only the pixels that are filled in both models remain.

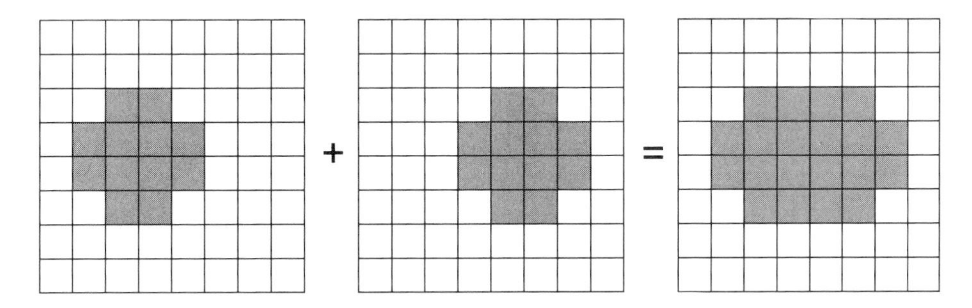

Figure 12.16 A Boolean unification. The result is all the pixels that are filled either in the first or second model.

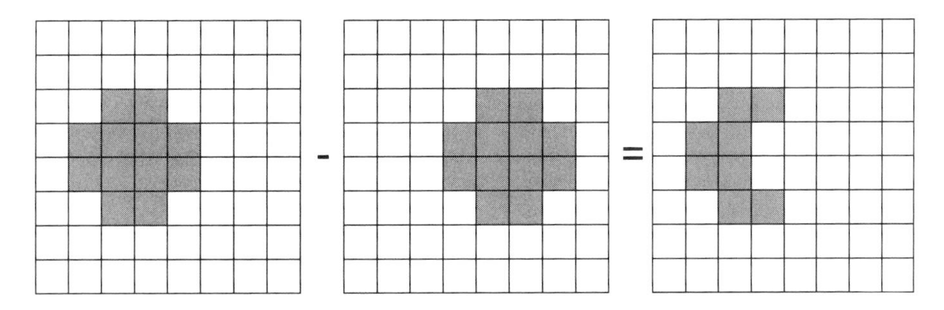

Figure 12.17 A Boolean subtraction. The result is the filled pixels of the first model that aren't covered by the filled pixels of the second model.

robustness nudge us toward building multistep workflows on voxel models. Let's look at a few such workflows.

12.3.5 *A few other uses of dilation and erosion*

Denoising isn't the only thing that dilation and erosion are good for. If you reverse the operation, doing dilation first and erosion later, you get an operation that patches all the thin cracks in your voxel model.

Knowing how Boolean works, you can also use the denoising operations as part of a thin-wall detector. In 3D printing, thin walls don't print well, so it's best to detect them early and report them to the users of the printing software before you start the printing process. To do that, you can run the denoising sequence on your voxel model first, eroding and dilating by the thin-wall threshold value, and then subtract the result from your source model. The rest will be all the features thinner than the predefined thickness threshold.

You can also detect thin cracks or check whether different models are placed too close on a platform by using the reverse operation: dilate, then erode, and do the same Boolean subtraction.

These operations are simple to perform on voxel models, but imagine doing denoising on a triangle mesh. This operation is still possible. Some noise shell detectors work with triangle meshes rather efficiently, but they're complicated and error-prone.

As with SDFs, sometimes it pays to switch to a new data representation, do some operations there, and switch back. The more data representations you have in your framework or development kit, the more freedom you have to choose the most efficient workflow possible.

In the next section, we'll learn how to turn a gray-value image into a set of smooth contours—effectively, how to convert data from an image to a boundary representation.

12.3.6 Section 12.3 summary

Dilation and erosion are basic operations on voxel models. They're analogous to the offset operation in SDF or boundary representation. Dilation is outer offset (the model gains volume), and erosion is inner offset (the model loses volume).

Boolean operations on voxels are fast and reliable as soon as all the operands share a voxel grid. This speed and reliability come in handy when we combine several operations in a single workflow to solve practical problems.

12.4 Practical example: Image vectorization

Since its inception, computer graphics has developed in two parallel ways. One approach, raster graphics, is about representing visual information as grids of colored squares—in other words, 2D images. The other approach, vector graphics, is about representing shapes as primitives—points, lines, and of course curves, or in other words, boundary representation (figure 12.18).

In the early days, when computer memory was expensive, vector graphics was prevalent. Games such as Space Wars (1977) or Asteroids (1979) had vector graphics. But when enough memory became available, the benefits of images started to kick in. In the modern world, almost all displays, except maybe some vintage oscilloscopes, are raster-based. Almost all computer graphics, even when displaying smooth mathematical objects, end up putting colors in square grid cells.

But we might still prefer a vector representation every now and then, if not for storing, for processing, such as when we want to make a satellite image into a map. Google

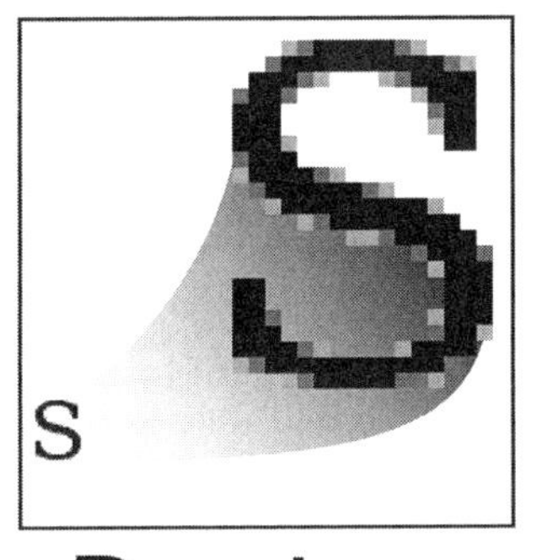

Figure 12.18 The difference between raster and vector graphics (Yug, modifications by 3247 [CC BY-SA 2.5])

Maps is a good example because when you zoom in on a satellite image, you start seeing pixels. When you zoom into a ready-made map, you don't. All the features scale up nicely; a straight road remains a straight road.

Another good example is blueprint vectorizing. Historically, a lot of old blueprints have been scanned. We can still use them to produce things, but we can't modify them efficiently with modern computer-aided design (CAD) software. To do so, we need to vectorize the old scans, turning them into a set of mathematically recognizable primitives such as lines, circles, and splines. The process is a bit like a reverse-engineering problem from chapter 11, only in 2D.

But how does vectorization work? How do you turn an image into a set of curves? Let's find out.

12.4.1 Input image

First, let's discuss our input image. The most common image data representation in regular use is 8-bit RGB, which means that an image is a grid of colored pixels in which every color is represented by 3 bytes. Each byte indicates a separate color channel. *R* stands for red, *G* for green, and *B* for blue. Every 8-bit channel may take one of $2^8 = 256$ different values. By mixing them together in different proportions, you get $256^3 = 16,777,216$ different colors—quite enough for pictures of cats and animated explosions, which covers pretty much all of the internet and gaming.

Medical images usually don't have colors. If a medical image is colored, the colors should have been added by the scanning software after the image was obtained. On the other hand, medical imagery needs more precision in terms of channel depth, because 256 different values aren't nearly enough to accurately depict all tissues and pathologies. Even the early CT scanners produced 12-bit images while memory was still expensive. Today, the common channel depth for CT and MRI is 16-bit, so the data comes as a grid of voxels, each hosting a single 16-bit value.

Images from satellites can have as many channels as they need. Satellites work in many wavelengths, so you can have several channels for visible light, plus separate channels for x-ray, ultraviolet, or infrared light. The depth is also limited to the capturing device's resolution and isn't guided by memory-saving constraints. The reason is economics: it's still rather expensive to put a satellite into orbit. Memory prices and even bandwidth costs dim by comparison.

To be honest, the algorithm I'm about to show you isn't meant for turning satellite imagery into road maps. It doesn't work well for turning blueprints into CAD files, either. But it's good at showing you the basic concepts of vectorization, so let's concentrate on that. Our image will be a simple 8-bit, 1-channel grid, something like figure 12.19.

0	0	0	0	0	0	0	0	0	0	0	0	0	0	0	0
0	77	125	38	0	0	0	0	0	0	0	0	0	0	0	0
0	120	255	254	203	144	96	3	0	0	0	0	0	0	0	0
0	34	253	255	255	255	255	230	154	94	8	0	0	0	0	0
0	0	196	255	255	255	255	255	255	252	241	139	83	6	0	0
0	0	149	255	255	255	255	255	255	255	255	255	250	213	80	0
0	0	98	255	255	255	255	255	255	255	255	255	255	224	58	0
0	0	2	224	255	255	255	255	255	255	255	242	152	4	0	0
0	0	0	145	255	255	255	255	255	255	255	154	1	0	0	0
0	0	0	82	251	255	255	255	255	255	255	253	156	1	0	0
0	0	0	6	237	255	255	255	255	255	255	255	252	146	3	0
0	0	0	0	149	255	255	243	149	252	255	255	255	240	21	0
0	0	0	0	69	249	255	152	1	150	252	255	238	71	0	0
0	0	0	0	0	211	224	4	0	1	137	240	86	0	0	0
0	0	0	0	0	73	57	0	0	0	2	20	0	0	0	0
0	0	0	0	0	0	0	0	0	0	0	0	0	0	0	0

Figure 12.19 An example of an 8-bit, 1-channel depth image. Each cell contains a number from 0 to 255. The lower the number is, the darker the pixel.

White pixels are full 255, the highest value that an 8-bit cell can host. The lower the cell value is, the darker the color the cell represents. Fully dark pixels are 0. Normally, those pixels would be black, but let's keep them lighter so we can see numbers and contours on top of them.

What we want is to calculate a line that, while being nice and smooth, separates dark-gray values from light-gray ones. We need to agree on a threshold number that separates light (high values up to 255) from darkness (low values starting from 0). Our threshold value could be 128, for example.

To reiterate: the input data for our algorithm is an 8-bit, 1-channel image and a light/dark threshold value, which is a number from 0 to 255.

SEE ALSO If you also want to know how images are compressed in practice, you can find a comprehensive interactive explanation of JPEG by Omar Shehata at http://mng.bz/zmAg.

12.4.2 Step 1: Obtain a contour

The first step looks a lot like what we did in chapter 11 when we got a contour out of an SDF by imposing a grid over it. Now we don't have an SDF, but we do have a grid, which comes automatically with the image. And because for the first step, we didn't use all the SDF, only the values in the grid-cell centers, we can pretend that our color values are also the values in the grid-cell centers and use the same logic as before.

Let's put a vertical segment between neighboring cells where the value in the left cell falls on one side of the light/dark threshold and the value in the right cell falls on the other. And let's put a horizontal segment between cells where color values in a top neighbor and in a bottom neighbor are also separated by the same light/dark threshold. In the end, we'll get a contour made of 1-pixel-long segments (figure 12.20).

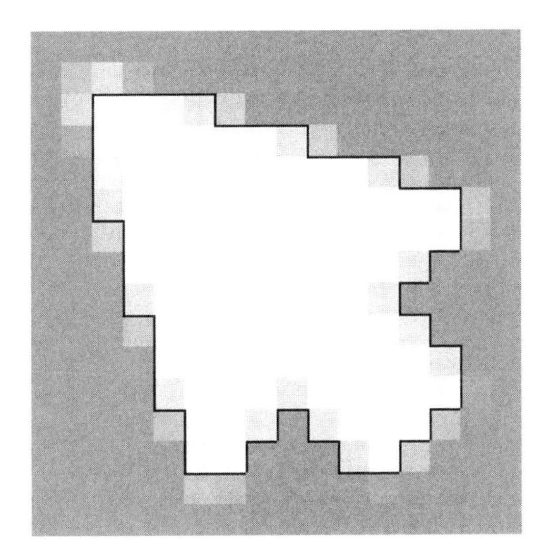

Figure 12.20 An initial contour is a border between pixels with values below the threshold and above.

Technically, this image is already a vectorization. Pragmatically, though, it's useless. As we zoom in, we want to see nice, smooth contours, not gigantic pixel borders. So let's get to the next step.

12.4.3 Step 2: Fit the contour

With smooth contouring from chapter 11, the next step would be fitting the contour by moving each vertex against the SDF's gradient for the distance that's exactly the SDF's value in the vertex. This time, however, we don't have an SDF. An image isn't a continuous function, so it doesn't have a gradient. It's a finite grid of values. If only we could turn a finite set of values into a continuous entity!

Well, hold on a minute. That's a job for an interpolation algorithm. In fact, we solved this exact problem in chapter 8 when we turned a pixel image into a continuous, smooth function of two variables. In section 12.4.5, we'll take a look at the specific implementation again, but for now, let's consider this problem to be solved.

When we have a continuous function, we can turn it into something like SDF. It wouldn't be signed, because all the values come from the [0, 255] range. But if we subtract our light/dark threshold value from the function we got by interpolating the image, we get exactly the signed function we want.

It won't be a true distance function, however! That's why we can't reduce this algorithm to the smooth contouring from chapter 11. We can't move every segment's vertices to the distance equal to the SDF value in them and be done with fitting, because our function doesn't reflect the distance to the nearest 0.

But what if we pretend that it does? What if we take the function's value at a vertex as a suggestion of a distance for the vertex translation and still use the gradient to find the direction? Well, moving vertices like that won't give us the correct contour right away. But their positions will become a little bit more correct each time we move them. Maybe eventually, they'll become correct enough to move on.

How do we know when their positions are good enough, though? Let's introduce another threshold value—the approximation threshold—and use it to compare with the signed function's value in a vertex. If we take 1.4 as this threshold and keep the light/dark threshold as 128, we'll consider the contour to be good enough if every vertex of it lies where the interpolated image has a value in the range (126.6, 129.4)—not exactly 128, but close enough.

Now let's remind ourselves how iterative algorithms (chapter 3) work. You need a starting position, a rule that brings you closer to the goal, and the definition of the goal: the predicate that says the result is good enough. The good news is that we already have all those things!

The boxy contour we got from the first step provides the initial positions for the vertices. The "going against the gradient by the function's value" rule is our iterative step. And "the new absolute function's value in vertex is lower than a given accuracy threshold" condition is our exit condition. It's like the case of the linear solver again: if we don't have a good direct way to get the solution immediately, we'll use an iterative algorithm and wait.

The approximation threshold is a lever that helps us balance the algorithm's accuracy and speed. The lower the threshold is, the more accurate the algorithm will be.

But it takes the algorithm more iterations to get to the result. With a larger threshold, we get fewer iterations and better speed but sloppier results (figure 12.21).

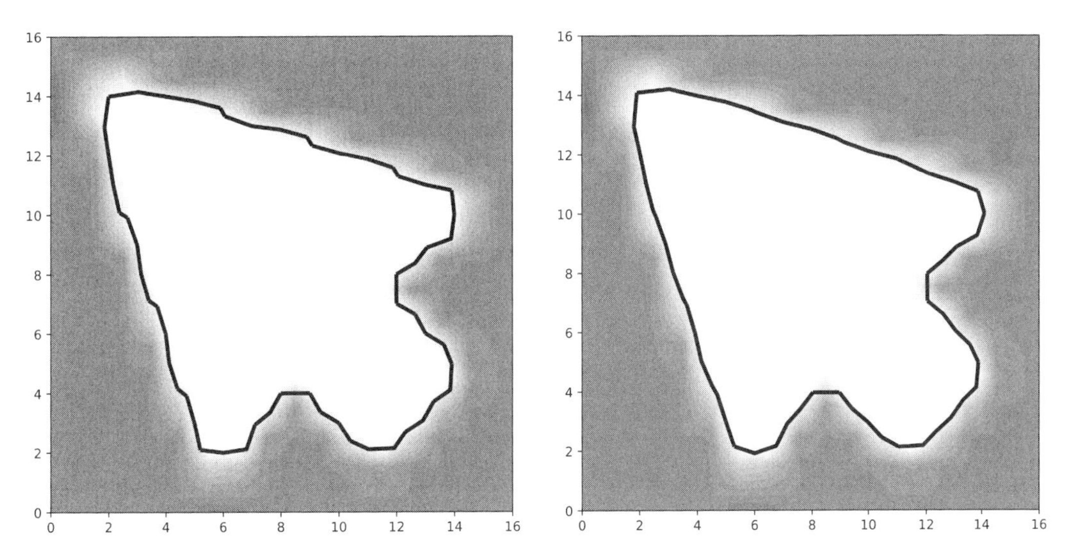

Figure 12.21 A fitted contour with an approximation threshold set to 20 (left) and 2 (right)

So we'll have a contour that lies close enough to the isoline we got by interpolating an image. It's still made of line segments, though. Let's turn it into a smooth contour made of Bezier pieces.

12.4.4 Step 3: Make the contour smooth

This step is almost exactly the same as smooth contouring (chapter 11). We use the gradient of our interpolated image to find possible tangent values, correct the tangent values so that they align with the segments, and then we add them to the "equation soup" along with vertex positions. With the equations we get, we can describe a cubic polynomial piece. Turning all our segments into cubics will make our contour smooth.

As before, a segment starts at (x_0, y_0) and ends at (x_1, y_1). Partial derivatives are dx_0 and dy_0. Partial derivatives in the second point are dx_1 and dy_1. Then the systems that hold our cubic curve's coefficients are

$$a_x t_0^3 + b_x t_0^2 + c_x t_0 + d_x = x_0$$

$$a_x t_1^3 + b_x t_1^2 + c_x t_1 + d_x = x_1$$

$$3a_x t_0^2 + 2b_x t_0 + c_x = dx_0$$

$$3a_x t_1^2 + 2b_x t_1 + c_x = dx_1$$

$$a_y t_0^3 + b_y t_0^2 + c_y t_0 + d_y = y_0$$

$$a_y t_1^3 + b_y t_1^2 + c_y t_1 + d_y = y_1$$

$$3a_y t_0^2 + 2b_y t_0 + c_y = dy_0$$

$$3a_y t_1^2 + 2b_y t_1 + c_y = dy_1$$

As we did in chapter 11, we can simplify the equations by choosing t_0 to be 0 and t_1 to be 1. Now the piece of the curve we're supposed to replace a linear segment with will be parametrized by t lying on [0, 1] interval. In $t = 0$, the curve segment will start in (x_0, y_0), and in $t = 1$, it will finish in (x_1, y_1):

$$d_x = x_0$$

$$a_x + b_x + c_x + d_x = x_1$$

$$c_x = dx_0$$

$$3a_x + 2b_x + c_x = dx_1$$

$$d_y = y_0$$

$$a_y + b_y + c_y + d_y = y_1$$

$$c_y = dy_0$$

$$3a_y + 2b_y + c_y = dy_1$$

The only trouble is that although the gradient of our quasi-SDF still shows the direction of the function's larger increment, the size of this vector is absolutely out of scale. Although a distance function of a point lying exactly 1 unit from the target surface is 1 by definition, our quasi-SDF at this point could easily be as high as 128. Exaggerated gradients result in exaggerated tangent vectors, and they spoil the contour by making cubics do the loops (figure 12.22).

The problem is easy to mitigate if we remember that a curve's tangent vector's direction at a point is determined by a ratio of partial derivatives. But the length of the tangent vector is determined by the magnitude of partial derivatives, not their ratio. $dx = 1$ and $dy = 2$ signify the same tangent line as $dx = 2$ and $dy = 4$, but the length of the tangent vector will be twice as large for the latter example.

To keep the whole curve smooth, the tangent vectors in the curve segment's ends don't have to be a specific length; an infinite set of codirected vectors sets the same tangent line. Therefore, we can choose the length we want. If our tangent vectors are disproportionate to the pixel size, let's make them proportionate. Let's normalize them. This approach does the trick and "tames" the contour well enough (figure 12.23). We also used this trick in chapter 7 to build a nice quadratic spline, remember?

Tangent vector normalization here is the only difference from the smoothing scheme we used in chapter 11.

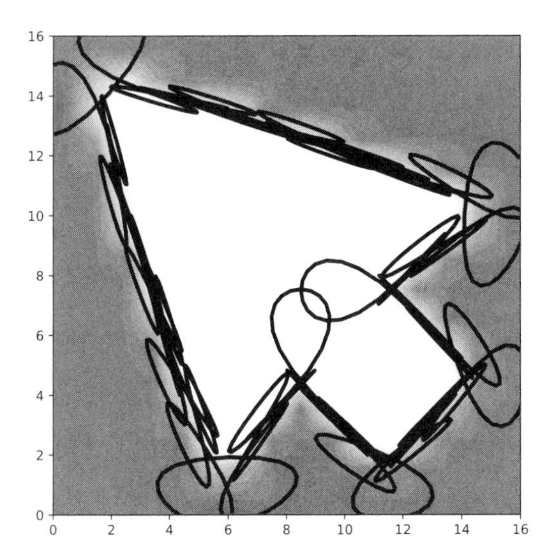

Figure 12.22 Exaggerated tangent vectors spoil the contour. It's still technically smooth and continuous, though.

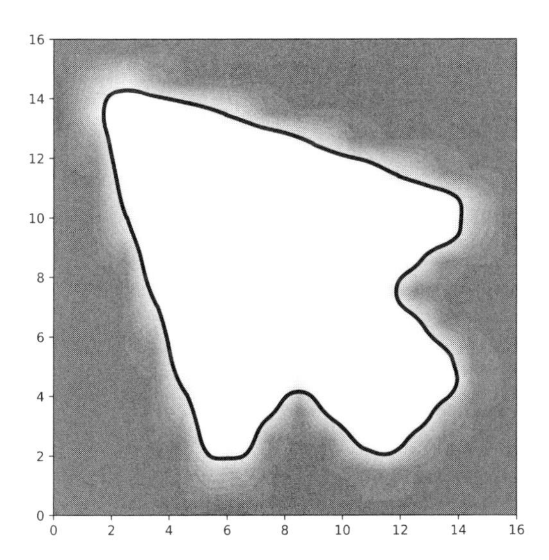

Figure 12.23 By normalizing tangent vectors, we keep the contour smooth, but the unwanted loops are gone.

12.4.5 Implementation

As always, the example is written in Python, but you're free to use any language you want. First, let's define an image:

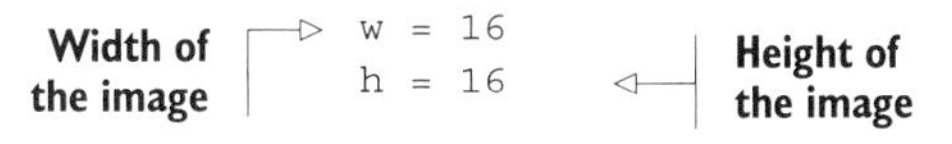

```
img = [
[0,  0,  0,  0,  0,  0,  0,  0,  0,  0,  0,  0,  0,  0,  0,  0],
[0,  77, 125,38, 0,  0,  0,  0,  0,  0,  0,  0,  0,  0,  0,  0],
[0,  120,255,254,203,144,96, 3,  0,  0,  0,  0,  0,  0,  0,  0],
[0,  34, 253,255,255,255,255,230,154,94, 8,  0,  0,  0,  0,  0],
[0,  0,  196,255,255,255,255,255,255,252,241,139,83, 6,  0,  0],
[0,  0,  149,255,255,255,255,255,255,255,255,255,250,213,80, 0],
[0,  0,  98, 255,255,255,255,255,255,255,255,255,255,224,58, 0],
[0,  0,  2,  224,255,255,255,255,255,255,255,242,152,4,  0,  0],
[0,  0,  0,  145,255,255,255,255,255,255,255,154,1,  0,  0,  0],
[0,  0,  0,  82, 251,255,255,255,255,255,255,253,156,1,  0,  0],
[0,  0,  0,  6,  237,255,255,255,255,255,255,255,252,146,3,  0],
[0,  0,  0,  0,  149,255,255,243,149,252,255,255,255,240,21, 0],
[0,  0,  0,  0,  69, 249,255,152,1,  150,252,255,238,71, 0,  0],
[0,  0,  0,  0,  0,  211,224,4,  0,  1,  137,240,86, 0,  0,  0],
[0,  0,  0,  0,  0,  73, 57, 0,  0,  0,  2,  20, 0,  0,  0,  0],
[0,  0,  0,  0,  0,  0,  0,  0,  0,  0,  0,  0,  0,  0,  0,  0]
]
```

The image is a 2D array of 1-channel, 8-bit depth "gray values."

Now let's define the thresholds:

```
distance_threshold = 128
accuracy_threshold = 2
```

With distance threshold, we govern where we want our isoline to pass. The number 128 implies that we expect a curve roughly in the middle between full light and full dark pixels.

With the accuracy threshold, we regulate how fast our fitting will exit. It's a lever to balance speed and accuracy.

Then let's present our image as a function. For now, let it be discontinuous:

```
@np.vectorize
def img_as_f(x, y):
    if x > 0 and x < w and y >= 0 and y < h:
        return img[h-1-int(y)][int(x)]
    return 0
```

Figure 12.24 shows the pixels represented as a function.

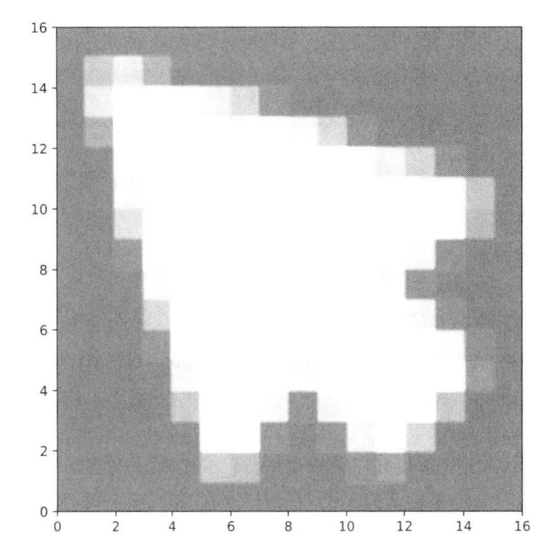

Figure 12.24 A function of pixel values. It isn't continuous.

Now let's turn the pixelated image into a continuous, smooth one. We'll use the same scheme as in chapter 8 (see section 8.3.3): the inverse weight interpolation. Refer to chapter 8 if you need a reminder of how this particular interpolation works:

```
@np.vectorize
def img_as_smooth_f(x, y):
    if x > 0 and x < w and y >= 0 and y < h:
        centered_x = x - 0.5
        centered_y = y - 0.5
        ix = int(centered_x)
        iy = int(centered_y)
        x0 = centered_x - ix
        x1 = 1 - x0
        y0 = centered_y - iy
        y1 = 1 - y0
        if x0 == 0 and y0 == 0:
            return img_as_f(ix, iy)
        if x0 == 0 and y1 == 0:
            return img_as_f(ix, iy + 1)
        if x1 == 0 and y0 == 0:
            return img_as_f(ix + 1, iy)
        if x1 == 0 and y1 == 0:
            return img_as_f(ix + 1, iy + 1)
        if x0 == 0:
            return ((img_as_f(ix, iy) / y0 + img_as_f(ix, iy + 1) / y1)
                   / (1/y0 + 1/y1))
        if x1 == 0:
            return ((img_as_f(ix+1, iy) / y0 + img_as_f(ix+1, iy+1) / y1)
                   / (1/y0 + 1/y1))
        if y0 == 0:
            return ((img_as_f(ix, iy) / x0 + img_as_f(ix+1, iy) / x1)
                   / (1/x0 + 1/x1))
        if y1 == 0:
            return ((img_as_f(ix, iy+1) / x0 + img_as_f(ix+1, iy+1) / x1)
                   / (1/x0 + 1/x1))
        return ((img_as_f(ix, iy) / (x0*y0) +
                img_as_f(ix, iy+1) / (x0*y1) +
                img_as_f(ix+1, iy) / (x1*y0) +
                img_as_f(ix+1, iy+1) / (x1*y1))
               / (1/(x0*y0) + 1/(x0*y1) + 1/(x1*y0) + 1/(x1*y1)))
    return 0
```

This is a trick to shift the whole function to (0.5, 0.5) vector. Without this shift, our pixel's colors are defined in the pixels' top-left corners. With this shift, the colors are defined in the pixel's centers.

These are the in-pixel distance coefficients. We take four of them—one for each pixel's side.

Literally the corner cases. With our (0.5, 0.5) shifts, the corners now coincide with the pixel's centers, so if we detect a corner case, we return the color value from the corresponding pixel.

Pixel's sides, special cases of inverse distance interpolation with two interpolated values

Finally, the general case of inverse distance interpolation, with all four corner values being mixed

Now the image becomes a smooth and continuous function (figure 12.25).

The next step is defining a quasi-SDF. As a reminder, an SDF is a signed distance function, and our function isn't really a distance function, but we pretend that it is—hence, the *quasi-* prefix.

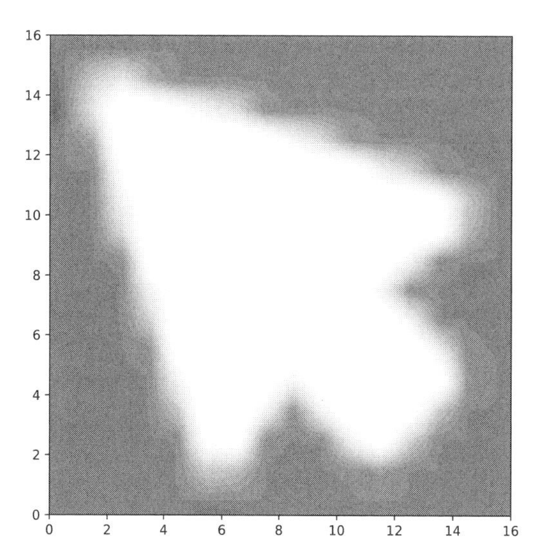

Figure 12.25 The continuous and smooth function that interpolates the pixel values in the pixels' centers

Our image is already a smooth, continuous function, so to turn it into a quasi-SDF, we only need to make it signed. After we do that, we can make the target polyline go through exactly this implicit curve:

```
SDF(x, y) = 0.

def SDF(x, y):
    return img_as_smooth_f(x, y) - distance_threshold
```

Now that we have a quasi-SDF, let's use the three-step algorithm that we used in chapter 11, but with the proper modifications. First, let's build an initial border with lots of corners:

```
border = []
for i in range(0, 16):
    for j in range(0, 16):
        if SDF(j - 0.5, i + 0.5) * SDF(j + 0.5, i + 0.5) < 0:
            border += [[(j, i), (j, i+1)]]
        if SDF(j + 0.5, i - 0.5) * SDF(j + 0.5, i + 0.5) < 0:
            border += [[(j, i), (j+1, i)]]
```

The border will be a list of line segments where each line segment is a two-piece list of 2D points. (points to `border = []`)

Adds a vertical line when needed (points to the first `if` block)

Adds a horizontal line where expected (points to the second `if` block)

An initial border looks like figure 12.26.

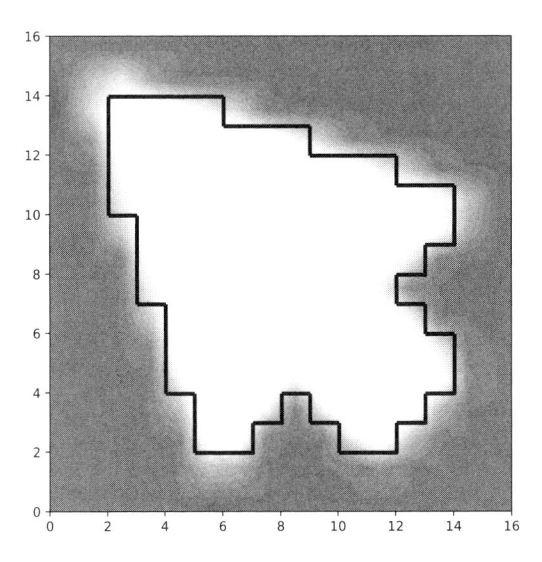

Figure 12.26 The initial contour made of the pixels' borders

The next step is fitting the border's vertices closer to the target implicit surface. We'll use an iterative algorithm, pretty much as we did for a linear solver in chapter 3:

```
epsilon = 0.1     <-- We approximate partial derivatives with differences. An epsilon is a different step size. Normally, the smaller the size, the better the approximation.

def dx(x, y):
    return (SDF(x+epsilon, y) - SDF(x-epsilon, y)) / (2*epsilon)
def dy(x, y):
    return (SDF(x, y+epsilon) - SDF(x, y-epsilon)) / (2*epsilon)
                                        Partial derivatives

contour = []
for segment in border:
    new_segment = []
    for i in range(2):
        x, y = segment[i]
        sdf_value = SDF(x, y)
        while abs(sdf_value) > accuracy_threshold:
            step_length = sdf_value / 256
            pdx = dx(x, y)
            pdy = dy(x, y)
            dxy_length = distance_p2p(pdx, pdy, 0, 0)
            x = x - pdx / dxy_length * step_length
            y = y - pdy / dxy_length * step_length
            sdf_value = SDF(x, y)
        new_segment += [(x, y)]
    contour += [new_segment]
```

Annotations: "We take the vertices from the initial border as a set of starting points." (on `for segment in border:` … `for i in range(2):`); "The exit condition is "stop when the SDF value is close enough to 0."" (on the `while` line); "The iteration itself" (on the `x = …` / `y = …` lines).

Now we have a fitted contour (figure 12.27).

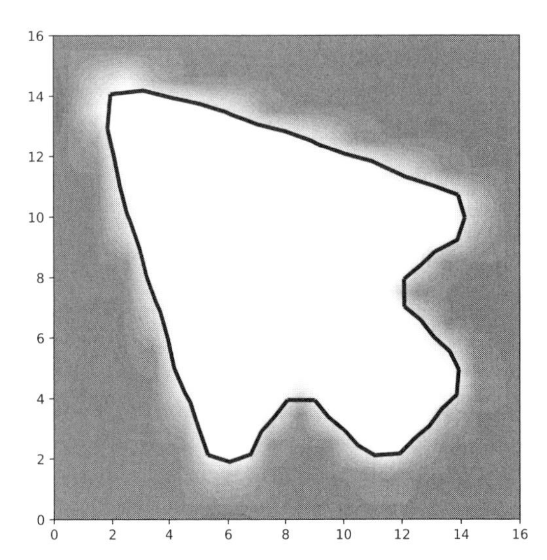

Figure 12.27 The fitted contour. It originates from the initial border, but now every vertex is fitted into the threshold isoline by means of an iterative algorithm.

This image is already starting to look good, but wait until we add curvature to our segments. To do so, we'll compute tangent vectors in each segment's vertices and substitute a cubic polynomial curve for each segment. We'll also split the curves into 16 little linear segments each so we can draw them easily, but that's a technicality. Representing curves as approximated sets of segments makes plotting simpler, but it isn't essential to the algorithm. Here's the code:

```
smoothened = []
for segment in contour:
    x0 = segment[0][0]
    y0 = segment[0][1]
    x1 = segment[1][0]
    y1 = segment[1][1]

    dx0 = dy(x0, y0)
    dy0 = -dx(x0, y0)
    dx1 = dy(x1, y1)
    dy1 = -dx(x1, y1)
    if dot2d(x1-x0, y1-y0, dx0, dy0) < 0:
        dx0 = -dx0
        dy0 = -dy0
    if dot2d(x1-x0, y1-y0, dx1, dy1) < 0:
        dx1 = -dx1
        dy1 = -dy1
    dxy_length0 = distance_p2p(dx0, dy0, 0, 0)
    dxy_length1 = distance_p2p(dx1, dy1, 0, 0)
    dx0 = dx0 / dxy_length0
    dy0 = dy0 / dxy_length0
    dx1 = dx1 / dxy_length1
    dy1 = dy1 / dxy_length1
```

Initial values for the cubics' ends come from the segments' vertices.

Initial tangent vectors come from partial derivatives.

We co-orient the tangent vectors with the segments' direction . . .

. . . and then we normalize them to get rid of the "loops."

```
a_x = dx0 + dx1 + 2*x0 - 2*x1
b_x = -2*dx0 - dx1 - 3*x0 + 3*x1
c_x = dx0
d_x = x0
a_y = dy0 + dy1 + 2*y0 - 2*y1
b_y = -2*dy0 - dy1 - 3*y0 + 3*y1
c_y = dy0
d_y = y0

for i in range(16):
    t0 = i / 16.
    t1 = t0 + 1. / 16.
    x0 = a_x*t0**3 + b_x*t0**2 + c_x*t0 + d_x
    y0 = a_y*t0**3 + b_y*t0**2 + c_y*t0 + d_y
    x1 = a_x*t1**3 + b_x*t1**2 + c_x*t1 + d_x
    y1 = a_y*t1**3 + b_y*t1**2 + c_y*t1 + d_y
    smoothened += [[(x0, y0), (x1, y1)]]
```

We get the cubics' coefficients by solving two systems of equations, ready-made solutions provided by SymPy (chapter 11).

A technicality. We turn cubics back into linear segments so we can draw them in the same way we drew the initial and fitted contours.

The resulting contour appears to be smooth (figure 12.28).

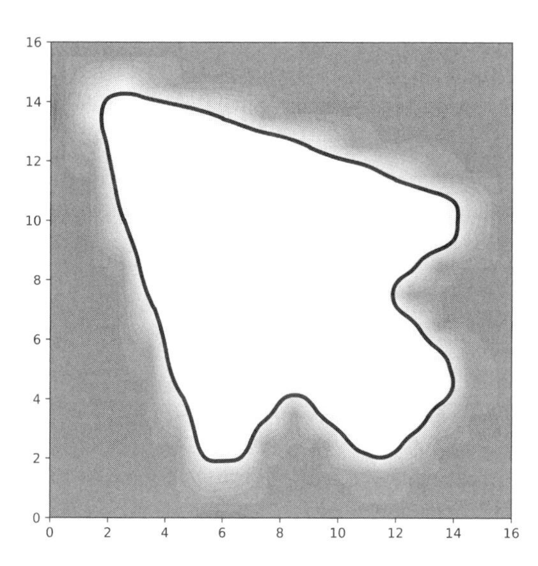

Figure 12.28 Smooth contour made of cubic polynomials

The full source code, with all the utility functions and plotting, is available at ch_12/image_vectorization.py in the book's source code.

12.4.6 Section 12.4 summary

The image vectorization algorithm we implemented may not be practical. Yes, we did turn a raster image of an arrow into a set of curves, and we did keep the formal criteria for smoothness and continuity. But an artist with a vector graphics editor will draw a much better arrow in a minute. For any practical use, we need to employ more elaborate algorithms.

I do think, however, that this algorithm is perfect for didactic purposes. Every other vectorization algorithm out there depends on the same basic mathematical concepts: SDFs, partial derivatives, tangent vectors, and so on. The algorithm presented here doesn't add much new to these concepts—no dynamic programming, no computer vision, no machine learning. The algorithm simply binds the geometric concepts together to show the power of geometry, not how far we can go with programming.

12.5 How voxels, triangles, parametric surfaces, and SDFs work together

We've looked at how geometric data is usually represented in computers. Obviously, if one ultimate representation outperformed all the others by any metric imaginable, this book would be at least two chapters shorter. All the representations and all data models we've learned about are used in programming, but in different contexts.

3D images traditionally come from medical imagery, but in recent years, CT is used in other industries as well. For instance, in transportation. Trains, ships, and airplanes work for decades with proper care and maintenance, and sometimes, this longevity allows them to outlive their own producers. A plane from the 1950s might still be in reasonably good condition, but the plant where its spare parts were produced might have disappeared a long time ago. Even the blueprints for the spare parts may have been lost.

Not to worry! With CT, you can scan a part that you want to replicate. Next, using a technique not unlike the one described in the previous section, you can make the part's boundary representation. Then you can mill the part from the model you get, or you can turn its parametric surfaces into triangle meshes and 3D-print the thing. By the way, you might still want to keep the smooth surfaces to postprocess the part with milling machines.

With this technology, or a whole bouquet of technologies, you can keep your business operational for decades to come. But even for this rather simple scenario, you need at least three different data representations.

Different representations have different strong points:

- *SDFs* give you fast, robust offsets and Boolean operations. They're also immune to problematic elements such as near-degenerate triangles and cracks between surfaces. A lot of model-fixing algorithms use SDFs underneath.
- *Boundary representation* with parametric surfaces—spheres, planes, cylinders, NURBS, and so on—isn't as fast and robust at operations, but it's precise. You need precise parametric surfaces in engineering and manufacturing.
- *Triangle mesh representation,* which is a special case of a boundary representation, is generally thought of as a faster, more robust alternative to engineering surfaces because the algorithms for processing triangles are usually simpler than the same algorithms working on NURBS. The choice boils down to the balance between the accuracy of your representation and the performance benefits you get to enjoy. A triangle mesh is only an approximation of a surface, after all.

- *Voxels* are fast and robust regarding erosion and dilation. Boolean operations and intersection detection are ultrafast and robust on voxels, too. In terms of approximation, however, balancing accuracy and speed for voxels is even harder than with triangles, because you get to keep not only a surface, but also the whole 3D model in memory.

The spare-part workflow described earlier isn't unique in its dependence on several representations at the same time. Almost any workflow that drives a thing from design to production requires jumping from one data model to another.

In medicine, producing any custom material object, such as an implant or a surgery guide, starts with obtaining 3D images. Next, the object is designed in parametric surfaces or (if the actual production is to be done with a 3D printer) triangle meshes. Then the model gets converted to a voxel representation so we can fit it into our images and double-check whether it works there.

Sometimes, simply converting one representation to another to pick the most fitting technology for the moment isn't enough anymore. In games and computer-generated images, we don't often design models the same way that we do in industrial design. We don't build a game character from planes and spheres. The preferred way to make highly detailed, highly realistic models is so-called *3D sculpting*.

With 3D sculpting, artists are free to create any form they choose and make it as intricate as they like. Do they have to worry about the model size or slow processing? Well, yes and no. The sculpting software usually works in a hybrid representation in which the model is based on voxels, but every boundary voxel contains a bit of surface data as well. It can contain an exact surface point and a normal vector, for example. So you get to enjoy fast operations and heightened precision of the representation, too.

This representation has limits as well. It does require a lot of memory, and it's relatively slow to render. You can't simply put your 3D sculptures into a game. But you can create several triangle mesh models out of a sculpture, making each more intricate than the one before. These models are called *levels of detail* (LODs). Having a good sculpture, you can make as many different LODs as you want out of it.

I'm certain that this isn't the last hybrid representation we'll hear about in the coming decades. As hardware evolves, more data models will become economically plausible and, hence, popular. Multicore CPUs, superscalar processing, heterogeneous computing, and cloud technology all affect the technological landscape we live in.

In the 1990s, there was no alternative to a triangle mesh in 3D printing; now almost every manufacturing startup brings in its own new data model which is either optimized for running on graphic cards or made inherently parallelizable to work well in a cloud. I'm sure that the same thing happens in industries I'm not aware of as well.

If you want to stay in programming for a few decades, you should be ready for changes. The data models will change; the algorithms will evolve; all the frameworks, all the new and shiny libraries that are now in their infancy, will mature and get obsolete eventually. New ones will come along to replace them.

The only thing that will survive this circle of life is geometry.

12.6 Exercises

Exercise 12.1 Let's say you want to restore an old valve lever. You can scan the original, edit its model in an editor of your choice, 3D-print the thing, and then polish it with an automatic milling machine. How many data conversions would you have to do in the process?

Exercise 12.2 We have two black-and-white images of the same size. Each pixel is completely black or completely white. The images are identical if a pixel at coordinates [i, j, k] from the first image is exactly the same as the pixel with the same coordinates from the second image. That makes sense, right?

But what if they're not identical? Most pixels are the same, but some are only similar. How do you assess the degree of sameness for these images?

Exercise 12.3 What if the images aren't strictly black-and-white, but based on gray values? Let's say that a pixel may have any value in the range of [0, 1]. How do you assess sameness then?

Exercise 12.4 We have a voxel model of several separate objects that don't touch. Voxels may only have two values: 1 for material and 0 for emptiness. We want to count the separate objects in the model. How do we do that?

Exercise 12.5 We've prepared a platform for 3D printing. We have several models that we want to print at the same time. For technological reasons, we can't put them too close to one another, so we have to ensure a certain gap between them. How do we check for this gap quickly enough, considering that we already have a detailed voxel model of the whole platform?

12.7 Solutions to exercises

Exercise 12.1 The answer really depends on specific technologies, but we need at least one, from a 3D image to a boundary representation for the 3D printer and the milling machine. A 3D printer might require a triangle mesh; then there are two conversions, and of course the editing phase may add several more. If you want to add thickness to some part of your lever, for example, you may want to convert the model to an SDF, put the offset there, and convert the model back.

Exercise 12.2 This question is an open question, of course. You're more than welcome to invent your own way. In practice, we often use the so-called Sørensen–Dice coefficient, sometimes abbreviated as Dice. It's a proportion of the number of pixels with the same values in both images to the total number of pixels. A pair of fully identical images has a Dice coefficient of 1.

Exercise 12.3 There's a Sørensen–Dice coefficient for continuous values, too.

SEE ALSO For more information, see "Continuous Dice Coefficient: A Method for Evaluating Probabilistic Segmentations," by Reuben R Shamir, Yuval Duchin, Jinyoung Kim, Guillermo Sapiro, and Noam Harel, at https://www.biorxiv.org/content/10.1101/306977v1.

Exercise 12.4 This question is another open question. Please feel free to invent your own algorithm. If you want a hint, though, use an image mask with a part number. Each filled voxel should either take the lowest part number of its neighbors or get a new part number if there are no filled neighbors with part numbers yet. You may want to repeat propagation iteratively if you choose to traverse the image directly.

Exercise 12.5 This question is yet another open question. I suggest counting the disjointed parts, dilating them all by half of the gap distance, and recounting them. If there will be fewer separate parts than before, some of them got glued together, so they were too close together to begin with.

Summary

- The concept of a 3D image is the generalization of a 2D raster picture made out of pixels to the 3D space.
- A voxel in 3D is the same as a pixel in 2D: a building block for images.
- Operations such as erosion and dilation are analogous to offsets in SDFs or boundary representations.
- Boolean operations on voxel models are fast and reliable as long as the operands share a voxel grid.
- Reliability and speed are the most welcome traits of voxel operations when we compose operations into practical workflows.
- With image vectorization, we have a full set of conversions. Now we know three ways of representing geometric data—SDFs, boundary representations, and voxels—and we can convert them all from and to one another.
- In practice, true efficiency in geometric modeling and processing often comes from picking the right data representation for the job, not from the quality of the algorithms' implementations.

appendix Sources, references, and further reading

Books

Applied Geometry for Computer Graphics and CAD, 2nd ed., by Duncan Marsh (Springer, 2005), is written for undergraduate math students, not programmers. It doesn't concern computational aspects such as memory consumption or code performance. Most importantly, it expects the reader to be a future mathematician, not a current programmer, so it uses the established language of math to explain more math.

Handbook of Geometry for Competitive Programmers, by Victor Lecomte (https://github.com/vlecomte/cp-geo), explains computational error along with geometric concepts. This fact alone makes it stand out from the majority of math books.

A Programmer's Geometry, by Adrian Bowyer and John Woodwark (Butterworth-Heinemann, 1983), contains detailed solutions for basic geometric problems. It was published in 1983, though, and has code samples in FORTRAN 77.

Mathematics for Computer Science, by Eric Lehman, F. Tom Leighton, and Albert R. Meyer (Samurai Media Limited, 2017), although not particularly a book on geometry, is an excellent example of approachable mathematical writing for nonmathematicians. It explains important mathematical concepts with relatable examples. It's even entertaining, as far as a math book can be.

A Brief Course in Modern Math for Programmers, by Vlad Partyshev (GUMROAD, 2019), is another good example of a "math for programmers" book. It's concise and doesn't have theorems and proofs, but instead focuses on important concepts from the realm of logic, algebra, and category theory. It doesn't cover geometry explicitly, although some of the ideas are universal throughout the whole body of applied mathematics.

Foundations of Game Engine Development, Volume 1: Mathematics, by Eric Lengyel (Terathon Software, 2016), is a more fitting example in terms of content. It's a a concise, practical book filled with C++ code examples.

Secure Coding in C and C++, 2nd ed., by Robert C. Seacord (Addison-Wesley Professional, 2013), surprisingly, has a few chapters on numeric programming from the standpoint of security.

Accuracy and Stability of Numerical Algorithms, 2nd ed., by Nicholas J. Higham (Society for Industrial and Applied Mathematics, 2002), is a classic work on computational stability and error.

Immersive Math, by Jacob Ström, Kalle Åström, and Tomas Akenine-Möller (http://immersivemath.com/ila/index.html), is the world's first linear algebra book with fully interactive figures.

Interactive Linear Algebra, by Dan Margalit and Joseph Rabinoff (https://textbooks.math.gatech.edu/ila/index.html), is a pretty standard textbook by instructors at the Georgia Institute of Technology, except that it's free, available online, and filled with interactive elements.

The Mathematics of Bézier Curves, by Mike Kamermans (https://pomax.github.io/bezierinfo/#explanation), is a free online book to check when you need to know how to do Bézier things.

Papers

"Scattered Data Interpolation in Three or More Variables," by Peter Alfeld. Available from https://www.math.utah.edu/~alfeld/papers.html. The paper contains a survey of interpolation methods.

"A Survey of Spatial Deformation from a User-Centered Perspective," by James Gain and Dominique Bechmann (https://pubs.cs.uct.ac.za/id/eprint/475/1/ACMTOGDef.pdf).

"How to Write Fast Numerical Code: A Small Introduction," by Srinivas Chellappa, Franz Franchetti, and Markus Püschel (https://www.spiral.net/doc/papers/FastNumericalCode.pdf). This paper is about numeric programming in the context of the current technological landscape.

"What Is a Good Linear Finite Element?: Interpolation, Conditioning, Anisotropy, and Quality Measures," by Jonathan Richard Shewchuk (https://people.eecs.berkeley.edu/~jrs/papers/elemj.pdf). The paper contains a survey on quality measures for triangles.

"Discrete Differential Geometry: An Applied Introduction," by Keenan Crane (https://www.cs.cmu.edu/~kmcrane/Projects/DDG/paper.pdf). This paper is more a book than an article. It explains how to make differential concepts such as smoothness and curvature work on triangle meshes.

"Algorithms for Accurate, Validated and Fast Polynomial Evaluation," by Stef Graillat, Philippe Langlois, and Nicolas Louvet (https://hal.archives-ouvertes.fr/hal-00285603/document). The paper is a survey on polynomial computation.

Websites

Better Explained (https://betterexplained.com). Clear, intuitive lessons about imaginary numbers, exponents, and more

Red Blob Games (https://www.redblobgames.com). Interactive visual explanations of math and algorithms, with motivating examples from computer games

Mathigon (https://mathigon.org/courses). Interactive courses on geometry and other topics

Visualize It (https://visualize-it.github.io). Interactive visualizations for various topics from physics and mathematics, including trigonometry, polynomial regression, and Bézier splines

Inigo Quilez: Articles (https://www.iquilezles.org/www/index.htm). Assorted tutorials on computer graphics, size coding, and math, including a mini-library of signed distance functions

Intuitive Math (https://intuitive-math.club). A collection of visualizations on linear algebra and geometry

Sublucid Geometry (https://zalo.github.io). A collection of interactive visualizations on somewhat less-known topics of 3D geometry

Image Processing Learning Resources (https://homepages.inf.ed.ac.uk/rbf/HIPR2/hipr_top.htm). A great free online resource if you want to learn more about image processing

Words and Buttons Online (https://wordsandbuttons.online), a shameless plug for my website. A growing collection of interactive tutorials, demos, and quizzes on math and programming

Interactive web pages

Curves and Surfaces (https://ciechanow.ski/curves-and-surfaces). An interactive explanation of curves and surfaces

What Is the Inverse of a Vector? (https://mattferraro.dev/posts/geometric-algebra). An introduction to geometric algebra, an invertible vector product, and the implication of having one

That Creative Code Page (https://www.notion.so/Creative-coding-algorithms-techniques-c5550ef2f7574126bdc77b09ed76651b). A visual overview of commonly used creative coding techniques and algorithms

Complex Analysis (https://complex-analysis.com). A visual and interactive introduction to complex analysis

An Interactive Introduction to Splines, by Evgeny Demidov (https://www.ibiblio.org/e-notes/Splines/Intro.htm). A collection of interactive pages explaining Bézier splines, B-splines, NURBS in particular, and more

An Interactive Introduction to Rotors from Geometric Algebra, by Marc ten Bosch (https://marctenbosch.com/quaternions). An explanation of how to program rotations without quaternions and why

Half-Edge Data Structures, by Jerry Yin and Jeffrey Goh (https://jerryyin.info/geometry-processing-algorithms/half-edge). An interactive explanation of half-edge data structures

Unraveling the JPEG, by Omar Shehata (https://parametric.press/issue-01/unraveling-the-jpeg). A comprehensive interactive explanation of the JPEG format

index

D

E

F

G

H

I

S

T

U

V

W